"This is a valuable resource for agritourism academics a[...] Africa who intend to create a sustainable competit[...] alignment to practice and theory makes it more contemporary and relevant."

Professor Dimitrios Buhalis, *Bournemouth University Business School, United Kingdom (UK), Editor-in-Chief,* Tourism Review

"This book provides critical insights for both agritourism scholars and practitioners in Africa. More importantly, it draws from the rich and diverse expertise of authors drawn from various parts of Africa. The book acts as a key resource as Africa strives to create sustainable and diversified economies".

Dr. George. T. Mudimu, *Senior Editor,* Cogent Social Sciences

Agritourism in Africa

This insightful, timely and multi-faceted book offers significant insight into the role and complex dynamics of agritourism in Africa.

Logically structured, data-led, and richly illustrated throughout, chapters provide theoretical, policy and practical implications on the successes and challenges of achieving sustainable agritourism destinations, with an emphasis on technology, that not only grows African economies, but offers work opportunities, increased social empowerment and diversity. Based on empirical research, the volume covers a wide range of topics relating to agritourism in Africa, elucidated through inclusion of case studies and examples from around the continent, including Ghana, Angola and Nigeria. Topics covered include discussion of the features required for a successful agritourism business, the impact of social media and digital marketing on new agritourism destinations.

This volume will be of pivotal interest to students, researchers, and scholars of Tourism, African Studies and Development Studies.

Brighton Nyagadza (CIM, PhD) is currently with the York St John University (YSJU), London Campus, United Kingdom, Full member of the Marketers Association of Zimbabwe (MAZ), an Associate of The Chartered Institute of Marketing (ACIM), United Kingdom, and Power Member of the Digital Marketing Institute (DMI), Dublin, Ireland. He has published several book chapters and sits on various international academic and national strategic boards, including the editorial board membership of *Journal of Digital Media & Policy*, The *Retail and Marketing Review* and *Mashonaland East Province Zimbabwe National Development Strategy (NDS) Committee* (2021–2025) – ICT & Human Capital Development cohort.

Farai Chigora (DBA) holds a Doctorate in Business Administration from the University of KwaZulu-Natal (South Africa) and is a Senior Lecturer in Business Science at the College of Business, Peace Leadership and Governance, Africa University in Zimbabwe. He is a branding specialist with interest in destination branding, strategic marketing, business research, and related business areas which he has authored in various refereed international journals.

Azizul Hassan (PhD) is a member of the Tourism Consultants Network of the UK Tourism Society. Dr. Hassan has been working for the tourism industry as a consultant, academic, and researcher. His research interest areas are technology-supported marketing for tourism and hospitality, immersive technology applications in the tourism and hospitality industry, and technology-influenced marketing suggestions for sustainable tourism and hospitality industry in developing countries. Dr. Hassan has authored articles and book chapters in leading tourism outlets. He is also part of the editorial team of book projects from Routledge and various other academic publishers. He is a regular reviewer of a number of international journals.

Contemporary Geographies of Leisure, Tourism and Mobility

Series Editor: C. Michael Hall
Professor at the Department of Management, College of Business and Economics, University of Canterbury, Christchurch, New Zealand

The aim of this series is to explore and communicate the intersections and relationships between leisure, tourism and human mobility within the social sciences.

It will incorporate both traditional and new perspectives on leisure and tourism from contemporary geography, e.g. notions of identity, representation and culture, while also providing for perspectives from cognate areas such as anthropology, cultural studies, gastronomy and food studies, marketing, policy studies and political economy, regional and urban planning, and sociology, within the development of an integrated field of leisure and tourism studies.

Also, increasingly, tourism and leisure are regarded as steps in a continuum of human mobility. Inclusion of mobility in the series offers the prospect to examine the relationship between tourism and migration, the sojourner, educational travel, and second home and retirement travel phenomena.

The series comprises two strands:

Contemporary Geographies of Leisure, Tourism and Mobility aims to address the needs of students and academics, and the titles will be published simultaneously in hardback and paperback.

Routledge Studies in Contemporary Geographies of Leisure, Tourism and Mobility is a forum for innovative new research intended for research students and academics, and the titles will initially be available in hardback only. Titles include:

Tourism Interventions
Making or Breaking Places
Edited by Rami K. Isaac, Jeroen Nawijn, Jelena Farkić and Jeroen Klijs

Agritourism in Africa
Edited by Brighton Nyagadza, Farai Chigora and Azizul Hassan

For more information about this series, please visit: www.routledge.com/ Contemporary-Geographies-of-Leisure-Tourism-and-Mobility/book-series/ SE0522

Agritourism in Africa

Edited by Brighton Nyagadza,
Farai Chigora and Azizul Hassan

Routledge
Taylor & Francis Group

LONDON AND NEW YORK

First published 2025
by Routledge
4 Park Square, Milton Park, Abingdon, Oxon OX14 4RN

and by Routledge
605 Third Avenue, New York, NY 10158

Routledge is an imprint of the Taylor & Francis Group, an informa business

British Library Cataloguing-in-Publication Data
A catalogue record for this book is available from the British Library

ISBN: 978-1-032-69617-1 (hbk)
ISBN: 978-1-032-69615-7 (pbk)
ISBN: 978-1-032-69618-8 (ebk)

DOI: 10.4324/9781032696188

Typeset in Times New Roman
by MPS Limited, Dehradun

To all agritourism enthusiasts dotted around the globe.

Contents

List of Figures   *xiii*
List of Tables   *xiv*
Editors   *xvi*
List of Contributors   *xvii*
Foreword by Prof. Lucyna Przezbórska-Skobiej   *xxi*
Foreword by Dr. Erisher Woyo   *xxiii*
Preface   *xxv*
List of Abbreviations   *xxxi*

1 Exploring the Nexus between Agritourism
   Development, Economic Factors, and Agricultural
   Trade Flows: A Case of Zimbabwe's Agritourism Sector   1
   ADMIRE CHAWARIKA AND TINASHE CHUCHU

2 Artificial Intelligence in Agritourism: Utilitarian
   Analysis of Opportunities, Challenges, and Ethical
   Considerations in the African Context   17
   VARAIDZO DENHERE AND DEO SHAO

3 Critical Success Factors for Agritourism
   Entrepreneurship in Africa   37
   MUFARO DZINGIRAI, RUKUDZO DZAPASI, AND
   THEOLOGIENCE MAZWIEMBIRI

4 Economic Potentials of Precision Agritourism in
   Alleviating Income Poverty among Farming
   Communities in Tanzania: The Cottage
   Farming Approach   54
   PROSCOVIA PASCHAL KAMUGISHA

5 Artificial Intelligence and Sustainable Agritourism
  for Human Development in Africa   77
  LESLIE WELLINGTON SIRORA AND MARTIN MUDUVA

6 Exploring the Ethical Dimensions of Sustainable
  Agritourism in Central Angola: Unveiling the Role
  of Local Actors in Fostering Economic Growth
  and Environmental Conservation   96
  JABULANI GARWI

7 African Tourism and Hospitality Experience:
  The Zimbabwean Perspective   119
  JULIUS TAPERA, PURITY HAMUNAKWADI, RAHABI MASHAPURE,
  AND ENET MUKURAZITA

8 The State of Agritourism Development in Southern
  Africa: A Regional Analysis   145
  ZIBANAI ZHOU

9 Bibliometric Analysis of Agritourism for
  Rural Transformation and Poverty Reduction
  Policy Implications   167
  FRANCIS ARON MWAIJANDE

10 Agritourism Digital and Social Media Marketing
   in Africa: The Case Study of Zimbabwe   191
   SINOTHANDO TSHUMA, MERCY DUBE, AND PHILLIP DANGAISO

11 Exploring How Disruptive Technological Innovation
   Can Positively Impact Agritourism in Developing
   Countries: Echoes from Zimbabwe   212
   SAMUEL MUSUNGWINI, EDWARD MUDZIMBA, MERCY DUBE, AND
   SINOTHANDO TSHUMA

12 Conclusion: The Future of Agritourism Development
   in Africa   240
   BRIGHTON NYAGADZA, FARAI CHIGORA, AND AZIZUL HASSAN

   Index   *245*

Figures

3.1	Critical success factors for agritourism entrepreneurship	50
5.1	A conceptual framework for the study	86
5.2	Word cloud of themes identified in the study	88
5.3	Market inefficiencies for agritourism in Africa	89
5.4	Distribution of major challenges associated with access to information on agritourism in Africa	90
5.5	Impact of AI-enabled sustainable agritourism practices on local communities in Africa	91
6.1	Triple Bottom Line (TBL) approach by John Elkington in 1994	103
9.1	Tourism arrivals in 000 and earnings in millions	172
9.2	Agritourism policy issues	177
9.3	Agriculture-tourism interface	178
9.4	Agritourism nexus for rural transformation	179
9.5	Mapping stakeholder by AIIM matrix	181
9.6	Visualization of key nodes for agritourism network in Tanzania	184
9.7	Visualization of key nodes for agritourism network in Tanzania	186
10.1	Model of agritourism transformation	196
10.2	Key enablers for tourism growth	197
10.3	Model for agritourism sustainable digital and social media marketing in Zimbabwe	201

Tables

1.1	Study results	10
2.1	Conceptualisation of utilitarianism theory of AI integration in agritourism	20
2.2	Agritourism income streams and their activities	23
2.3	Digital tools potential for KT in agritourism	27
2.4	Associated challenges and ethical considerations of AI fusion in agritourism	30
4.1	Literature review selection criteria	59
4.2	Status of agritourism industry in Tanzania	64
4.3	Potential economics of adopting precision agritourism in Tanzania	69
Appendix 4.1	Agritourism book chapter matrix of reference publishing databases	76
5.1	Market inefficiencies in the agritourism industry in Africa	89
5.2	Challenges related to limited access to information in agritourism in Africa	90
6.1	Interview participants' profile	105
7.1	Major tourist attractions in Zimbabwe by Province	133
7.2	Major Zimbabwean hotels by city/tourism destination	134
7.3	UNESCO World Heritage Site	134
9.1	Linkage effects of tourism in the Tanzanian economy	170
9.2	Local impact index	172
9.3	Examples of agritourism in South Africa	173
9.4	Tourism and rural employment	176
9.5	Agritourism policy issues	178
9.6	List of abbreviations for organizations	187
10.1	List of scientific papers on enhancing consumer experience in agritourism through digital marketing in Zimbabwe (2013–2023)	200

10.2	List of scientific papers on Social media marketing for sustainable agritourism (2012–2023)	202
10.3	List of scientific papers on challenges of social media marketing in agritourism (2013–2023)	204
10.4	List of scientific papers on opportunities for social media marketing in agritourism (2013–2023)	207
11.1	Demographic data of research participants	217

Editors

Brighton Nyagadza (CIM, PhD) is currently with the York St John University (YSJU), London Campus, United Kingdom, Full member of the Marketers Association of Zimbabwe (MAZ), an Associate of The Chartered Institute of Marketing (ACIM), United Kingdom, and Power Member of the Digital Marketing Institute (DMI), Dublin, Ireland. He has published several book chapters and sits on various international academic and national strategic boards, including the editorial board membership of *Journal of Digital Media & Policy*, The *Retail and Marketing Review* and *Mashonaland East Province Zimbabwe National Development Strategy (NDS) Committee* (2021–2025) – ICT & Human Capital Development cohort.

Farai Chigora (DBA) holds a Doctorate in Business Administration from the University of KwaZulu-Natal (South Africa) and is a Senior Lecturer in Business Science at the College of Business, Peace Leadership and Governance, Africa University in Zimbabwe. He is a branding specialist with an interest in destination branding, strategic marketing, business research and related business areas which he has authored in various refereed international journals.

Azizul Hassan (PhD) is a member of the Tourism Consultants Network of the UK Tourism Society. Dr. Hassan has been working for the tourism industry as a consultant, academic, and researcher for over 20 years. His research interest areas are technology-supported marketing for tourism and hospitality, immersive technology applications in the tourism and hospitality industry, and technology-influenced marketing suggestions for sustainable tourism and hospitality industry in developing countries. Dr. Hassan has authored over 150 articles and book chapters in leading tourism outlets. He is also part of the editorial team of 25 book projects. He is a regular reviewer of a number of international journals.

Contributors

Admire Chawarika is a Trade Policy Analyst and Agricultural Economist. He is currently a PhD candidate in Agricultural Economics at the University of South Africa. He has experience in research including trade policy analysis, agricultural policy and environmental assessments. He has published multiple journal papers in well-revered journal publishing outlets, such as Taylor & Francis. He is currently involved in monitoring and evaluating several trade and agricultural projects and combines both quantitative and qualitative research to proffer solutions for the transformation of trade policy initiatives. He combines theoretical and practical aspects of research to tackle contemporary issues in international trade and agricultural policy issues.

Tinashe Chuchu (PhD) is a Marketing and Tourism academic and researcher. He holds a Doctorate in Marketing from the University of the Witwatersrand. Currently, he works as a Senior Lecturer in the Marketing Division of the School of Business Sciences at the University of the Witwatersrand, South Africa. He is a consumer behaviour and tourism scholar who has published numerous studies in these fields in top journals and has presented at international conferences. Dr. Chuchu has published and reviewed for major publishing. He is a member of The Academy of Business and Retail Management Conferences based in the United Kingdom. He sits on the editorial board of the Retail and Marketing Review as well as the African Journal of Business and Economic Research.

Phillip Dangaiso is a Lecturer at the School of Entrepreneurship and Business Sciences, Department of International Marketing, Chinhoyi University of Technology (CUT), Zimbabwe. He is a multi-award winner and researcher. His research interests are in social marketing, sustainable and inclusive development, services marketing, higher educational technology, customer experience management, digital service scapes and relationship marketing.

Varaidzo Denhere (PhD) is a Senior Postdoctoral Research Fellow at the University of Johannesburg, in the Johannesburg Business School. Varaidzo is a multi-disciplinary researcher who has published articles in cutting-edge fields, particularly AI applications in accredited peer-reviewed journals. She has been the conference secretary for the International Conference of Accounting and Business (I-CAB) hosted by the University of Johannesburg from 2021 to date.

Mercy Dube is a Lecturer in the Department of Information and Marketing Science at Midlands State University (MSU), Faculty of Business Sciences, Gweru, Zimbabwe. Her research interests are in digital marketing, social marketing, consumer buying behaviour, and relationship marketing. She is involved in lecturing courses in digital marketing, e-commerce, consumer behaviour and neuro-marketing aspects of branding.

Rukudzo Dzapasi is a student at Midlands State University currently studying for a Bachelor of Commerce in Business Management Honours degree. His research interests are strategy, management, development, finance, and business intelligence.

Mufaro Dzingirai (PhD) is a Lecturer in the Department of Business Management, Midlands State University. He is also currently a Senior Fellow at the Nexus Think Tank (Zarawi Trust).

Jabulani Garwi is an academic researcher and a development practitioner with interdisciplinary interests and specialization in sustainable development and African rural studies. He completed a BA(Hon.) in Economic History from the University of Zimbabwe, MA in Development Studies from the Midlands State University, and a PhD in Sustainable Agriculture from University of the Free State, South Africa. His research areas of interest include sustainability, rural development, rural livelihoods, poverty analysis, agrarian studies, food security, climate change and contract farming. His articles appear in journals such as *Southern Journal for Contemporary History*, *The Dyke*, *Journal of Economic Impact* and *IOSR Journal of Humanities and Social Science (IOSR-JHSS)*. He is currently a diplomat at the Zimbabwe Embassy in Angola.

Purity Hamunakwadi (PhD) is a Postdoctoral Research Fellow at Nelson Mandela University in the Department of Building and Human Settlements Development. She has a Master's in Development Studies and holds a PhD at North-West University (Doctor of Philosophy in Social Science (Social Transformation and Management). Her research interests are in sustainable livelihoods, human settlements, Local Economic Development, and entrepreneurship. She was appointed as one of the reviewers of the Journal of Inclusive Cities and Built Environment (JICBE). Currently, she has co-supervised 3 Honours students to completion.

Proscovia Paschal Kamugisha (PhD) is a Senior Lecturer in the Department of Economics under the Faculty of Social Sciences and is an Economist majoring in the agricultural sector. Her research interests are in economics of crops, livestock, natural resource, development and supply chain management. She has published a number of articles and book chapters. She has served as Head of the Department of Administrative Studies during 2020–2023 during that time and served as Acting College Principal severally over that period.

Rahabhi Mashapure is a full-time Lecturer in the Department of Entrepreneurship and Business Management at Chinhoyi University of Technology (CUT). She holds a Master's degree of Philosophy (MPhil) in Business Management and Entrepreneurship (Women Entrepreneurship and Sustainable Rural Livelihoods) (CUT), and a BSc Honours Degree in Business Management and Entrepreneurship (CUT). She is currently pursuing a Doctor of Philosophy (D.Phil) in Business Management and Entrepreneurship (Entrepreneurship Digitalization for Women Entrepreneurship Enhancement) at the same university.

Theologience Mazwiembiri is a student at Midlands State University, currently studying for a Bachelor of Commerce, Business Management Honours degree. His research interests are strategy, management, development, finance, and business intelligence.

Francis Aron Mwaijande (PhD) is a Fulbright Scholar holding a PhD in Public Policy from the University of Arkansas (USA), MA Communications from the University of Wolverhampton, UK and BA Education (Hons) from the University of Dar Es Salaam, Tanzania. He is a Senior Lecturer of Public Policy, M&E at Mzumbe University, Tanzania. His publications include Case Studies in Public Policy, Handbook on Evaluation Research Methods, Managing Blue Economy, and Digitalization of Agricultural Policy.

Martin Muduva is a PhD candidate in Data Science and Lecturer at Midlands State University (MSU), Zimbabwe. With an MSc, in Big Data Analytics and a Master of Leadership and Corporate Governance, he is an esteemed academic and a professional in the field of big data analytics.

Edward Mudzimba is a Lecturer in the Department of Information and Marketing Sciences at Midlands State University, Zimbabwe. His research interests are social, digital and green marketing and brand management. He has published book chapters and research articles in these areas. He lectures courses, including Brand Ethics and Sustainability, Consumer Intelligence, Digital Consumer Intelligence and Neuropreneurship.

Enet Mukurazita is a PhD student in the Organizational, Leadership, Policy, and Development Department at the College of Education and Human Development, University of Minnesota. Her research interests are in African Development, Indigenous Knowledge, Entrepreneurship Training, and Women's education.

Samuel Musungwini (PhD) is a Lecturer at Midlands State University, in the Department of Information and Marketing Sciences. He holds a PhD in Information Systems from the University of South Africa (UNISA), a Post-Graduate Diploma in Tertiary education from Midlands State University, a Bachelor of Science Honours Degree in Information Systems from Midlands State University, a Master Science Degree in Information Systems Management from Midlands State University. He is a founding member of the Zimbabwe Conference on Information Technology (ZCICT). He is a member of the ZIMCHE physical assessment for facilities and resources checklist for academic programs. He is a published author who has written research articles in Scopus indexed journals and non-Scopus indexed journals. He has also attended and presented at a number of international conferences.

Deo Shao (PhD) is currently a Post-Doctoral Research Fellow in the fields of Digital Transformation and Technology Governance at the Business School, University of Johannesburg. He has authored numerous articles in peer-reviewed accredited journals, participated in various workshops and conferences as a keynote speaker and presenter, and has written multiple book chapters. In addition, he is a peer reviewer for several journals. He is also a Senior Lecturer at the University of Dodoma. Deo has also supervised and examined numerous PhDs and Master's degrees to completion.

Leslie Wellington Sirora is an Artificial Intelligence Engineer and Researcher with 11 years of experience. He holds Master's degrees in Software Engineering, as well as Mechatronics and Artificial Intelligence, along with other qualifications in Law, Physics and Cybersecurity. Leslie's research interests lie at the intersection of artificial intelligence and law.

Julius Tapera (PhD) is the Assistant to the Vice-Chancellor at Lupane State University (LSU), Zimbabwe. He holds a PhD in Strategic Management (CUT), a Master's in Business Administration (NUST), and a Bachelor of Commerce Honours Degree in Management (NUST). His research interests are in strategy, leadership, corporate governance, higher education management, entrepreneurship, quality assurance, tourism and hospitality. He lectures Strategic Management in the Master's programmes and supervises some postgraduate research projects. He has published articles in referred journals, books and book chapters.

Sinothando Tshuma is a Lecturer and Program Coordinator in the Department of Marketing Management, Zimbabwe Open University. Her research interests are digital, green, and international marketing, and she has published articles on these topics. Her main responsibility is lecturing courses that include Principles and Practices of Direct Marketing, Digital Marketing, and Industrial Marketing.

Zibanai Zhou is a Lecturer in the Department of Tourism, Hospitality and Leisure Sciences at Midlands State University, Zimbabwe. His main research interests are tourism development, event tourism, and local communities.

Foreword

Agriculture and tourism are important economic sectors for many countries in Africa, employing significant parts of labour and accounting for significant parts of GDP. During the past years agriculture sector's annual growth has maintained a relatively low rate, while tourism has been one of the fastest-growing economic sectors and has been considered a driver of economic development, diversification of economic activities and economy, a catalyst for employment creation at the community level and an engine for sustainable development. Global trends indicate that the tourism market is increasingly becoming more inclined towards alternative forms of tourism and special-interest tourism such as agritourism or ecotourism. In African countries, agritourism can be described as a diverse, innovative, and considerably rapidly advancing industry. The touristic uniqueness of Africa due to the diversity and multitudes of environments and landscapes, with most of these attractions spread in rural areas is perceived as an engine of rural economic development. Other non-economic benefits attributable to tourism include the impact on the culture and goodwill of a people, and to elimination of social bias.

Nevertheless, despite the promising growth of the tourism sector in many developing African countries, macroeconomists argue that the revenue receipts from the tourism sector do not effectively reach the marginalized poor rural citizens, who often constitute a significant proportion of the population in developing countries. The presence of the agritourism industry in a rural economy also stimulates entrepreneurship among the rural population; this drives wealth creation and can tackle poverty, inequality, and social vices. In light of this, any efforts to link the two sectors and collaboration between them are a prerequisite step to alleviate poverty and sustainable rural development.

Readers of this 11-chapter book will benefit from novel interdisciplinary findings, and insights arising from transitional issues that have brought great paradigm shifts in agritourism in Africa, based on a combination of academic and practical insights. This special approach separates itself from other publications that tend to explore these issues independently and selectively, denying the reader a panoramic view of the confluence of these concepts. This book fills a gap in the growing interest in agritourism in African countries, offering a multidisciplinary, multi-focus approach to agritourism and underlying the role it plays in economic and non-economic development, and exploration of rural areas.

Prof. Lucyna Przezbórska-Skobiej,
Department of Economics and Economic Policy in Agribusiness,
Faculty of Economics, Poznań University of Life Sciences, Poland

Foreword

The relationship between tourism development and agriculture in developing countries, particularly in Africa, is complex. It spans from situations of conflict, where tourism development competes for resources, to symbiosis, where tourism enhances local agriculture through food supply and agritourism. This book emphasises the potential for local agricultural development through tourism, which benefits both sectors and fosters symbiosis. The benefits of a closer relationship include decreased linkages through imports which is critical in reducing tourism revenue leakages, improvement in tourism industry food supplies, increased tourist access to local foods and improved environmental and social sustainability for tourism, not least through alleviation of poverty.

Agritourism in Africa highlights the pressing need for discussions on the intersection of tourism development and agriculture in Africa. This edited volume fills critical theoretical gaps in our understanding of agritourism and tourism development on the continent, where authentic experiences are in high demand. Although tourism has long been seen as a pro-poor development strategy, studies examining agriculture-tourism linkages are scarce.

This book addresses the role of agritourism in diversification and community empowerment across different African contexts. It raises critical questions for practitioners and policymakers, showing that agritourism is more than a buzzword; it is a dynamic force connecting travellers to African agriculture's roots while empowering local communities. Case studies from Angola, Tanzania, Zimbabwe, and several other African countries demonstrate its potential to rejuvenate rural economies, preserve culture, and promote entrepreneurship and sustainability.

The insights presented here are essential for policymakers and practitioners to minimise import-related revenue leakage and invest in agritourism initiatives. These measures, such as farm-based accommodation, agricultural festivals, and farm tours, are critical for sustainable tourism in African destinations that seek to diversify their rich natural and man-made attractions. This edited volume contributes significantly to the literature on agritourism, offering fresh perspectives on leveraging African agricultural heritage for tourism development, job creation, and rural community improvement. With its diverse perspectives, it provides a solid theoretical foundation on agritourism and tourism development in developing African countries.

I am confident that this book will serve as a valuable reference for years to come. As you delve into *Agritourism in Africa*, I encourage you to absorb the case studies and insights within these pages. They bear testament to the resilience, creativity, and potential of agritourism in shaping a brighter future for Africa and its communities.

Dr. Erisher Woyo,
Manchester Metropolitan University,
United Kingdom (UK). Extraordinary Research Scientist,
North-West University, South Africa (SA)

Preface

Readers of this edited volume benefit from novel interdisciplinary findings, and insights arising from transitional issues that have brought great paradigm shifts in agritourism in Africa, based on a combination of academic and practical insights. Each chapter provides some theoretical, policy and practical implications for policy makers, educational practitioners, students and business communities on how agritourism in Africa strategies could be of benefit to them. It takes and considers macro, meso and firm-level analyses relating to how agritourism digital and social media marketing could be harnessed to fuel tourism corporate brands development in under-researched African contexts. Its empirical case studies provide scientific evidence-based approaches that inform understanding on the contribution of these antecedents to the realisation of tourism corporate brands development. This publication is a significant addition. The proposed edited volume balances theory and experimentation, providing a thorough explication of the tools and techniques of agritourism relevant to the tourism and agricultural organisations in the African continent and the whole globe at large.

In Chapter 1, Chawarika and Chuchu explore complex interactions between Zimbabwe's agritourism growth, economic factors, and their effects on agricultural trade flows in a number of select countries across the globe. The dynamic interactions among agritourism activities, economic indicators (such as GDP Deflator and Exchange Rates), geographic and trade-related factors (such as Distance, Southern African Development Community (SADC) trade agreement membership, and EPA participation), and agricultural trade volumes are examined. Secondary data from multiple sources were utilised for analysis. Panel data on bilateral agricultural trade flows between Zimbabwe and 11 other countries spanning 18 years from 2001 to 2018 was sourced from the World Integrated Trade Solutions (WITS) database. The chapter makes an attempt to close a gap in the literature by supplying empirical proof of the methods by which agritourism affects trade dynamics, providing a useful tool for academics, practitioners, and policymakers alike. This chapter presents an understanding into the potential role of agritourism as a driver of economic growth and the improvement of agricultural commerce through meticulous analysis of panel data from a wide range of countries. It is hoped that insights presented in this chapter will give rise to the continuing discussion on effective trade policies in a globalised world and sustainable rural development by providing light on these intricate relationships.

Denhere and Deo in Chapter 2 attempted to explore the phenomenon through the lenses of the utilitarian theory in the African context. It explored various etiquettes in which AI could improve the economic, educational, environmental, and cultural aspects of agritourism. Furthermore, the chapter explained ethical considerations such as data management, equitable distribution, and sustainability in the course of integrating AI in the subsector. The ultimate goal was to provide a comprehensive examination of the opportunities and drawbacks of AI in the agritourism sector within the African context based on existing literature. To achieve this objective, a critical literature review strategy was followed. Following the introduction, the chapter covered the following subtopics: the role of AI in diversifying income streams within the agritourism subsector; the potential of AI in knowledge transfer and training in the agritourism subsector; resource optimisation through AI to bolster sustainable farming practices; the cultural implications of AI-enhanced agritourism in preserving indigenous traditions and fostering interactions; and challenges and ethical considerations arising from the integration of AI in agritourism. This chapter is significant because it aimed to explain the relationship between emerging AI technologies and agritourism, a subsector that is critical for many African economies. The outcomes of this research could help stakeholders to strategically leverage the potential of AI and address underlying challenges.

Dzingirai, Dzapasi, and Muziwembiri succinctly explain the critical success factors for agritourism entrepreneurship in Africa in Chapter 3. Their study adopted a structured literature review methodology whereby secondary data was collected from the Scopus database, Google Scholar, and other organizational sources. Six critical success factors for agritourism entrepreneurship, namely, central government support, adult entrepreneurship education, conducive legal framework and robust policy, financial resources, rural infrastructure development, and agritourism entrepreneurship awareness campaigns were established in this study. The results from this study will benefit policymakers, tourism professionals, agricultural practitioners, entrepreneurs, and researchers.

In Chapter 4, Kamugisha tackles economic potentials of precision agritourism in alleviating income poverty among farming communities in Tanzania, through the cottage farming approach. Tanzania's agriculture employs more than two-thirds of the national labour force and contributes over a quarter of the national GDP. The farming communities are characterized by high poverty levels, making it challenging for them to afford basic needs. The poverty is aggravated by the seasonal nature of the incomes earned, price volatility of farm produce and risks involved in the production process. Agritourism has been deployed to some parts of the country in an effort to alleviate poverty by diversifying income sources. Despite these efforts, agritourism practiced in Tanzania depends on rain-fed systems that are prone to climate change vagaries. Precision agritourism is anticipated to offer a promising solution to mitigate the effects of climate change in the agritourism sector. This study utilized a narrative review approach and Roger's diffusion theory to estimate the economic potentials of precision agritourism in reducing income poverty among Tanzanian farming communities. Five search queries namely: Agritourism, Precision agriculture, poverty levels among farming communities, Tanzania and tourism regulatory framework were used to solicit information from Google Scholar, African Journals Online (AJOL), IEEE Xplore, Emerald, Research gate, Taylor & Francis, MDPI, Springer, EBSCO, Elsevier and Routledge search engines within 1997 – 2023-time period. The results suggest that deployment of precision agritourism can be facilitated through farmers' collective action, where a group of individuals invests in precision agritourism, and the resulting profits are distributed among the members.

Further to this, Sirora and Muduva their work in Chapter 5, delves into the potential of artificial intelligence (AI) to facilitate sustainable agritourism in Africa. It begins by providing an overview of the current state of agritourism in Africa, highlighting the challenges it encounters. The concept of sustainable agritourism and its contribution to human development in Africa are then discussed. The chapter further explores the potential of AI in addressing the challenges faced by sustainable agritourism, namely, limited access to information, market inefficiencies, and environmental sustainability. Lastly, the chapter concludes by discussing the key areas for future research. To steer this research endeavour, several key research questions were formulated. Firstly, the study aims to identify specific market inefficiencies within the agritourism industry in Africa and explore how AI can effectively mitigate these inefficiencies. Secondly, the research examines how AI technologies and data-driven approaches can contribute to enhancing environmental sustainability in agritourism practices in Africa. Additionally, the study investigates the challenges associated with limited access to information in the context of agritourism in Africa and explores how AI-based solutions can improve information dissemination and accessibility. Furthermore, the research investigates the potential applications of AI in improving the overall efficiency and effectiveness of agritourism operations in Africa, with a focus on fostering economic growth and human development. Lastly, the study explores how the adoption of AI-enabled sustainable agritourism practices contributes to the empowerment and socio-economic development of local communities in Africa. This research holds significant importance as it explores the potential of AI in promoting sustainable agritourism in Africa. The findings may offer valuable insights and recommendations for policymakers. By addressing market inefficiencies, environmental sustainability concerns, and limited access to information, AI can contribute to sustainable agritourism practices that preserve natural resources and promote responsible tourism.

Chapter 6 by Garwi discusses the realm of sustainable agritourism in the municipality of Caála, Huambo Province, Central Angola, with particular emphasis on its ethical dimensions and the vital role local actors play in advancing both economic growth and environmental conservation. As agritourism gains momentum as a vehicle for achieving economic prosperity and environmental stewardship, a critical examination of the inherent ethical considerations within this industry is paramount, especially within the socio-cultural and environmental context of Huambo. Employing a comprehensive research approach comprising qualitative interviews and a thorough analysis of local policies and practices, this study explores the ethical facets of sustainable agritourism in Caala Municipality. Engaging a diverse sample of 21 participants representing various stakeholder groups, including local farmers, community leaders, tourism operators, government authorities, non-governmental organizations (NGOs), environmental conservation groups, and educational institutions, the study sheds light on the intricate ethical issues entwined with land use, resource management, cultural preservation, and community engagement. The study particularly underscores the complex interplay between the economic aspirations of agritourism entrepreneurs and the imperative to safeguard the region's cultural heritage and delicate ecosystems. Furthermore, this research highlights the pivotal role of local actors, such as farmers, indigenous communities, and government authorities, in shaping the ethical contours of sustainable agritourism. It illuminates their agency in championing responsible tourism practices, ensuring equitable revenue distribution, and conserving biodiversity. The study elucidates the ethical dilemmas faced by these actors, including the challenge of balancing economic gains with environmental conservation and preserving cultural authenticity while meeting the demands of tourists. The investigation into the ethical aspects in this case study provides valuable insights for policymakers, entrepreneurs, and communities engaged in sustainable agritourism in Caala Municipality and other regions with similar aspirations.

Tapera, Hamunakwadi, Mashapure, and Mukurazita in Chapter 7 articulate the wealth and uniqueness of the African tourism and hospitality experience within the Zimbabwean context. A qualitative inquiry is employed to explore the Zimbabwean tourism and hospitality landscape, from the perspective of local and foreign tourists, industry players, regulators and policy makers. This study was based on data collected through qualitative analysis of published academic articles and unpublished documents, including websites, media reports, documentaries and global tourism and hospitality reports by leading global institutions like the World Tourism Authority (WTA). Based on the findings, recommendations are proffered with regard to how the African tourism and hospitality experience can be continuously improved to retain its pole position within the global space, and how sustainability could be achieved from both policy and operational points of view. Related areas of further study are also identified and recommended as part of supporting the continuous growth and development of the African tourism and hospitality industry.

In Chapter 8, Zhou's study offers an alternative narrative by addressing the following research questions: i) What is the state of agritourism development in southern Africa and what are the prospects of the region to become an agritourism destination? ii) What is the spatial distribution of agritourism resources in southern Africa and how has it affected the development of agritourism at regional level? iii) Which are the specific marketing and promotional strategies that have been deployed to increase market awareness about southern Africa's agritourism product? A qualitative research approach was adopted. Unstructured interviews were conducted with tourism stakeholders selected through purposive and snow ball sampling procedures. Findings show that overall the state of agritourism development in southern Africa is uninspiring, despite significant regional variations; however, there is huge potential for further development. Furthermore, findings revealed an uneven spatial distribution of agritourism resources across the region. Additionally, findings unravelled numerous agritourism marketing and promotional strategies which have been initiated at regional level such as agritourism fairs, to market the region's agritourism product to the wider global tourism market. Uncoordinated and exclusive current tourism policy frameworks were identified as major factors which undermine agritourism development in southern African countries. The study is significant as it amplifies an alternative tourism product often overlooked in southern Africa's traditional tourism packages at the global tourism stage. Additionally, the study provides insights on the need to reform regional agritourism policy frameworks, and regular reviews of agritourism marketing and promotional strategies at regional level. The study recommends the establishment of a regional agritourism body aimed at coordinating the marketing of the region's agritourism product.

Mwaijande in Chapter 9 examines how agritourism is conceptualized as a framework for rural development whose intervention integrates agriculture and tourism. This approach is well developed in the Western world, but has received limited attention in the African region. The purpose of Mwaijande's chapter is to put agritourism in rural development policy space as an agenda for serious consideration that has potential for rural transformation and poverty reduction. Secondary data were collected from review of documents, interview and bibliometric analysis for identifying the forward-backward benefits, challenges and the scientific publications in Africa to inform agritourism policy decision making. We found that agritourism is an emerging phenomenon in Africa, but is gaining importance in rural development because it contributes to farm diversified economy in countries such as South Africa, Kenya, Nigeria, and Tunisia. In addition, agritourism create touristic jobs on farms. On the other hand, the tourism sector benefits from agritourism by having new tourist destinations and attractions. The main findings indicate that agritourism can bring economic benefits both at the farm and national level. However, developing agritourism is constrained by inadequate entrepreneurial education, rural infrastructure and government enabling environment. The chapter sets the agritourism policy agenda for the rural transformation.

Tshuma, Dube, and Dangaiso in Chapter 10 identify the digital and social media channels that are available for marketing agritourism products in Africa, analysing the challenges being faced by marketers in agritourism digital and social media marketing and finally proffering solutions to the highlighted challenges. The research will be literature-based through the use of recent relevant journals in gathering literature based on the objectives. The study's framework is based on key elements from Zvavahera and Chigora (2023) needed to transform the agritourism sector in Zimbabwe. The methodology will also include the use of secondary data to validate the findings. The need by economies to meet their development goals has spearheaded the rise in the recognition of the long-forgotten agritourism sector as one of the anchors of economic development. This is based on the validation that the development of the sector can curb rural-to-urban migration through employment creation and increasing the flow of money to the rural sphere and in turn increasing the Gross Domestic Product (GDP) of a nation. The need by many farmers to be sustainable has also led to the growth of the sector as many farmers seek the need to increase profit margins and gain relevance to acquire state funding. It is also argued that agritourism requires little investment yet high long-term returns, which creates a gate pass of business growth for many farmers. The recognition of the sector by farmers as a lucrative investment is likely going to intensify competition which can only be curbed through effective digital and social media marketing.

Lastly, in Chapter 11, Samuel Musungwini, Edward Mudzimba, Mercy Dube, and Sinothando Tshuma explored how disruptive technological innovation can positively impact agritourism in developing countries, with views from Zimbabwe. This chapter aimed to analyse the literature on tourism and opportunities presented by ICT usage with more thrust on emerging technologies. The chapter used a qualitative approach to data collection and analysis by reviewing available literature to inform our understanding. Empirical data was collected from 12 key informants to situate the research in a developing context. The chapter established that the Zimbabwean tourism sector can benefit immensely from the adoption and increased usage of disruptive technologies if the prevailing challenges are addressed.

By Editors
Brighton Nyagadza
York St John University, London Campus,
United Kingdom (UK)
Farai Chigora
College of Business, Peace Leadership and Governance,
Africa University (AU), Zimbabwe
Azizul Hassan
Tourism Consultants Network of the Tourism Society,
United Kingdom (UK)

Abbreviations

ACT	Agricultural Council of Tanzania
AfCFTA	Africa Continental Free Trade Area
AI	Artificial Intelligence
ASEAN	Association of Southeast Asian Nations
ASLMs	Agricultural Sector Development Lead Ministries
AWF	African Wildlife Foundation
COMESA	Common Market of Eastern and Southern Africa
COVID-19	Coronavirus Disease of 2019
CWMAC	Community Wildlife Management Areas Consortium
EAC	East African Community
FZS	Frankfurt Zoological Society
GDP	Gross Domestic Product
HAT	Hotel Association of Tanzania
HAZ	Hospitality Association of Zimbabwe
KT	Knowledge Transfer
LATF	Lusaka Agreement Task Force
LOCALCOMM	Local Community
ML	Machine Learning
MNRT	Ministry of Natural Resources and Tourism
MNTDIV	Ministry of Natural Resources Tourism Division
NCAA	Ngorongoro Conservation Area Authority
NCAA	Ngorongoro Conservation Areas Authority
NEMC	National Environment Management Council
PARKS	Parks Congress
PO-RALG	President's Office, Regional Administration and Local Governments
RBT	Resource Based Theory
RTTZ	Responsible Tourism United Republic of Tanzania
SADC	Southern African Development Community
SDGs	Sustainable Development Goals
TACTO	Tanzania Association of Cultural Tourism Organizers
TAHA	Tanzania Horticultural Association

TAHOA	Tanzania Hunting Operators Association
TANAPA	Tanzania National Parks Authority
TANAPA	Tanzania National Parks
TATO	Tanzania Association of Tourism Operators
TAWIRI	Tanzania Wild Life Research Institute
TCT	Tourism Confederation of the United Republic of Tanzania
TFTA	Tripartite Free Trade Area
TIC	Tanzania Investment Center
TPDF	Tanzania Peoples Defense Force
TTB	Tanzania Tourist Board
TZPOLICE	Tanzania Police Force
UNCTAD	United Nations Conference on Trade and Development
UNDP	United Nations Development Programme
UNEP	United Nations Environmental Programme
UNESCO	United Nations Educational, Scientific and Cultural Organization
UNWTO	United Nations World Tourism Organisation
USDA	United States Department of Agriculture
WITS	World Integrated Trade Solutions
WTO	World Tourism Organization
WTTC	World Travel and Tourism Council
WWF	World Wildlife Fund
ZCT	Zimbabwe Council of Tourism
ZITF	Zimbabwe International Trade Fair
ZTA	Zimbabwe Tourism Authority
ZTDC	Zimbabwe Tourism Development Corporation

1 Exploring the Nexus between Agritourism Development, Economic Factors, and Agricultural Trade Flows

A Case of Zimbabwe's Agritourism Sector

Admire Chawarika and Tinashe Chuchu

Introduction

Agritourism has been defined as a form of tourism involving farms used for leisure, recreation or educational activities (Ammirato et al., 2020; Roman and Grudzień, 2021; Santeramo and Barbieri, 2017). Agritourism has been studied in varying contexts and it can be debated, however, research has yet to present a conclusive comprehension of the facets that underpin and define agritourism (Phillip et al., 2010). Agritourism initiatives are being viewed as an important diversification strategy for agricultural entrepreneurs which has the potential to contribute to the development of the rural economy (Canovi and Lyon, 2020). Agritourism is a diverse, innovative, and considerably rapidly advancing industry (DeLay et al., 2019). The following section discusses and justifies the theoretical underpinning which formed the basis for this research. Agritourism generates an array of economic incentives, which include increased income and marketing possibilities as well as economic incentives such as the betterment of farmers' livelihoods (LaPan and Barbieri, 2014; McGehee and Kim, 2004; Schilling et al., 2012). The following section explores the theoretical development that underpins this study.

Theoretical Development

This research was grounded in five frameworks: the gravity model of trade, Ricardo's competitive advantage theory, the Heckscher-Ohlin theory, the Marshall Lerner condition and the J-Curve. Detailed descriptions and their application in the research in question are presented below.

Gravity Model of Trade

The gravity model of trade is an established framework for analysing international economics and trade patterns of countries since its conception by Tinbergen (1962) (Mohmand et al., 2015). The basic premise of the gravity model of trade is that the volume of trade between two countries (or entities) is proportional to the product of their sizes (often GDP) and inversely

DOI: 10.4324/9781032696188-1

proportional to the distance between them. The gravity model of trade is viewed as a rigorous approach to partial analysis of international trade flows, which has become established as the principal tool for measuring the effect of trade policies at the cumulative, sectoral and product levels (Yotov, 2022). This model was deemed appropriate within the context of this research and was therefore adapted as illustrated in the methodology section. Effective multilateral trade agreements depend on having regional trade agreements and a multilateral trading framework relies on these regional trade agreements (Chawarika et al., 2022). According to Piermartini and Teh (2005), the gravity model provides the theoretical foundation for the consequences of regional integration. Therefore, theoretical justifications ought to align with these models. Ricardo's theory provides the theoretical foundation for using the gravity model to assess how regional integration affects trade flows. Costinot and Donaldson (2012) state that trade can benefit all nations if they focus on areas where they have a comparative advantage. This is in line with Ricardo's idea. This is a significant problem for the analysis of agricultural trade flows in Zimbabwe. By investing in economic activities, particularly agricultural production in industries where they have a comparative advantage, countries in the SADC-FTA and EPA can gain. This comparative advantage relates to low production costs brought about by relative pricing differences, with technology playing a significant role. It's crucial to remember, though, that the Ricardo theory's ability to explain trade flows is seriously flawed. These include the presumptions of constant costs, the absence of demand, and the static character of the model, which ignores the growth's dynamic effect.

The Heckscher-Ohlin Theory

The Heckscher-Ohlin theory addresses some of the previously mentioned shortcomings of the Ricardo hypothesis. According to Bajona and Kehoe (2010), the Heckscher-Ohlin theory provides significant insights into the characteristics of trade flows that nations encounter, specifically in the context of Zimbabwe's agricultural trade flows. According to the Heckscher-Ohlin theory, nations will export commodities made with highly abundant but intensively used components. In this regard, varying factor endowments become crucial. In light of this, it is anticipated that nations with strong agricultural production endowments will export a greater volume of their products. Relative factor endowment is a key component of trade flows that the theory highlights. This can aid in the explanation of certain patterns and characteristics of the agricultural trade flows in nations with high relative factor endowments, such as labour and land. One major shortcoming of the Heckscher-Ohlin theory was pointed out by Leontief. Leontief claims that a nation with a high relative capital abundance can export more items requiring a significant amount of labour. As a result, the Leontief paradox occurs when it is not always possible to utilise the variable factor endowments to explain the excessive trade flows between countries.

According to Krugman, low transportation costs are a major factor in the economic benefits of regional integration between neighbouring countries. In light of this, trading among close regions plays a crucial role in shaping trade flows. Therefore, it is anticipated that neighbouring SADC-FTA and EPA nations will have a bigger influence on Zimbabwe's agricultural trade flows. Nonetheless, it is crucial to remember that owing to economic size (GDP), a country can be far away from another and yet trade more with it. This demonstrates how different variables, such as GDP, distance, and other economic factors, have varying effects on trade flows. Krugman offers a significant perspective on scale economics, which directs trade between nations and private enterprises. Thus, in order to have a competitive advantage in a regional trade agreement, countries need to become more cost-efficient. According to Bergstrand (1989), Krugman pointed out that the gravity model is predicated on monopolistic rivalry. Products that set each nation apart were identified as the "push factors" that allowed for international trade.

The Marshall Lerner Condition and the J-Curve

The Marshall Lerner condition and the J-Curve are important theories in elucidating the connection between trade balance and exchange rates, claim Auboin and Ruta (2013). According to the Marshall Lerner condition, real exchange rate deprivation improves trade balance when the total of import and export price elasticity is more than one. Therefore, quantity exported will increase more than a comparable fall in export price of the same items if the price of exported goods is price elastic. This provides a crucial summary of how price elasticity explains the value of exports for a range of different items. But it's important to remember that the Marshall Lerner condition is based on a static analysis. The J-Curve is a crucial illustration of the nature of exchange rate volatility. The dynamic approach to the Marshall Lerner condition on the link between trade balance and exchange rate is highlighted by the J-Curve.

According to Berthou (2008), the J-curve illustrates a nation's trade balance trend in reaction to exchange rate deprivation. The short-term decline in the trade balance is linked to the real exchange rate depreciation. The main cause is that, according to contractual agreements, the majority of imports and exports are scheduled several months in advance. Over time, the country's export volumes will climb in response to declining export prices, creating trade surpluses. Due to consumers looking for alternatives realising the adjustment time period, transaction volume will not change after real exchange rate depreciation. Over an extended period, the depreciation of the real exchange rate might yield positive effects on the current account balance by making exports more affordable than imports (Auboin and Ruta, 2013). In this regard, we anticipate that Zimbabwe's exchange rates will have a significant long-term impact on agricultural trade flows. This is due to the J-Curve theory's explanation of the long-term impacts of exchange rates.

The movement of exchange rates is also highlighted in a significant way by the purchasing power parity theory. According to the hypothesis, prices for the same items in various markets will be the same (Prodan, 2008). Therefore, shifts in the relative costs of products and services in various nations will cause the exchange rate to fluctuate. As a result, the theory provides a crucial summary of how actual exchange rates move. The absolute and relative buying power parities are the two components of purchasing power parity theory. While the relative power parity idea accounts for inflation rates, absolute purchasing power parity aims to level the playing fields in terms of price levels between nations. The notion of relative power parity is essential to understanding the fluctuations in Zimbabwe's exchange rate, especially during the period of hyper-inflation. Demand-side variables, in Linder's opinion, are crucial in influencing international trade. The Heckscher-Ohlin and other supply-side theoretical frameworks have disregarded the demand component. In light of this context, Linder suggests that international trade is determined by demand. As a result, we would anticipate that nations with larger populations and per capita incomes would have higher demand for goods. In this regard, trade flows would emphasise strong trade figures between more demand-driven nations. The literature development informed by prior research in agritourism is presented in the section that follows.

Literature Development

Prior research on agritourism explored looked at agritourism from a sustainability perspective (Ammirato et al., 2020; Barbieri et al., 2019; Canovi and Lyon, 2020), while others investigated its importance and resulting profit potential during the COVID-19 global pandemic (Roman and Grudzień, 2021). Agritourism, which involves inviting the public onto a farm, ranch, or other agricultural setting to participate in various activities and experiences, offers a range of opportunities and challenges.

Agritourism Opportunities

Diversified Income

Agritourism can provide an additional revenue stream for farmers, helping to stabilise income, especially during off-seasons or downturns in commodity prices. Due to the constant decline in farm revenues, farmers have sought to make use of alternative revenue-generating channels to boost their earnings, usually through agritourism (LaPan and Barbieri, 2014). Agritourism presents an important diversified production and earning model which has generated income for over 30 years (Schmitt, 2010). Small-scale farmers who are primarily producers of beef stand to benefit from diversification into agritourism as a risk mitigation strategy (Pitrova et al., 2020).

Education

Agritourism has been hailed as an important educational tool by Petroman et al. (2016). Agritourism provides a platform for farmers to educate the public about farming practices, the source of their food, and the importance of agriculture. Furthermore, agritourism is a form of educational tourism that is viewed as a strategic option to mass tourism in the rural areas where it cannot be conducted due to the negative impact on the human communities and the environment (Petroman et al., 2016).

Cultural Preservation and Rural Development

It can help in preserving rural traditions, lifestyles, and crafts that might otherwise be lost in the face of modernisation. Barbieri et al. (2019) posit that agriculture tourism contributes to the preservation of natural resources through sustainable farming engagement and thus should be encouraged. Through attracting tourists, agritourism can boost local economies, create employment opportunities, and revitalise rural communities (Bwana et al., 2015). Farmers have managed to preserve their farmlands' heritage through agritourism initiatives and the agritourists were mainly motivated by intrinsic motives to preserve the nation's rural heritage (LaPan and Barbieri, 2014).

Agritourism Challenges

Regulation and Skill Requirements

Running a successful agritourism operation requires a mix of farming and business skills, including marketing, customer service, and event management. Farmers need to navigate regulatory environments, including health and safety standards. Liability issues, such as a visitor getting injured on the property, can be a concern. DeLay, Chouinard and Wandschneider (2019) concluded in their study that state regulations and zoning laws are significant barriers to the expansion of agritourism in the United States.

Cost and Seasonality

Setting up an agritourism operation might require significant investment in infrastructure and staff training continuously development. Just as agriculture can be seasonal, agritourism attractions may also experience fluctuations, with certain times of the year being more popular than others. Agritourism is expensive to operate (DeLay et al., 2019), and the challenge brought about by seasonality could be mitigated through charging different prices for products depending on the stage of the season at which tourists visit (Matei, 2015).

Agricultural Trade Flows

Agricultural trade is an economic exchange process which involves the production, importation and exportation of food and other various resources across nations (Hoekman and Nelson, 2020). The desire to advance intra-Africa trade is a theme that dominates discussions in many settings across numerous public and private sector institutions, and plays a key role in various continental and regional-level economic agreements in Africa (Chibira and Moyana, 2017). The present study adopted the gravity model as one of its frameworks to analyse agriculture trade flows which has been used in Asia to analyse total and agriculture trade flows between countries which include Indonesia, Malaysia, The Philippines and Thailand (Zhou and Tong, 2022). In addition, the gravity model approach in analysis of trade flows has also been utilised by Yang and Martinez-Zarzoso (2014) to investigate the relationship between the Association of Southeast Asian Nations (ASEAN) and China's trade treaties (Uwakata and Aregbeshola, 2023). Stemming from an African perspective, as far as agricultural trade is concerned, South Africa plays a key role in the continent's trade activities leading to the nation's ability to sustain and increase its agricultural exports globally (International Trade Center, 2021; Seti, 2023). The research in question analyses the agriculture trade flow between Zimbabwe and 11 other nations, namely: South Africa, Kenya, Zambia, Egypt, Mozambique, China, the USA, the Netherlands, Belgium, the UK, and Botswana. The selection criteria of the abovementioned nations were based on the availability of trade data on agricultural trade flows between Zimbabwe and the nations. For instance, South Africa is central to Zimbabwe's agriculture sector as it accounts for 77% of all agriculture exports from Zimbabwe and 56% of agriculture imports into Zimbabwe with only 2% from US agricultural trade (United States Department of Agriculture (USDA), 2015). The other African nations, namely Botswana, Zambia, Mozambique, and Kenya are part of the Tripartite Free Trade Area (TFTA) (Qoto, 2018). The TFTA is a regional economic agreement which makes efforts towards encouraging intra-regional trade between SADC, the Common Market of Eastern and Southern Africa (COMESA) and the East African Community (EAC) (Qoto, 2018).

The selected nations might span a range of geographical areas, enabling a thorough examination of how trade agreements affect global agricultural trade flows. This regional diversity guarantees a more comprehensive and diverse dataset, enhancing the study's external validity. It's possible that the goal of including nations like Zimbabwe, South Africa, Kenya, Zambia, Egypt, Mozambique, and Botswana is to offer a thorough picture of the Southern African continent. This makes it possible to investigate the ways in which agricultural trade flows are influenced by regional dynamics, such as trade agreements and economic alliances.

Furthermore, the selected nations may participate in trade agreements, such as the SADC-FTA and EPA, to varying degrees. An in-depth examination of how the length and conditions of agreements affect trade flows is made possible by evaluating the effects on agricultural commerce with nations at various phases of participation in these accords. Some examples of agritourism practices in Africa are discussed in the section that follows.

Examples of Agritourism in Africa

Wine Tourism in Stellenbosch, South Africa

Context: The well-known South African wine region Stellenbosch has effectively incorporated agritourism into its wine business.

Agritourism Activities: Wine tastings, vineyard tours, and grape harvesting are available to visitors.

Economic Impact: By drawing foreign visitors to the region's distinctive blend of viticulture and agriculture, agritourism in Stellenbosch has helped the local economy flourish.

South Africa (Stellenbosch): Situated in the Western Cape, Stellenbosch is a premier example of agritourism perfection. Primarily concentrated on the well-known wine business, the area takes advantage of its rich soil and pleasant weather to provide a variety of wine-related activities. International tourists are drawn to the area by wine tasting events, vineyard tours, and grape harvesting activities, all of which have a major positive impact on the local economy. The agritourism model of Stellenbosch combines tourism with the rich viticulture traditions of South Africa, offering guests an educational tour of the wine-making process. However, there might be issues with how the wine business affects the environment, which would need sustainable viticulture methods. However, the region's success comes from turning its agricultural strength into a valuable cultural and commercial asset that illustrates the complex inter-relationship between agriculture and tourism.

Ghana: Kumasi's Cocoa Farm Experiences

Background: To provide insights into the cocoa industry. Kumasi, a major cocoa-producing region in Ghana, has integrated agritourism.

Agritourism Activities: Activities related to agritourism include visiting cocoa plantations, discovering how chocolate is made, and interacting with regional farmers.

Impact on Education: This example shows how agritourism can help tourists learn about the challenges involved in farming and develop a respect for locally produced goods.

Ghana (Kumasi): Situated in a well-known cocoa-producing area, Kumasi is a shining example of agritourism that emphasises the cocoa sector. Agritourism activities in the area include visiting cocoa plantations, creating chocolate, and interacting with nearby farmers. Ghana's economy benefits greatly from Kumasi's support of regional farmers and provision of educational opportunities for tourists. The Kumasi agritourism approach educates guests about the subtleties of cocoa growing and its cultural significance while honouring Ghana's rich cocoa legacy. While maintaining sustainable cocoa farming methods may provide difficulties, the region's prospects rest in showcasing Ghana's cocoa tradition internationally. In the centre of West Africa, agriculture, culture, and economic development are inextricably linked, as seen by Kumasi's agritourism initiatives. The research methodology adopted for the study is presented below.

Research Methodology

A mechanism for implementing the set goals is developed in this section. As a result, a methodological framework is outlined, showing how population, exchange rates, economic scale, and agritourism agreements affect commerce. A significant amount of focus is placed on the data sources, the analytical methods used, and the empirical model formulation. In order to quantify the influence of the dummy variables linked with agritourism agreements, we will also describe how to transform them and apply the gravity model with panel data in the analysis.

Research Context

In as much as the research initially discussed agritourism from a broader sense, this is an attempt to provide some background to the topic under investigation. This research is specifically concerned with investigating agritourism in South Africa, Kenya, Zambia, Egypt, Mozambique, China, the USA, the Netherlands, Belgium, the UK, and Botswana and thus agritourism data associated with these regions was retrieved.

Data Sources and Description

The analysis carried out by the research is based on secondary data from multiple sources. The World Integrated Trade Solutions (WITS) database provided panel data for the 18-year period (2001–2018) trade between Zimbabwe and 11 other nations. The following nations: South Africa, Kenya, Zambia, Egypt, Mozambique, China, the USA, the Netherlands, Belgium, the UK, and Botswana are included in this research as separate agritourism destinations. The World Development Indicators database provided the GDP and population estimates, while the Center for Prospective Studies and International Information (CEPII) provided information on the distances between large cities. The Reserve Bank of Zimbabwe and the IMF international financial statistics database were the sources of exchange rate data. Independent variables pertaining to agritourism and regional trade agreements were included in the study. The gravity model was used for the analysis, and STATA was the main statistical programme used.

Gravity Model Specification

The gravity model to determine the impact of agritourism agreements and other independent variables on trade, is specified as highlighted:

$$\ln T_{ijt} = \beta_0 + \beta_1 \ln Agritourism_{ij} + \beta_2 \ln Y_{it} + \beta_2 \ln Y_{jt} + \beta_3 \ln N_{it}$$
$$+ \beta_4 \ln N_{jt} + \beta_5 \ln D_{ij} + \beta_6 Ex_{ijt} + \varepsilon_{ijt} \tag{1.1}$$

where

i: is Zimbabwe country

j: is partner countries 1, ... , n

T_{ijt}: Trade with country j in time t (USD\$ monetary term)

$Agritourism_{ij}$: Dummy variable for Agri-tourism agreement (0 = No agreement and 1 = Presence of agreement)

Y_{it}: Zimbabwe's GDP in time t

Y_{jt}: GDP of partner j in time t

Ex_{ijt}: Ehange rate between Zimbabwe and country j in time t

N_{it}: Population of country i (Zimbabwe) in time t

N_{ij}: Population of partner country j in year t

D_{ij}: Distance between capital of Zimbabwe and country j

ε_{ijt}: Error term

Variable Description and Measurement

The method for measuring the independent and dependent variables is described in this section.

Dependent Variable

The dependent variable, which represents the bilateral trade activity between Zimbabwe and its various trading partners, serves as the main focus of our analysis. To measure this dependent variable, the total value of imports and exports expressed in US dollars (USD\$) will be the main indicator. We have omitted forestry and fisheries items from our research in compliance with the World Trade Organization's Harmonised System because they fall into different categories. Zimbabwe's trade destinations are represented by the list of trading partners that we analysed in our analysis, which includes South Africa, Kenya, Zambia, Egypt, Mozambique, China, the USA, the Netherlands, Belgium, the UK, and Botswana. Hence, the dependent variable will measure whether an agritourism agreement has an impact on the quantum of trade between Zimbabwe and its trading partners.

Independent Variables

• **Agritourism agreements**

The agritourism trade agreement dummy variable is used in the examination of how regional trade agreements affect Zimbabwe's trade flows. The dummy variable only indicates whether the agritourism agreement has a positive influence on Zimbabwe's trade flows.

- *Exchange Rates*
  Exchange rate swings are a major source of information about the effects of monetary policy on Zimbabwe's agritourism. Nominal exchange rates are used as an explanatory variable in this study. Our analysis of the impact of exchange rates on trade flows is based on the Marshall Lerner condition and previous empirical research. Nominal exchange rates are frequently examined because they provide information about how different macro-economic indicators within an economy are performing. As a result, in the economic framework, exchange rates take on a crucial function as independent variables.
- *Gross Domestic Product*
  The Gross Domestic Product (GDP) will be used as the gravity model's indicator of economic size. GDP is significant because it is a key indicator of how a nation's economic success affects the movements of tourists across its borders. We shall utilise Zimbabwe's and its trading partners' nominal GDP values, expressed in US dollars, for this research.
- *Distance*
  Distance is a proxy for transaction ease and transportation expenses. The variable's measurement will be the length of time, in kilometres, that separates Zimbabwe's capital city (Harare) from its partner nations.
- *Population*
  Population is a separate variable that was extracted from the original Newtonian gravity model and is used as a gauge of a nation's market size. The idea behind this is that countries with higher population sizes typically have larger marketplaces, which raises the demand for agritourism products and services. As a result, one would expect the population variable to have a positive coefficient. It is crucial to remember that other factors have a greater overall impact on trade between nations; therefore, having a larger population does not always translate into higher trade flows (Table 1.1).

Table 1.1 Study results

Independent variables	Coefficients	t-Ratio	P-Value
Agritourism Agreement	0.5750*	3.23	0.001
Zimbabwe Gross Domestic Product	0.02868	0.06	0.951
Trading Partner Gross Domestic Product	2.8001**	2.33	0.020
Zimbabwe Population	−0.1602	−1.01	0.315
Trading Partner Population	−0.8867**	−2.17	0.030
Distance	−0.0202	−0.31	0.757
Exchange Rates	0.2196	0.48	0.631
R^2 Within = 40.15			

Source: Synthesised by the authors.

*Significant at 1% (*), significant at 5% (**).

Results and Discussion

Agritourism Agreements

The estimated coefficient for agritourism trade agreement from the empirical results in the table is 0.5750 significant at 1%. It is important to emphasise that being a part of agritourism agreements has a favourable influence on Zimbabwe's trade activities. Enhancements in trade facilitation, encompassing improvements in infrastructure and the efficiency of customs clearance, further contribute to the beneficial effects of agritourism agreements on Zimbabwe's trade activities. These elements collectively exert a positive influence on Zimbabwe's trade flows. Hence, agritourism agreements are pivotal to Zimbabwe's trade flows. Agreements pertaining to agritourism sometimes entail selling and promoting a nation's agricultural goods to visitors. The demand for these goods may rise as a result of the exposure, increasing their export to the countries of origin of the travellers. Agritourism contracts promote export market diversity. A nation's agricultural sector can increase international trade by drawing visitors from other nations, which can broaden its consumer base and lessen its dependence on a single market.

Trading Partner Gross Domestic Product

The coefficient associated with the Gross Domestic Product of the trading partner is noteworthy, as it holds statistical significance at the 5% level. This coefficient can be directly interpreted as an elasticity due to GDP's continuous nature. Thus, a 1% rise in the trading partner's GDP is expected to result in a 2.8001% increase in trade flows. This is attributed to the trade creation in agritourism as a result of the increase in the Gross Domestic Product of Zimbabwe's trading partners. Hence, the economic performance of Zimbabwe's trading partners is pivotal to the growth of trade between the countries. GDP growth is usually a sign of a more robust and prosperous economy in a given nation. Customers and businesses are more likely to purchase products and services, particularly those from outside markets, as a result of increasing incomes and greater spending power. This may increase the desire for imported goods and boost global trade.

Trading Partner Population

The coefficient of the trading partner population is −0.8867 and significant at 5%. Therefore, as the trading partner's population increases, there is a decrease in trade between Zimbabwe and its trading partners. A country's resources may be strained by a growing population, making it more difficult to generate surplus items for export. In these situations, it's possible that a sizable percentage of the resources will be used for domestic purposes rather than trade. Furthermore, Trade patterns can be impacted by government policies, such as currency controls or protectionist measures. Countries with large populations may adopt policies that prioritise trade above self-sufficiency.

Implications

Gravity Model of Trade

The application of the Gravity Model of Trade within the context of the research implies that the observed impacts of Zimbabwe's membership in EPA and SADC-FTA may be influenced by the economic size and distance between trading partners. According to the Gravity Model, two important factors influencing bilateral trade flows are the partner countries' economic size and geographic proximity. Analysing how these variables match the results seen in Zimbabwe's agricultural trade may shed light on the Gravity Model's wider applicability and the degree to which its predictions agree with the actual data.

Marshall Lerner Condition and J-Curve

A consideration of the Marshall Lerner condition and the J-Curve is implied by the research investigation of exchange rates and their detrimental effects on agricultural trade flows. The negative effects of exchange rate changes on trade flows, as found in the empirical findings, should stimulate an evaluation of whether currency depreciation, as expected by the Marshall Lerner condition, leads to an anticipated improvement in the trade balance over the long term. Furthermore, Zimbabwe's agricultural commerce may benefit from the application of the J-Curve idea to explain short-term obstacles and possible long-term advantages.

Heckscher-Ohlin Theory

Beyond the Heckscher-Ohlin theory, the theoretical implications highlight the importance of this theory in comprehending the relationship among agritourism development, economic considerations, and agricultural trade flows. The application of the Heckscher-Ohlin theory implies that the comparative advantage achieved through factor endowments, such as labour and capital, plays a vital role in determining Zimbabwe's trade dynamics under EPA and SADC-FTA. The hypothesis sheds insight on the underlying mechanisms causing positive effects and helps to explain why particular agricultural goods and industries see higher trade flows.

Contributions to Agritourism in Africa

The empirical findings in this chapter provide insightful information about how Zimbabwe's trade activities are affected by agritourism agreements, trading partner GDP, and trading partner population. These findings shed light on variables that can form and impact the agritourism environment in the area, which has important ramifications for the larger African setting.

First off, the trade agreements related to agritourism have a positive and statistically significant coefficient, which highlights their critical function in augmenting trade flows in Zimbabwe. This has wider ramifications for the continent of Africa since it implies that signing up for agritourism agreements may play a significant role in boosting global trade. By entering into such partnerships, African countries can use the exposure to foreign tourists as a tool to market and sell their agricultural products. This lessens reliance on a single market by diversifying export markets in addition to increasing demand. This study supports the strategic significance of encouraging and engaging in agritourism agreements as a way to increase trade in general and the agriculture sector in particular in African nations.

Secondly, the significant influence that trading partners' GDPs have on trade flows is directly related to the economic conditions in Africa. The correlation between the GDP growth of trading partners and the rise in trade flows indicates that there is room for additional trade activity when the trading partners of African nations enjoy economic development. This is significant because it highlights how important it is for African nations to take trading partners' economic health into account when navigating international trade ties. According to the findings, African countries should strategically align with economically prospering trading partners in order to capitalise on the positive link between greater commerce and wealth in order to pursue economic expansion through agritourism.

But the population of trading partners coefficient adds a more complex viewpoint. Trade flows and trading partner population have a negative connection, which implies that trade between Zimbabwe and its trading partners declines as trading partner populations rise. This research suggests possible difficulties in producing excess goods for export, which should worry populous African countries. It also emphasises how government policies affect trade patterns, showing that populous nations could put commerce secondary to self-sufficiency. This knowledge could be used by African policymakers to create policies that strike a compromise between goals for foreign trade and home needs, particularly in the context of agritourism.

To sum up, the empirical findings presented in this chapter make a substantial contribution to our understanding of the effects of trade agreements, GDP, and population of trading partners on Zimbabwe's trade operations. The results offer significant perspectives for African countries, directing tactical choices in cultivating global trade connections, advancing economic expansion via agritourism, and manoeuvring through the intricacies of population-associated obstacles in commerce.

Contribution to Knowledge

The research expands the theoretical framework and offers a more thorough explanation of the variables impacting Zimbabwe's agricultural trade flows by combining insights from the Gravity Model of Trade, Marshall Lerner condition, and J-Curve. This extended theoretical framework highlights the applicability of many economic theories in understanding the dynamics of trade in the real world and adds to the continuing discussion on trade agreements and their consequences. By focusing on the Heckscher-Ohlin theory and integrating various theoretical viewpoints to thoroughly analyse the intricate interactions between economic agreements, agricultural trade, and economic development in Zimbabwe and, by extension, Africa, the study expands our understanding of the subject.

Study Limitations

There might be issues with the study's ability to prove a link between trade agreements and agricultural trade flows. To support causal inference, more sophisticated econometric techniques or quasi-experimental designs should be used in future studies. Furthermore, taking into account extraneous elements like political stability, technical improvements, and climate change might offer a more thorough comprehension of the complex effects on agricultural trade. The generalisability of the findings may be a constraint due to the focus on Zimbabwe. Subsequent research endeavours ought to evaluate the suitability of the findings for other nations possessing analogous trade agreements and economic environments. To do this, cross-national comparisons must be made in order to determine the variables that affect how differently trade agreements affect agricultural trade in various geographical areas.

References

Ammirato, S., Felicetti, A. M., Raso, C., Pansera, B. A., and Violi, A. (2020). Agritourism and sustainability: What we can learn from a systematic literature review. *Sustainability*, 12(22), 1–18.

Auboin, M., and Ruta, M. (2013). *The Relationship Between Exchange Rates and International Trade: A Review of Economic Literature*. WTO Staff Working Paper, No. ERSD-2011-17, World Trade Organization (WTO), Geneva, 10.30875/13e835 62-en

Bajona, C., and Kehoe, T. J. (2010). Trade, growth, and convergence in a dynamic Heckscher–Ohlin model. *Review of Economic Dynamics*, 13(3), 487–513.

Barbieri, C., Sotomayor, S., and Aguilar, F. X. (2019). Perceived benefits of agricultural lands offering agritourism. *Tourism Planning & Development*, 16(1), 43–60.

Bergstrand, J. H. (1989). The generalized gravity equation, monopolistic competition, and the factor-proportions theory in international trade. *The Review of Economics and Statistics*, 143–153.

Berthou, A. (2008). An investigation on the effect of real exchange rate movements on OECD bilateral exports. Retrieved May 6, 2019, from http://papers.ssrn.com/sol3/papers.cfm?abstract_id=1157776

Bwana, M. A., Olima, W. H., Andika, D., Agong, S. G., and Hayombe, P. (2015). Agritourism: Potential socio-economic impacts in Kisumu County. *Journal of Humanities and Social Science*, 20(3), 78–88.

Canovi, M., and Lyon, A. (2020). Family-centred motivations for agritourism diversification: The case of the Langhe Region, *Italy. Tourism Planning & Development*, 17(6), 591–610.

Costinot, A., and Donaldson, D. (2012). Ricardo's theory of comparative advantage: Old idea, new evidence. *American Economic Review*, 102(3), 453–458.

Chawarika, A., Madzokere, F., and Murimbika, A. (2022). Regional trade agreements and agricultural trade: An analysis of Zimbabwe's agricultural trade flows. *Cogent Economics & Finance*, 10(1), 2048482.

Chibira, E., and Moyana, H. (2017). Enhancing intra-Africa trade: The need to go beyond hard infrastructure investment, border management reform and customs processes enhancement. The 36th Southern African Transport Conference, CSIR International Convention Centre, Pretoria, South Africa on 10–13 July 2017.

DeLay, N. D., Chouinard, H. H., and Wandschneider, P. R. (2019). Categorizing agritourism operations and identifying barriers to success. *Journal of Agribusiness*, 37(1), 1–22.

Hoekman, B., and Nelson, D. (2020). Rethinking international subsidy rules. *The World Economy*, 43(12), 3104–3132.

International Trade Center (2021). *Trade Statistics for International Business Development*. Retrieved January 15, 2024, from http://www.trademap.org/ Country_SelProductCountry_TS.aspx.

LaPan, C., and Barbieri, C. (2014). The role of agritourism in heritage preservation. *Current Issues in Tourism*, 17(8), 666–673.

Matei, F. D. (2015). Study on the evolution of seasonality in agritourism by regions of Romania. *Romanian Economic Journal*, 18(55), 149–162.

McGehee, N., and Kim, K. (2004). Motivation for agri-tourism entrepreneurship. *Journal of Travel Research*, 43(2), 161–170.

Mohmand, Y. T., Salman, A., Mughal, K. S., Imran, M., and Makarevic, N. (2015). Export potentials of Pakistan: Evidence from the gravity model of trade. *European Journal of Economic Studies*, 14(4), 212–220.

Petroman, I., Varga, M., Constantin, E. C., Petroman, C., Momir, B., Turc, B., and Merce, I. (2016). Agritourism: An educational tool for the students with agro-food profile. *Procedia Economics and Finance*, 39, 83–87. 10.1016/S2212-5671(16)30244-1

Phillip, S., Hunter, C., and Blackstock, K. (2010). A typology for defining agritourism. *Tourism management*, 31(6), 754–758.

Piermartini, R., and Teh, R. (2005). *Demystifying modelling methods for trade policy*. No. 10. WTO Discussion Paper. Retrieved November 3, 2023, from https://www. econstor.eu/bitstream/10419/107045/1/wto-discussion-paper_10.pdf

Pitrova, J., Krejčí, I., Pilar, L., Moulis, P., Rydval, J., Hlavatý, R., Horáková, T., and Ticha, I. (2020). The economic impact of diversification into agritourism. *International Food and Agribusiness Management Review*, 23(5), 713–734.

Prodan, R. (2008). Potential pitfalls in determining multiple structural changes with an application to purchasing power parity. *Journal of Business & Economic Statistics*, 26(1), 50–65.

Qoto, L. (2018). The COMESA-SADC-EAC free trade area: Rules of origin–An impediment to regional trade and economic integration (Doctoral dissertation). University of KwaZulu-Natal.

Roman, M., and Grudzień, P. (2021). The essence of agritourism and its profitability during the coronavirus (COVID-19) pandemic. *Agriculture*, 11(5), 1–25. 10.3390/ agriculture11050458

Santeramo, F. G., and Barbieri, C. (2017). On the demand for agritourism: A cursory review of methodologies and practice. *Tourism Planning & Development*, 14(1), 139–148.

Schilling, B., Sullivan, K., and Komar, S. (2012). Examining the economic benefits of agritourism: The case of New Jersey. *Journal of Agriculture, Food Systems, and Community Development*, 3(1), 199–214.

Schmitt, M. (2010). Agritourism–From additional income to livelihood strategy and rural development. *The Open Social Science Journal*, 3(1), 41–50.

Seti, T. M. (2023). Determinants of South African agricultural exports to African markets. *Journal of Economic and Financial Sciences*, 16(1), 1–8.

Tinbergen, J. (1962). An analysis of world trade flows. *Shaping the World Economy*, 3, 1–117.

United States Department of Agriculture (2015). Zimbabwe: Zimbabwe Agricultural Economic Fact Sheet, October 6, 2015. Retrieved January 15, 2024, from https://fas.usda.gov/data/zimbabwe-zimbabwe-agricultural-economic-fact-sheet

Uwakata, O. Y., and Aregbeshola, R. A. (2023). The impact of multilateral trade agreements on intra-regional Trade: A case of the economic community of West African States (ECOWAS). *Journal of African Union Studies*, 12(2), 121–143.

Yang, S., and Martinez-Zarzoso, I., (2014). A panel data analysis of trade creation and trade diversion effects: The case of ASEAN-China Free Trade Area. *China Economic Review*, 2(1), 138–151.

Yotov, Y. V. (2022). On the role of domestic trade flows for estimating the gravity model of trade. *Contemporary Economic Policy*, 40(3), 526–540.

Zhou, L., and Tong, G. (2022). Structural evolution and sustainability of agricultural trade between China and countries along the "Belt and Road". *Sustainability*, 14(15), 1–24.

2 Artificial Intelligence in Agritourism

Utilitarian Analysis of Opportunities, Challenges, and Ethical Considerations in the African Context

Varaidzo Denhere and Deo Shao

Introduction

The concept of agritourism was established at the global level in the early 20th century (Wagner and Engel, 2018) with the United States and Italy being the first countries to have some form of agritourism in the world (van Zyl and van der Merwe, 2021). According to Lak and Khairabadi (2022), the past 25 years have seen the term agritourism becoming noticeable in literature. Agritourism resulted from the integration of agriculture and tourism. Extant literature shows that agritourism can also be called tourist farm, rural tourism, or farm-based tourism. These various names stem from the lack of both a common definition of the term and a lack of consensus on agritourism activities (Pérez-Olmos and Aguilar-Rivera, 2021). Lak and Khairabadi (2022, p. 2) describe agritourism as "farming activities performed on the farm or in other agricultural environments for entertainment or education". Ecotourism World (2020) also defines agritourism as "the symbiotic relationship between tourism and agriculture". The lack of both consensus on agritourism activities and standard definition has hampered the advancement of effectual policies in the sector (Pérez-Olmos and Aguilar-Rivera, 2021). In general, agritourism could be defined as an emerging industry that integrates agricultural practices with tourism, providing a distinctive fusion of agricultural activities with recreational pursuits. Agritourism enables tourists to engage in local agricultural practices, traditions, and cultural activities, thereby offering an authentic experience of rural life across different regions of the continent.

Agritourism has emerged as a potentially viable means of generating sustainable income in the diverse rural regions of Africa (Ciolac et al., 2019). Before integrating agriculture with tourism, agriculture was the cornerstone of many African economies with 90% of the African rural population depending on it as the main source of income. Back then, the agricultural sector employed about 60% of the African population and yet yielded relatively low income (Kanu et al., 2014). To promote development in rural areas as well as reducing the urban-rural affluence gap, various governments initiated agritourism to help farmers diversify their entrepreneurial portfolio (Wang et al., 2023). Extant literature suggested a boost in

DOI: 10.4324/9781032696188-2

income through the fusion of agriculture and tourism. Furthermore, an observation was made that agritourism has transformed its focus, with an increasing emphasis on the promotion of the inherent worth of rural and natural surroundings (Sgroi et al., 2018). Despite the integration of agriculture with tourism, the income levels remained relatively low mainly because the mode of production in the sector remained largely traditional (Wang et al., 2023). Therefore, the infusion of technology into agritourism would boost both productivity and income.

Amongst various technologies on the market, AI has been used by different business sectors to improve productivity. Generally African economies have been lagging in adopting AI compared to developed economies. However, an upsurge in AI adoption has been witnessed during the COVID-19 pandemic. Literally, AI offers a lot of solutions that can potentially transform many business sectors among them agritourism (Perifanis and Kitsios, 2023). John McCarthy coined the term AI in 1955 and defined it as, "the science and engineering of making intelligent machines, especially intelligent computer programs" (Rajaraman, 2014). According to Perifanis and Kitsios (2023, p. 1), AI refers to "a wide range of cutting-edge analytics, applications, and logic-based methods that imitate human behaviour, decision-making, and processes, including learning and problem-solving". This suggests that AI has numerous tools that can be used for different purposes to enhance tasks and processes for the benefit of society. Consequently, the infusion of AI in agritourism is envisaged to diversify and improve income levels. Indeed, the integration of AI in agritourism could offer significant potential for transformation, as it can improve the tourist experience, streamline operational processes, and provide data-driven insights to optimise agritourism enterprises. Furthermore, it should be noted that the sustainability of agritourism expansion is not uniform and exhibits distinct characteristics across various regions. This implies that the employment of AI has the potential for the customisation of strategies to cater to specific requirements (Belliggiano et al., 2020).

The infusion of AI in agritourism presents both untapped opportunities and challenges for sustainable income generation. Certainly, the sector can strategically leverage AI to shift from traditional practices to digitally transformed practices. This will see the growth of income in the African economies that adopt the AI tools. Therefore, this chapter sought to explore the intersection of AI and agritourism in the lenses of utilitarian theory in the African context. It will explore various etiquettes in which AI can improve the economic, educational, environmental, and cultural aspects of agritourism. Furthermore, the chapter will explain ethical considerations such as data management, equitable distribution, and sustainability in the course of integrating AI into the subsector. The ultimate goal is to provide a comprehensive examination of the opportunities and drawbacks of AI in the agritourism sector within the African context, based on existing literature.

Methodology

The study conducted a critical literature review on the challenges, opportunities and ethical issues of applying AI tools in the agritourism subsector. The Oates (2006) literature review strategy was followed. Using keywords such as "artificial intelligence," "agritourism," and "AI", we searched and retrieved from scholarly research databases, such as Scopus, Science Direct and Google Scholar using keywords such as "artificial intelligence," "agritourism," and "AI".

The review process involved analysis of the findings and identification of research gaps. The findings were explained using the utilitarian theory analytical framework. The review analysis and synthesis of the literature offered a deep understanding of the potential, challenges and ethical considerations of AI technologies to improve efficiency in agritourism and to identify areas that need more research.

Theoretical Framework

The Utilitarian theory is a consequentialist theory that aims to maximise benefits and minimise drawbacks (Mitov, 2021). The adoption of AI in agritourism offers opportunities and challenges that require a thorough examination. This study employed Utilitarian theory to examine opportunities and challenges while ensuring an approach that is grounded in ethical principles. The investigation focuses on exploring whether the use of AI leads to increased benefits and reduced drawbacks for all stakeholders involved. Utilitarianism is relevant to this study because of its focus on consequences. Table 2.1 describes how utilitarianism theory conceptualises the integration of AI in the agritourism subsector.

Literature Review

Artificial Intelligence in Agritourism

Agritourism was described by Rauniyar et al. (2021) as a slowly emerging distinct available option of recreational activities at the disposal of tourists as they decide on tourism destinations. Due to the ongoing rapid technological development, agritourism business operators mainly in developed economies have integrated it with AI (Xie and He, 2022). In developing economies, the growth and technological innovation of agritourism is critical as it has the potential to reduce poverty and promote rural development but these economies lag behind the developed economies. Bhatta and Ohe (2020) argued that the farmers' willingness and capacity to put up the necessary AI infrastructure together with tourists' intrigue to visit the agritourism sights determines agritourism innovation and development. This scenario creates complexity in the integration of AI in agritourism because the whole process depends on the willingness and capacity of the business owner. This might have cost implications and the attitude of the business owner on the adoption of AI technologies for their business. This then justifies an

Table 2.1 Conceptualisation of utilitarianism theory of AI integration in agritourism

Construct	Item	Description
Maximising utility	Economic and operational enhancement	The integration of AI in agritourism operations has the potential to maximise utility by improving the economic and operational aspects of the industry. This includes analysing the impact of AI-driven optimisations in resource management and production processes on the overall economic development and well-being of stakeholders.
	Sustainability and environmental considerations	AI can be leveraged to promote agritourism sustainable practices, exploring whether the implementation of AI technologies, such as precision farming and resource optimisation impact the environment and livelihoods positively.
	Tourists' experience and satisfaction	The deployment of AI to provide personalised experiences for tourists can lead to increased satisfaction and happiness, ultimately maximising utility. As a result, the utilisation of AI to enhance tourist experiences is a valuable contribution to overall satisfaction.
Minimising disutility	Ethical and privacy concerns	The assessment of ethical issues, particularly about data protection and user agreement, is crucial in determining whether the potential drawbacks arising from ethical concerns are surpassed by the advantages offered by personalised and enhanced agritourism experiences.
	Socio-economic and employment implications	The potential socioeconomic implications of implementing AI, such as job displacement, must be thoroughly evaluated to assess whether the benefits, such as increased operational efficiency and profitability, outweigh the potential negative consequences.
	Technological challenges and accessibility	It is imperative to examine the technological challenges that arise from the integration of AI to determine whether the perceived advantages of AI adoption, such as improved operational efficiency, justify the disutility caused by technological challenges and accessibility issues.
Cultural implications	Cultural preservation and community involvement	The application of AI in the agritourism subsector should be evaluated in terms of its impact on cultural preservation and community involvement to determine whether it can be harnessed in a manner that optimises its benefits by promoting and enhancing local cultures and engaging communities in a meaningful way.

Source: Authors' construct.

observation by Suanpang and Pothipasa (2021) that the technological revolution has culminated in a vast improvement in the tourism industry, but AI-related solutions are still restricted. Extant literature indicates that big organisations in agritourism have the financial capacity to adopt AI technologies to provide intelligent solutions to their businesses, unlike small operators. This promotes the technological divide between agritourism businesses.

According to UNWTO (2023), agritourism is a kind of tourism and tourism in general as a service sector is amongst the top four global export earners which currently provides a tenth of jobs globally. Following the promulgation of sustainable development goals (SDGs), it is crucial for this sector to also contribute towards the relevant SDGs. Ghidouche et al. (2021) underscored the broadening of innovation in agritourism as well as the emphasis of sustainability of agritourism. The use of AI technologies in agritourism develops an original agritourism product and services and involves infrastructure, human management, and procurement processes among other things thus leading to improved competitiveness (Xie and He, 2022). Furthermore, AI-enhanced agritourism could potentially promote certain SDGs, such as SDG10, reduction of inequalities; SDG2, zero hunger, through food security and sustainable production; SDG6, clean water and sanitation; and SDG8, decent work and economic growth (Ghidouche et al., 2021; Santos, 2023). The integration of AI in agritourism can boost employment opportunities and reduce poverty in African economies.

Case Studies of AI in Agritourism across the African continent

As already indicated in this study, a technological revolution has made a vast change in the tourism industry, but AI-related solutions are still limited, particularly in developing economies. Agritourism is a type of tourism and therefore it is not an exception to this position. Similarly, AI is increasingly used in agriculture in several African countries. This results in the improvement of productivity in terms of agricultural products. However, the aspect of tourism which integrates with agriculture to form agritourism remains lagging in terms of employing intelligent solutions through the integration of AI in agritourism.

Grillini et al. (2022) conducted a study to compare supports and policies of agritourism in the USA, Italy, and South Africa. The study focused on these countries because of their strength in agritourism as well as their diversified socio-cultural characteristics. However, in this study, looking at South Africa as an African country, no information was provided on AI in agritourism. This shows a paucity of empirical literature on successful cases of AI in agritourism in Africa. If implemented in agritourism, the different AI tools described in this study are envisaged to improve the agritourism subsector in Africa by providing intelligence.

The Role of AI in Diversifying Income Streams within the Agritourism Subsector

Agritourism Activities and Income Streams

Agritourism involves visiting working farms by tourists for learning and entertainment. To create both educational and economic value, many farms have diversified into direct-to-consumer sales, food systems, vacation rentals, as well as farm tours (Quella et al., 2021). For example, van Zyl and van der Merwe (2021b) established that in South Africa, the agritourism activities and attractions on demand from the nine provinces were hunting, accommodation, hiking, nature trails, photography and wildlife viewing. Fleischer et al. (2018) classified agritourism activities into two categories, namely farm or rural-based activities, and activities indirectly related to farm areas. The farm or rural-based activities include accommodation, attractions, and food services, among others. Those that are directly related to the farm include arts and crafts stores, galleries, and restaurants.

Based on Quella et al.'s (2021) categorisation of agritourism framework, agritourism activities are organised into core and peripheral activities: where the core activities are closely linked to agricultural production happening on a working farm, while peripheral activities are not closely linked to agricultural production but also happening on a working farm. However, classifying peripheral activities as agritourism remains controversial; hence, Chase et al. (2018) further classified both core and peripheral under five categories, namely, education, direct sales, hospitality, entertainment, and outdoor recreation.

This chapter uses information from both Chase et al.'s (2018) framework and extant literature to identify the common agritourism activities and the income streams within the agritourism subsector. Table 2.2 illustrates the agritourism income streams and activities thereof.

AI and Agritourism Income Streams

There is a potential agritourism drawback resulting from farmers spending more time serving guests at the expense of agricultural production, a potentially counterproductive situation in terms of global food contribution (Fischer, 2019) rendering agriculture and tourism partly in conflict. When agritourism emerged, it was perceived to be an expedient of improving farm income streams as well as creating new on-farm income streams, diversifying and upgrading farm operations, raising awareness of local agricultural products, and promoting sustainability of farm businesses (Bernardo et al., 2004; Popescu et al., 2023). AI tools could be employed to reduce the potential drawback and enhance the perceived expediency role of agritourism, thereby offering a significant potential for transformation, improvement of income streams and sustainability.

Agritourism business owners must understand that farm events require careful planning and coordination to showcase the farm offerings, manage

Table 2.2 Agritourism income streams and their activities

Agritourism income streams	Agritourism activities
Education	• School/classes and tours • Farm-to-table and tastings • Agri museums off-farm • Agri fairs off-farm • Education and training • Cannery tours and cooking classes • Garden/nursery tours • Winery tours and tasting • Agricultural technical tours • Historical agricultural exhibits
Direct sales	• Farm/roadside stands • U-pick/cut operations • Farm-to-table and tastings • Tours • Farmers' markets/on-farm sales • Agricultural-related art and crafts/gifts
Hospitality	• Overnight accommodation/farm stays • Farm-to-table and tastings • Outfitter services on-farm • Restaurants/eateries • Guided tours • Bed & Breakfast • Farm/ranch vacations • Guest ranch
Entertainment	• Harvest festivals/fairs • Corn maze/hayrides • Weddings on-farm • Concerts on-farm/music, barn dances • Photography • Special events on-farm • Hunting/working dog trials/training
Outdoor recreation	• Schools/classes and tours • Horseback ridings • Hiking/nature trails • Art and photography • Fishing • Hunting • Wildlife viewing and photography • Landscape beauty viewing • Camping and picnicking • Wagon/sleigh rides • Cross-country skiing

Source: Authors' construct.

logistics, and provide engaging activities as well as the common customer experience (Giller, 2023). Consequently, agritourism operations should be streamlined with specialised AI technology to improve efficiency and attract more customers. This will improve the agritourism income streams as customers will get value for their money. For example, to manage farm events, a centralised dedicated event management AI tool can be employed to streamline registration, ticketing, communication, and agritourist management. Such a tool manages all facets of the event and allows smooth coordination. Being very good with repetitive tasks, AI could be applied for tasks that have to do with checking data entry, validation of information and processing of information into graphs and charts for easy interpretation (Tjoe, 2023).

Some of the farmers could still be using the traditional manual bookings, payments and ticketing processes which are time-consuming. AI tools could be implemented to allow for online booking and secure payment systems as well as an automated ticketing process. Such a system would allow customers to buy tickets before their travel and enable the host to know how many tourists to expect, and plan and prepare accordingly. This will be convenient for the intending tourists to make their bookings in advance while reducing administrative work for the host to allow more time for agricultural production.

AI for agritourism can also be in the form of specialised mobile apps focusing on the agritourism market. The app can have digital guides that provide the tourists with real-time access to the scheduled events, activity descriptions, maps, and updates. Such tools allow tourists to have the most of their farm visit experiences through enhanced engagement and improved navigation. Income streams can also be improved through digital marketing using AI tools and social media platforms that can be leveraged by farmers to reach a wider audience in marketing agritourism activities. Traditional marketing requires more manpower and material resources usually yielding unsatisfactory results for farm owners (Xie and He, 2022). An AI-based marketing strategy can efficiently integrate farm resources, establish customers' needs and provide the most suitable customers' envisaged experience. Furthermore, a farm should design an impressive website to create a lasting first impression for its potential customers. The ideal website should be designed in such a way that it fascinates users with its mesmeric visuals, refined aesthetics, and smooth navigation. Giller (2023) posits that every click on the website should reveal a world of picturesque landscapes, sterling farm experiences, and catchy farm moments. The website can also be enhanced by chatbots to enable customers to get assistance instantly (Tjoe, 2023).

Educational agritourism is aimed at helping tourists better understand the agricultural production and product distribution processes, the rural traditional culture, and the preservation of the environment, among other things (Petroman et al., 2016). These activities can be enhanced with relevant AI tools. For example, gamification can be introduced to support traditional teaching and learning methods at the farm. This could be done through mobile apps. The farmer could also employ QR code scavenger hunts hung around the farm. The use of such tools brings excitement as well as friendly competition among tourists during farm events. The educational games come with prizes and rewards making the whole exercise enjoyable. This will attract more customers and increase the agritourism income. Sometimes tourists can be full of questions and an AI Virtual Assistant can save the situation by answering customers' questions when the farmer is busy working in the field. An AI Virtual Assistant relies on location information, customer data, and prior dialogues from the farm's knowledge repository and integrates analytics and cognitive computing to generate customised responses.

AI tools, specifically chatbots, can be used to provide the best customer service. AI-powered Chatbots can be available for customers around the clock to provide the needed information by customers where humans are unavailable. AI tools can also be used for surveys to gauge customer satisfaction, a process which helps in improving customer services. These tools can also be leveraged to allow farmers to gather customer preferences and enable personalised experiences (Suanpang and Pothipasa, 2021). Customers will always have varied preferences and catering to these preferences would boost agritourism income streams. For example, farmers can use AI tools to create a sense of exclusivity and cater to individual interests by offering customised itineraries, and exclusive access to certain areas of the farm, and boost agritourism incomes.

The Potential of AI in Knowledge Transfer and Training in the Agritourism Subsector

The term knowledge transfer (KT) is defined by Hassan et al. (2017, p. 751) as "a process through which knowledge moves between a root and a recipient and where knowledge is given and practised". Gallemard (2023) also defines knowledge transfer as a process of sharing skills, experiences, ideas and information by employers or employees in a business through training, coaching, mentoring, and available communication channels. It can further be described as a way of managing knowledge in business organisations, the education sector, and social circles. Knowledge can be regarded as a long-term asset (Hassan et al., 2017). Consequently, the goal of knowledge transfer is the retention of critical information within a society, an organisation or a team. Gallemard (2023) posited that KT through sharing skills and experience helps individuals avoid costly mistakes committed by others, leading to improved performance.

Van Niekerk et al. (2015) highlight the existence of a wealth of knowledge valuable for sustainable commercial farming within family farmers and that this knowledge continues to evolve within the family structure. This knowledge has been transmitted from generation to generation for preservation as well as improvement. Knowledge from older generations is augmented by new knowledge that is generated through technological advancement. The main objective of transferring knowledge from one farming generation to the next is to perpetuate control of a vibrant and improved business to the next generation (Van Niekerk et al., 2015). KT in agritourism also known as advisory services at farm level also occurs between farmers and farm advisors (Cawley et al., 2023). This can be achieved through peer-to-peer learning using participatory activities during farm events or group discussions. For many years, agricultural advisory services employed various traditional ways and activities to transfer knowledge to farmers to shape farmer behaviour, and assist them in decision making (Cawley et al., 2023).

Ingram and Morris (2007) highlighted the existence of various forms of knowledge in general and acknowledged the existence of such a variety of knowledge in farming, with transferring this knowledge being the major challenge. According to Ingram and Morris (2007), the transfer of knowledge can be done through communication from one business unit to another in the form of instructions, absorbed through following routines, or standard verbal reports. Galindo (2007) asserted that the agricultural sector is diverse and fragmented; hence, the KT services in this sector differ from those employed in other business organisations. Consequently, KT in this sector is more likely through participatory learning, apprenticeship, one-on-one during an activity, or group discussion (Galindo, 2007). However, Cawley et al. (2023) argue that KT depends on participants' absorptive capacity and their learning mode preferences and yet the most relevant and most impactful type of activities and content on farm performance are not well understood.

The Use of AI Tools in Knowledge Transfer and Training

AI technology can be used to leverage KT in the agritourism subsector just like any other business organisation. The use of AI for KT promotes easier and faster communication among participants, increased accessibility of knowledge, reduced costs and risks of knowledge loss, and well-supported knowledge management and reuse of knowledge, among others (Taherdoost and Madanchian, 2023). Furthermore, AI tools can assist in connecting different locations, teams, and departments within an organisation. Meanwhile, the advantages of employing AI in knowledge management include easy problem solving, knowledge based on personal preferences, enhanced flexibility of knowledge representation, and enhanced knowledge sharing, usage, and capturing (Taherdoost and Madanchian, 2023). Gallemard (2023) cited collaborative software, knowledge management systems and learning management systems as examples of technologies used in KT. The AI tools that could be employed for KT in agritourism include blogs, podcasts, wikis, webinars, and social media platforms illustrated in Table 2.3.

Resource Optimisation through AI to Bolster Sustainable Farming Practices

The use of AI holds promises in augmenting sustainable practices within agritourism by optimising resource allocation (Linaza et al., 2021). Through a systematic analysis of data about the use of resources and the generation of waste, AI technology can discern opportunities for reducing waste and

Table 2.3 Digital tools potential for KT in agritourism

Tools	Description
AI supported blogs	These are online journals that allow users to post, comment and subscribe to content. These are convenient for sharing tacit knowledge such as insights, opinions, as well as stories. AI tools are used to generate blog posts through natural language processing (NLP), and then AI algorithms analyse content from multiple sources and generate ideas to produce fresh content. AI can also analyse already written text and suggest improvements in the grammar, consistency, clarity, and accuracy.
Podcasts	These are useful for distributing and learning embedded knowledge. AI can be used to create content for podcasts. These allow users to listen to and download content.
Wiki	This is a website or database supported by generative AI tools. It allows users to create, edit, and share content collaboratively. A wiki is handy for capturing and documenting knowledge, such as facts, procedures, and policies.
Webinars	These are online seminars that allow users to interact with and learn from experts or peers. Webinars are supported by AI virtual assistants which can automate repetitive tasks, assist with scheduling, manage recordings as well as provide real-time support during live sessions.
Social media	These are AI-enhanced platforms and can be used to build and maintain relationships among knowledge workers and stakeholders. AI tools enhance social media activities extensively across numerous cases, among them text and visual content creation which is crucial for KT.

Source: Authors' construct.

enhancing the efficient allocation of resources, thereby potentially leading to the implementation of sustainable practices and the attainment of optimal operational outcomes (Nishant et al., 2020).

In addition, AI technology offers the possibility of advancing the implementation of sustainable production approaches through data analysis data of resource consumption and environmental impacts. This enables the identification of potential ways to reduce environmental effects and enhance resource efficiency (Lakshmi and Corbett, 2020). AI has the potential to enhance the implementation of a circular economy in agritourism enterprises (D'Amore et al., 2022).

The use of AI plays a pivotal role in the field of precision farming, facilitating a data-centric approach to all stages of agricultural operations, ranging from seed sowing to crop harvesting (Van Wynsberghe, 2021). Through the analysis of extensive data derived from various sources, including weather forecasts, soil sensors, and satellite imagery, AI algorithms provide farmers with precise guidance to optimal planting schedules, appropriate irrigation quantities, and suitable periods for harvesting (Liao and Yao, 2021). Furthermore, AI enables the real-time analysis of diverse variables, including meteorological circumstances, temperature fluctuations, water consumption, and soil characteristics. This assists agritourism practitioners in making well-informed decisions based on data, and enhances planning strategies to achieve higher crop yields through the farming of appropriate crop types and optimising the utilisation of resources (Sgroi et al., 2018).

Moreover, AI plays a crucial role in the progression and implementation of sustainable agricultural methods through its ability to analyse and optimise the utilisation of resources (Sætra, 2021). For instance, the utilisation of AI to analyse data, which is subsequently stored and processed in cloud-based systems, enables farmers to enhance resource allocation, reduce inefficiencies, and enhance crop yields (Lutfi et al., 2022). In addition, AI has the potential to assist farmers in reducing their reliance on pesticides and herbicides by offering guidance on optimal application schedules and amounts, taking into account the current weather and soil conditions (Ciolac et al., 2021). Consequently, the agritourism industry experiences enhanced efficiency and profitability, accompanied by diminished waste and intensified crop yields (Drăgoi et al., 2017).

The Cultural Implications of AI-Enhanced Agritourism in Preserving Indigenous Traditions and Fostering Interactions

In addition to promoting income streams and profitability in agritourism, the subsector also plays a critical role in preserving indigenous traditions that have been handed down through generations (UN, 2019). Communities that engage in agritourism usually have unique, genuine and authentic resources which need to be preserved in their original form (Popescu et al., 2023). Examples include agricultural practices, traditional medicines, music, art, natural resources, traditional lifestyle, specific crafts, traditional languages, food products, and cultural events. However, these traditions are at high risk of being lost due to globalisation. It has always been a challenge to preserve these traditions in this rapidly evolving world to the extent that in some places these are fast disappearing due to a lack of better ways to preserve them. Employing various AI tools could successfully safeguard and preserve these indigenous traditions.

Frąckiewicz (2023) asserted that the capability of AI to process huge amounts of data and decipher patterns enables it to assist in the safeguarding, preservation, and transmission of indigenous traditions in agritourism. AI can be employed to preserve these traditions through digitising and cataloguing them to make them easily accessible to future generations. AI tools could be employed to create digital libraries that would store indigenous traditions for preservation. NLP, an AI branch that enables computers to comprehend, generate, and manipulate human language could be used in preserving indigenous languages, through scanning and analysing texts, images, and videos, in the process of identifying and categorising information about indigenous languages. Furthermore, for indigenous languages under threat, speech recognition technology could be employed to develop a language pronunciation app. The app uses a machine learning model to assess the language pronunciation as well as provide an oral experience. This model would preserve the language in its original form and assist in the learning of the language by future generations. AI can also assist in the indigenous language translation through the use of machine learning (ML) algorithms.

Traditional art and music can also be preserved through the use of AI. ML can be employed in preserving art and music as it is capable of analysing patterns and styles in indigenous art and music. Therefore, AI can assist in creating digital reproductions of the art and the music, thus preserving them in their original forms for future generations.

Traditional practices tend to be linked to specific locations on the land hence communities would have a deep connection with their land. The land in question would need to be preserved. Therefore, AI can be used to delineate and monitor such locations. AI and satellite imagery can be used to monitor any changes that occur in these locations and indicate potential threats thereby protecting the land in question and preserving the traditions associated therewith. Therefore, AI can be employed to preserve the rich indigenous traditions and foster interactions in agritourism. However, there are ethical issues to consider when adopting and implementing these technologies.

Challenges and Ethical Considerations Arising from the Integration of AI in Agritourism

The adoption of AI in the agritourism subsector is not free from challenges. It is imperative to undertake ethical considerations, including privacy concerns, bias and fairness, a lack of accountability, job displacement, and environmental impact. Agritourism firms should address these issues, ensuring that their AI systems are ethical, transparent, and sustainable. Table 2.4 describes the challenges and ethical considerations of AI adoption in agritourism.

Implications

This section describes both practical and theoretical implications of the current study.

Practical Implications

Currently, there is a literature dearth on AI in agritourism in the African context. The current study has highlighted what AI tools are capable of doing when adopted in the agritourism sub-sector. Extant literature has indicated that the world is currently going through an era of digitalisation, which was

Table 2.4 Associated challenges and ethical considerations of AI fusion in agritourism

Challenges	Ethical considerations	References
Privacy concerns	The integration of AI in agritourism necessitates the collection and analysis of substantial data, raising pivotal privacy concerns. It is imperative for farmers and agritourism businesses to uphold transparency regarding data collection and usage, secure consent from users prior to data collection, and ensure secure data storage	(Smith, 2020; Sparrow et al., 2021; Whittlestone et al., 2019)
Bias and fairness	AI systems can inherently possess biases, potentially leading to unjust outcomes. For instance, AI systems trained on biased data may generate results that are prejudiced against certain groups. It is crucial for agritourism businesses to ensure that AI systems are developed to be equitable and unbiased	(Ho et al., 2019; Mark, 2019; Smith, 2020; Sparrow et al., 2021)
Lack of accountability	The complexity and opacity of AI systems can hinder accountability for their decisions. Agritourism businesses must ensure that their AI systems are transparent and that users comprehend their functionality and decision-making processes	(Smith, 2020; Taddeo and Floridi, 2021)
Job displacement	The deployment of AI in agritourism may result in job displacement as AI systems can automate tasks previously performed by humans. It is vital for agritourism businesses to ensure that human workers are not supplanted by AI systems without being provided alternative employment opportunities	(Kerr et al., 2020; Ryan, 2022)
Environmental impact	The utilisation of AI in agritourism can have environmental repercussions, potentially leading to augmented resource usage and waste generation. Agritourism businesses must ensure that their AI systems are designed sustainably, optimising resource usage, and minimising waste	(Dara et al., 2022; Ryan, 2022; Smith, 2020; Taddeo and Floridi, 2021)

Source: Authors' own construct.

accelerated by the COVID-19 pandemic. This saw most businesses agri-tourism going online. Digitalisation refers to the adoption of digital technologies by businesses to change the ways of doing business through increased efficiency and productivity. This study has indicated possible opportunities where AI can be harnessed in agritourism. AI can be used to both diversify and enhance agritourism income streams. This increases profitability in the subsector. When adopted in agritourism in the African context, AI can also be used to diversify modes of knowledge transfer, which is critical in agritourism.

Agritourism is rich in indigenous knowledge that is passed from one generation to another. Hence, this study has indicated the importance of such knowledge as well as the possible ways of preserving the crucial knowledge with the aid of AI tools. African governments might use this information to formulate relevant policies that promote the use of AI in preserving indigenous agritourism knowledge. Tourism authorities in various African countries could also harness the information from this study to come up with awareness programmes that promote the adoption of AI solutions, especially by small operators that need support in the current rapid digital transformation environment in which they find themselves operating.

Furthermore, tourism as a service sector with agritourism as a subsector is among the top four global export earners. This huge economic contribution warrants active and positive contributions towards the attainment of SDGs. This study has shown that agritourism can significantly contribute towards SDG8, decent work and economic growth, as it employs one in ten jobs globally (Buhalis et al., 2023; UNWTO, 2023). Consequently, AI-enhanced agritourism will increase the opportunities for decent work for the employable population. The integration of AI in agritourism might have a ripple effect on tourism value chains thereby prompting the formulation of policies that bring about positive socio-economic impact. This study has also indicated that there are some ethical issues surrounding the use of AI. African governments could use this information to formulate policies around the responsible use of AI in agritourism.

Theoretical Implications

Hitherto, the utilitarian theory has mostly been used in social science fields. This study breaks ground by applying the theory to the agritourism subsector. The utilitarian theory offers an understanding of maximising usefulness while minimising harm. It also provides an approach to analyse the ethical socio-operational aspects of implementing AI in African agritourism.

The study offers knowledge for decision-makers and policymakers to assess the opportunities and challenges of AI applications in agritourism. It contributes to the theoretical understanding of ethical considerations for responsible integration of AI technologies.

In addition, the study explores the implications of AI in shaping tourist experiences and satisfaction. Theorisation of AI for enhancing tourist experiences suggests that overall satisfaction and happiness can be maximised, contributing to discussions on advancements in the tourism sector. This is in tandem with the evaluation of the consequences of AI adoption, particularly in terms of ethics, socio-economic factors and technology. Utilitarianism offers an approach to assess these trade-offs involved in integrating AI into agritourism.

Last but not least, through the utilitarianism lens, the study enriches the knowledge on how AI can be utilised to optimise benefits while respecting cultures within the agritourism subsector.

Study Limitations

This study solely relied on existing literature some of which could be outdated. This would deny the audience of this study the current information on the agritourism subsector in Africa. Furthermore, in searching for literature, researchers might not have been exhaustive, failing to cover all aspects and complexities of applying AI in agritourism within diverse and multifaceted African contexts. The study recommends empirical study in the African context where data is collected from persons who are operating agritourism businesses. Future research should empirically focus on tackling the ethical dilemmas associated with integrating AI into agritourism. Supplementary research should also focus on the establishment of frameworks for data management and the promotion of responsible development. Furthermore, empirical studies that investigate the direct effects of AI on livelihoods as well as their intersection with indigenous knowledge and traditions are important. Moreover, future research must explore the scalability and affordability of AI technologies in the context of African agritourism.

Conclusion

The infusion of AI into agritourism gives rise to a range of opportunities and challenges. The implementation of AI can expand the scope of agritourism offerings, improve income levels, facilitate knowledge transfer and training, optimise resources for sustainable agricultural practices, and culturally enhance agritourism while simultaneously preserving indigenous traditions. Nevertheless, it is imperative to concede and address noteworthy ethical concerns related to the management of data, allocation of resources, and assurance of long-term viability. The integration of AI into the agritourism subsector has substantial implications for the economic, educational, environmental, and cultural dimensions within the African context. From an economic perspective, the implementation of AI can enhance income levels and foster income diversification within the agritourism sector. It also has the potential to enhance sustainable farming practices through the optimisation of resources, thereby contributing to environmental preservation. From a cultural standpoint, AI assumes a pivotal function in the preservation of indigenous traditions and facilitation of interactions within the agritourism industry. The chapter demonstrated that the integration of AI within the agritourism subsector in Africa from the utilitarianism perspective presents both a plethora of potential opportunities and challenges.

Funding Declaration

The authors declare that the research was not funded.

References

Belliggiano, A., Garcia, E. C., Labianca, M., Valverde, F. N., and De Rubertis, S. (2020). The "eco-effectiveness" of agritourism dynamics in Italy and Spain: A tool for evaluating regional sustainability. *Sustainability*, *12*(17), 7080.

Bernardo, D., Valentin, L., and Leatherman, J. (2004). Agritourism: If we build it, will they come? *Risk and Profit Conference, Manhattan, KS*, 19–20.

Bhatta, K., and Ohe, Y. (2020). A review of quantitative studies in agritourism: The implications for developing countries. *Tourism and Hospitality*, *1*(1), 23–40.

Buhalis, D., Leung, X. Y., Fan, D., Darcy, S., Chen, G., Xu, F., Wei-Han Tan, G., Nunkoo, R., and Farmaki, A. (2023). Tourism 2030 and the contribution to the sustainable development goals: The tourism review viewpoint. *Tourism Review*, *78*(2), 293–313.

Cawley, A., Heanue, K., Hilliard, R., O'Donoghue, C., and Sheehan, M. (2023). How knowledge transfer impact happens at the farm level: Insights from advisers and farmers in the Irish agricultural sector. *Sustainability*, *15*(4), 3226.

Chase, L. C., Stewart, M., Schilling, B., Smith, B., and Walk, M. (2018). Agritourism: Toward a conceptual framework for industry analysis. *Journal of Agriculture, Food Systems, and Community Development*, *8*(1), 13–19.

Ciolac, R., Adamov, T., Iancu, T., Popescu, G., Lile, R., Rujescu, C., and Marin, D. (2019). Agritourism–A sustainable development factor for improving the 'health' of rural settlements. Case study Apuseni mountains area. *Sustainability*, *11*(5), 1467.

Ciolac, R., Iancu, T., Brad, I., Adamov, T., and Mateoc-Sîrb, N. (2021). Agritourism— A business reality of the moment for Romanian rural area's sustainability. *Sustainability*, *13*(11), 6313.

D'Amore, G., Di Vaio, A., Balsalobre-Lorente, D., and Boccia, F. (2022). Artificial intelligence in the water–energy–food model: A holistic approach towards sustainable development goals. *Sustainability*, *14*(2), 867.

Dara, R., Hazrati Fard, S. M., and Kaur, J. (2022). Recommendations for ethical and responsible use of artificial intelligence in digital agriculture. *Frontiers in Artificial Intelligence*, *5*, 884192.

Drăgoi, M. C., Iamandi, I.-E., Munteanu, S. M., Ciobanu, R., Țarțavulea, R. I., and Lădaru, R. G. (2017). Incentives for developing resilient agritourism entrepreneurship in rural communities in Romania in a European context. *Sustainability*, *9*(12), 2205.

Fischer, C. (2019). Agriculture and tourism sector linkages: Global relevance and local evidence for the case of South Tyrol. *Open Agriculture*, *4*(1), 544–553.

Fleischer, A., Tchetchik, A., Bar-Nahum, Z., and Talev, E. (2018). Is agriculture important to agritourism? The agritourism attraction market in Israel. *European Review of Agricultural Economics*, *45*(2), 273–296.

Frąckiewicz, M. (2023, September 9). Navigating traditions: AI in preserving indigenous practices. *TS2 SPACE*. https://ts2.space/en/navigating-traditions-ai-in-preserving-indigenous-practices/

Galindo, I. M. (2007). Regional development through knowledge creation in organic agriculture. *Journal of Knowledge Management, 11*(5), 87–97.

Gallemard, J. (2023). *The Basics of Knowledge Transfer: A Beginner's Guide (2023)*. https://blog.smart-tribune.com/en/knowledge-transfer

Ghidouche, K., Nechoud, L., and Ghidouche, F. (2021). Achieving sustainable development goals through agritourism in Algeria. *Worldwide Hospitality and Tourism Themes, 13*(1), 63–80. 10.1108/whatt-08-2020-0092

Giller, M. (2023, July 9). Agritourism software. *Farmlike*. https://farmlike.io/agritourism-software/

Grillini, G., Sacchi, G., Chase, L., Taylor, J., van Zyl, C. C., van Der Merwe, P., Streifeneder, T., and Fischer, C. (2022). Qualitative assessment of agritourism development support schemes in Italy, the USA and South Africa. *Sustainability, 14*(13), 7903.

Hassan, N., Noor, M. N. M., Hussin, N., and others. (2017). Knowledge transfer practice in organization. *International Journal of Academic Research in Business and Social Sciences, 7*(8), 2222–6990.

Ho, C., Soon, D., Caals, K., and Kapur, J. (2019). Governance of automated image analysis and artificial intelligence analytics in healthcare. *Clinical Radiology, 74*(5), 329–337.

Ingram, J., and Morris, C. (2007). The knowledge challenge within the transition towards sustainable soil management: An analysis of agricultural advisors in England. *Land Use Policy, 24*(1), 100–117.

Kanu, B. S., Salami, A. O., and Numasawa, K. (2014). Inclusive growth: An imperative for African agriculture. *African Journal Food Agriculture Nutrition Development, 14*(3).

Kerr, A., Barry, M., and Kelleher, J. D. (2020). Expectations of artificial intelligence and the performativity of ethics: Implications for communication governance. *Big Data & Society, 7*(1), 2053951720915939.

Lak, A., and Khairabadi, O. (2022). Leveraging agritourism in rural areas in developing countries: The case of Iran. *Frontiers in Sustainable Cities, 4*, 863385.

Lakshmi, V., and Corbett, J. (2020). How artificial intelligence improves agricultural productivity and sustainability: A global thematic analysis. *Proceedings of the Annual Hawaii International Conference on System Sciences*. 10.24251/hicss.2020.639

Liao, M., and Yao, Y. (2021). Applications of artificial intelligence-based modeling for bioenergy systems: A review. *GCB Bioenergy, 13*(5), 774–802.

Linaza, M. T., Posada, J., Bund, J., Eisert, P., Quartulli, M., Döllner, J., Pagani, A., G. Olaizola, I., Barriguinha, A., Moysiadis, T., and others. (2021). Data-driven artificial intelligence applications for sustainable precision agriculture. *Agronomy, 11*(6), 1227.

Lutfi, A., Al-Khasawneh, A. L., Almaiah, M. A., Alsyouf, A., and Alrawad, M. (2022). Business sustainability of small and medium enterprises during the COVID-19 pandemic: The role of AIS implementation. *Sustainability, 14*(9), 5362.

Mark, R. (2019). Ethics of using AI and big data in agriculture: The case of a large agriculture multinational. *The ORBIT Journal, 2*(2), 1–27.

Mitov, A. (2021). *Ethical use of artificial intelligence through the Utilitarianism perspective [B.S. thesis]*. University of Twente.

Nishant, R., Kennedy, M., and Corbett, J. (2020). Artificial intelligence for sustainability: Challenges, opportunities, and a research agenda. *International Journal of Information Management, 53*, 102104. 10.1016/j.ijinfomgt.2020.102104

Oates, B. J. (2006). *Researching Information Systems and Computing*. Sage Publications Ltd.

Pérez-Olmos, K. N., and Aguilar-Rivera, N. (2021). Agritourism and sustainable local development in Mexico: A systematic review. *Environment, Development and Sustainability*, *23*(12), 17180–17200.

Perifanis, N.-A., and Kitsios, F. (2023). Investigating the influence of artificial intelligence on business value in the digital era of strategy: A literature review. *Information*, *14*(2), 85.

Petroman, I., Varga, M., Constantin, E. C., Petroman, C., Momir, B., Turc, B., and Merce, I. (2016). Agritourism: An educational tool for the students with agro-food profile. *Procedia Economics and Finance*, *39*, 83–87.

Popescu, C. A., Iancu, T., Popescu, G., Adamov, T., and Ciolac, R. (2023). The impact of agritourism activity on the rural environment: findings from an authentic agritourist area—Bukovina, Romania. *Sustainability*, *15*(13), 10294. 10.3390/su151310294

Quella, L., Chase, L., Wang, W., Conner, D., Hollas, C., Leff, P., and others. (2021). *Agritourism and On-Farm direct sales interviews: Report of qualitative findings.* University of Vermont Extension; *2021*.

Rajaraman, V. (2014). JohnMcCarthy—Father of artificial intelligence. *Resonance*, *19*, 198–207.

Rauniyar, S., Awasthi, M. K., Kapoor, S., and Mishra, A. K. (2021). Agritourism: Structured literature review and bibliometric analysis. *Tourism Recreation Research*, *46*(1), 52–70.

Ryan, M. (2022). The social and ethical impacts of artificial intelligence in agriculture: Mapping the agricultural AI literature. *AI & SOCIETY*, 1–13.

Sætra, H. S. (2021). AI in context and the sustainable development goals: Factoring in the unsustainability of the sociotechnical system. *Sustainability*, *13*(4), 1738.

Santos, E. (2023). From neglect to progress: Assessing social sustainability and decent work in the tourism sector. *Sustainability*, *15*(13), 10329.

Sgroi, F., Donia, E., and Mineo, A. M. (2018). Agritourism and local development: A methodology for assessing the role of public contributions in the creation of competitive advantage. *Land Use Policy*, *77*, 676–682.

Smith, M. J. (2020). Getting value from artificial intelligence in agriculture. *Animal Production Science*, *60*(1), 46. 10.1071/an18522

Sparrow, R., Howard, M., and Degeling, C. (2021). Managing the risks of artificial intelligence in agriculture. *NJAS: Impact in Agricultural and Life Sciences*, *93*(1), 172–196.

Suanpang, P., and Pothipasa, P. (2021). AI recommended agrotourism supporting community-based tourism post Covid-19. *Journal of Management Information and Decision Sciences*, *24*(1), 1–10.

Taddeo, M., and Floridi, L. (2021). How AI can be a force for good–An ethical framework to harness the potential of AI while keeping humans in control. In *Ethics, Governance, and Policies in Artificial Intelligence* (pp. 91–96). Springer.

Taherdoost, H., and Madanchian, M. (2023). Artificial intelligence and knowledge management: Impacts, benefits, and implementation. *Computers*, *12*(4), 72.

Tjoe, K. (2023, April 27). *6 Use Cases of AI in Travel and Tourism*. Rezdy. https://rezdy.com/blog/use-cases-of-ai-in-travel-tourism/

UN. (2019). *Indigenous People's Traditional Knowledge Must Be Preserved, Valued Globally, Speakers Stress as Permanent Forum Opens Annual Session*. UN Press. https://press.un.org/en/2019/hr5431.doc.htm

UNWTO. (2023). *Tourism for SDGs*. https://tourism4sdgs.org/sdg-8-decent-work-economic-growth/

Van Niekerk, J., Mahlobogoane, M., and Tirivanhu, P. (2015). The transfer of intergenerational family knowledge for sustainable commercial farming in Mpumalanga Province of South Africa: Lessons for extension. *South African Journal of Agricultural Extension*, *43*(1), 66–77.

Van Wynsberghe, A. (2021). Sustainable AI: AI for sustainability and the sustainability of AI. *AI and Ethics, 1*(3), 213–218.

van Zyl, C. C., and van der Merwe, P. (2021). The motives of South African farmers for offering agri-tourism. *Open Agriculture, 6*(1), 537–548.

Wagner, N., and Engel, W. (2018). *Agritourism routes: Shaping Sustainability. Agricultural Economics Services: Macro and Resource Economics*. Western Cape Government Agriculture. https://www.elsenburg.com/wp-content/uploads/2022/03/2018-Agritourism-Routes-for-Sustainabilty.pdf

Wang, J., Zhou, F., Chen, C., and Luo, Z. (2023). Does the integration of agriculture and tourism promote agricultural green total factor productivity?—Province-level evidence from China. *Frontiers in Environmental Science, 11*, 404.

Whittlestone, J., Nyrup, R., Alexandrova, A., & Cave, S. (2019). The role and limits of principles in AI ethics: Towards a focus on tensions. *Proceedings of the 2019 AAAI/ACM Conference on AI, Ethics, and Society*, 195–200.

World, E. (2020, April 15). What is Agritourism? *Ecotourism World*. https://ecotourism-world.com/what-is-agritourism/

Xie, D., and He, Y. (2022). Marketing strategy of rural tourism based on big data and artificial intelligence. *Mobile Information Systems, 2022*.

3 Critical Success Factors for Agritourism Entrepreneurship in Africa

Mufaro Dzingirai, Rukudzo Dzapasi, and Theologience Mazwiembiri

Introduction

It is worth mentioning that agritourism entrepreneurship has gained substantial attention in recent years worldwide. The concept of agritourism entrepreneurship is related to how tourists learn and entertain themselves as they visit farms. As such, it is widely regarded as a strategic option to develop rural and agricultural communities, especially in Africa. It is a fact that agritourism started in the United States of America, Canada, and Italy in the early 1920s (Grillini et al., 2022; van Zyl and van der Merwe, 2021). As this concept of agritourism entrepreneurship cuts across disciplines like agriculture, tourism, and entrepreneurship, it is not surprising to note that it has a great potential to unlock economic value in many African economies. The economies of many African countries are anchored on these three aspects of agritourism entrepreneurship (Baipai et al., 2023). Therefore, the need to interrogate the preparedness of the African economies when it comes to agritourism entrepreneurship is the need of the hour. It is a fact that agritourism entrepreneurship relies on a plethora of factors that include accessibility to farms, availability of resources in rural areas, marketing strategies, government support, and management strategies. In this regard, many African countries perceive agritourism as an ideal opportunity for rural sustainability.

While agritourism entrepreneurship is a powerful strategy to promote rural sustainability in African economies, agritourism entrepreneurs play an instrumental role in the areas of organizing rural festivals and events, bed and breakfast on farms, farm-to-table dining experiences, and guided-based tours of farms (Grillini et al., 2022; van Zyl and van der Merwe, 2021). It is worth mentioning that agritourism entrepreneurship appears to be lucrative in contemporary tourism. As such, the success of agritourism entrepreneurs depends on the possession of robust knowledge related to the agricultural, tourism and hospitality sector, and entrepreneurship. It is within this context that agritourism entrepreneurship has been hailed by various agencies as a viable and strategic option to foster sustainable development in agricultural communities and rural economies.

DOI: 10.4324/9781032696188-3

In the context of South Africa, the concept of agritourism was witnessed in the early 1950s when tourists were paying a visit to ostrich farms and then wine tourism gained popularity in the early 1970s (Grillini et al., 2022; van Zyl and van der Merwe, 2021). However, limited policies focus on agritourism development in South Africa while most of the legislation frameworks and policies are biased towards wildlife conservation (Grillini et al., 2022). Despite the fact that rural areas in South Africa witness slow growth of economic activities, agritourism is gaining popularity in areas like Eastern Cape (Cheteni and Umejesi, 2023). This implies that the rural people can complement income by engaging in other economic activities like agritourism entrepreneurship.

Although Zimbabwe is an agro-based economy, it is surprising to observe that agritourism has not been fully adopted and promoted as compared to other African countries like South Africa, Tanzania, and Kenya. It is a fact that Zimbabwe embarked on land redistribution in 2000, but there have been limited concerted efforts to foster agritourism entrepreneurship among farmers (Baipai et al., 2023). It is of utmost importance to note that agritourism entrepreneurship is relevant to the Zimbabwean economy in the sense that it creates employment, promotes rural development, fosters the establishment of tourist enterprises in rural areas and then adds value to the gross domestic product. Despite the relevance of agritourism entrepreneurship, most of the studies on agritourism were carried out in developed countries whilst limited is known in the context of developing economies, especially in African economies (Baipai et al., 2023; Ciolac et al., 2019; Zvavahera and Chigora, 2023). More worryingly, the current literature is skewed towards agritourism without capturing the concept of entrepreneurship. To bridge these worrisome literature gaps, this chapter aims to investigate the critical success factors for agritourism entrepreneurship in Africa.

Conceptualization of Agritourism Entrepreneurship

Conceptualization of agritourism entrepreneurship is a contested issue in the current tourism discourse. This suggests that there is no agreement among scholars on the definition of agritourism entrepreneurship. This could be attributed to the fact that agritourism entrepreneurship encompasses three unique disciplines that are agriculture, tourism, and entrepreneurship. Generally, agritourism entrepreneurship refers to the development of a new and unique business venture that is dedicated to the provision of high-quality agritourism services to both local and international tourists. According to Pérez-Olmos et al. (2023), agritourism entrepreneurship refers to the process of setting up a new business venture that offers tourism-related experiences that are anchored on agriculture-based activities and rural experiences. In line with this definition, agritourism entrepreneurship encompasses farm-driven activities, rural experiences, and agricultural activities that provide entertainment and leisure to tourists. From an activity-based perspective, Zvavahera and Chigora (2023) described agritourism entrepreneurship as the provision of farm-based experiences to tourists like nature walks, vineyard visits, farm stays, educational programs, fruit picking, local culture exposure, animal husbandry experiences, hunting, photography, wildlife viewing as well as rural festivals. Deducing from this definition, agritourism entrepreneurship involves leveraging and capitalizing on the natural assets of a rural or agricultural community to create a tourism business that benefits both local economies and visitors.

In light of the above, agritourism entrepreneurs can easily establish a business in areas like the organization of rural festivals and events, bed and breakfast on farms, farm-to-farm-table dining experiences, and guided-based tours of farms. While agritourism entrepreneurship appears to be lucrative in the contemporary tourism sector, it is worth mentioning that the success of agritourism entrepreneurs depends on the possession of robust knowledge related to agriculture, tourism and hospitality sector, and entrepreneurship (Khazami and Lakner, 2022). As such, a need for knowledge and skills concerning marketing, customer service, customer relationship management, entrepreneurship, business management, and sustainable agritourism. The entrepreneurship site should be designed to showcase the unique offerings of a particular agricultural region and provide visitors with an immersive experience that allows them to learn about local farming practices, sample fresh produce, and engage in outdoor activities (Leonelli et al., 2022). To this end, agritourism entrepreneurs must understand the local culture and environment to create a successful business and should be knowledgeable about the tourism industry, local regulations, food safety, and land management.

The conceptualization of agritourism in this study was informed by the opportunity-based theory of entrepreneurship. This advocates for the utilization of opportunities that lie in the external environment by setting up a business with the main purpose of making a profit. With this in mind, the authors argue that agritourism has many untapped opportunities that can be utilized by farmers in rural areas in order to earn extra income. More interestingly, the smallholder farmers can provide entertainment and leisure activities on farms at a low cost which goes a long way in profit maximization and rural development.

Agritourism Entrepreneurship and Sustainable Development Goals (SDGs)

It is worth mentioning to interrogate the connection between agritourism and SDGs within the context of Africa. Despite the fact that many African countries are facing several political, economic, technological, and social challenges, it is imperative to note that agritourism has a great potential to foster socio-economic transformation as captured in the United Nations' SDGs (Baipai et al., 2023; Jean et al., 2023). African governments are prioritizing development agenda in an attempt to incorporate all the under-served, marginalized, and neglected members of society. In this regard, there is a dire need to promote rural development through agritourism entrepreneurship. Agritourism entrepreneurship improves the living standards of rural people, creates employment, eliminates poverty, and promotes economic growth (Lovelock et al., 2023; Zvavahera et al., 2023).

In light of the above analysis, it is worth mentioning that SDG 8 states that "promote sustained, inclusive and sustainable economic growth, full and productive employment and decent work for all". It is within this context that agritourism entrepreneurship feeds into this goal through rural employment and the promotion of sustainable economic development and growth in rural areas. More interestingly, SDG 8 target number 3 states that "promote development-oriented policies that support productive activities, decent job creation, entrepreneurship, creativity and innovation, and encourage the formalization and growth of micro-, small- and medium-sized enterprises, including through access to financial services". This implies that agritourism entrepreneurship can play a significant role in decent work and economic growth through the establishment of small and medium enterprises in African economies.

Agritourism Entrepreneurship: A Global Perspective

Agritourism entrepreneurship from a global perspective refers to the development and management of agricultural businesses that offer tourism experiences to visitors. It involves attracting tourists to rural areas by showcasing agricultural activities, products, and the local way of life. The aim is to diversify income for farmers, stimulate rural development, and provide authentic and educational experiences for tourists. Agritourism entrepreneurship opportunities vary across different countries and regions due to variations in agricultural practices, cultural heritage, and natural resources. For example, some countries may focus on vineyard tours, farm stays, or farm-to-table experiences, while others may emphasize cultural heritage, traditional food production, or rural crafts.

With the growing interest in agritourism entrepreneurship in recent years, it is deemed necessary to highlight that agritourism emerged in Italy and the United States of America in the early 1920s. More interestingly, Italy is widely regarded as a country that effectively implemented and embraced agritourism (van Zyl and van der Merwe, 2021). In this regard, it passed the Agriturismo law in 1985 as a strategic move to foster farm stay in its rural areas. Going forward, there is a plethora of research studies on agritourism in the context of the United States of America (Chiodo et al., 2019; Giaccio et al., 2018).

Agritourism entrepreneurship in developed countries has gained traction as a way to diversify agricultural businesses, stimulate rural economies, and provide unique experiences for both domestic and international tourists

(Grillini et al., 2022). It is a fact that the urgent need for economic diversification necessitated to development and growth of agritourism in developed countries like Canada and the United States of America. In developed countries with well-established agricultural sectors, it is salient to observe that agritourism provides an opportunity for farmers to generate additional income by capitalizing on the demand for rural experiences. By offering farm tours, farm stays, on-site dining, or selling local products, farmers can diversify their revenue streams. For example, countries like the USA are now catering luxurious accommodations on farms for tourists. They are making bookings online of farm tours and people seem to be enjoying it.

Moving forward, the urgent calls for supporting sustainable rural development have justified the need for supporting and funding agritourism in many developed countries. It is an open secret that agritourism entrepreneurship can contribute to the economic sustainability of rural areas in developed countries. It helps counteract the decline of traditional agricultural practices, rejuvenate rural communities, and create employment opportunities (Grillini et al., 2022). In addition, agritourism has been hailed for preserving cultural heritage. With this in mind, agritourism allows visitors to experience and learn about local traditions, customs, and traditional agricultural practices. It helps preserve cultural heritage by showcasing traditional farming methods, food production, and rural crafts (Chiodo et al., 2019; Giaccio et al., 2018).

There is a general consensus that agritourism entrepreneurship in developed countries fosters farm-to-table experiences among domestic and international tourists. With the principles of farm-to-table tourism experiences, visitors can participate in farm activities, pick fruits or vegetables, and learn about organic farming practices. They can also enjoy fresh, locally sourced meals prepared on-site. Italy is one of the countries that gives the best experience of fruit picking to the stage of wine tasting at the same farm (Grillini et al., 2022).

Moreover, agritourism opens an opportunity for promoting education and awareness within the context of rural setup. This is a situation whereby agritourism provides a platform for educating visitors about the importance of agriculture, sustainable food production, and environmental conservation. It raises awareness about the challenges facing rural communities and fosters connections between urban and rural areas. To this end, developed countries like Italy, Canada, and the United States of America often have well-developed infrastructure, established tourism networks, and marketing resources that can support the growth of agritourism entrepreneurship. Additionally, government policies and funding initiatives may exist to encourage and support the development of agritourism ventures.

In the context of the European Union, there is a myriad of funding programs that are generally dedicated to fostering tourism in rural areas covering the period from 2021 to 2027. The European Regional Development Fund (ERDF), European Agricultural Fund for Rural Development (EAFRD), and Cohesion Fund are some of the most relevant European

Union funding schemes towards setting up, promoting, and extending agritourism (Grillini et al., 2022). Within the context of Italy, agritourism is flourishing given the existence of robust legal frameworks that the development of agritourism (Grillini et al., 2022).

From a global perspective, agritourism entrepreneurship plays a vital role in promoting sustainable rural development, preserving local traditions, and fostering agricultural education. It can contribute to economic growth, job creation, and the preservation of rural landscapes and ecosystems. Agritourism entrepreneurs must understand the local culture and environment to create a successful business and should be knowledgeable about the tourism industry, local regulations, food safety, and land management.

Importance of Agritourism Entrepreneurship

Agritourism entrepreneurship has been increasing its importance in recent years. As such, it is justified to interrogate the merits of agritourism, especially from a sustainable development perspective. The greatness of it can be seen in different sectors namely: economically, socially, culturally, technically, and politically. First of all, farmers, especially those in rural areas, can make money outside of their regular farming operations by hosting tourists on their farms or agricultural assets. Through agritourism, they can generate new revenue streams from their land, buildings, and resources (Grillini et al., 2022). A range of revenue streams can be employed, including entry charges, farm stays, guided tours, educational seminars, on-site meals, retail sales of farm goods and crafts, and organizing special events like festivals or weddings. Farmers may be able to lessen their reliance on income from crops alone with the aid of these alternative revenue streams (Boone and Duim, 2017; Winna et al., 2019).

In addition, agritourism entrepreneurship provides value addition to agricultural outputs (Koens et al., 2018). Value-added goods and services with an agricultural focus are a common aspect of agritourism. By creating and marketing goods resulting from their farming operations, such as artisanal food items, specialty crops, or handcrafted goods, farmers can diversify their sources of revenue. In addition, they can offer services like agricultural workshops, cooking classes, and farm-to-table dining experiences. These value-added products have the potential to bring in more money and attract higher prices.

Rural experience is improved through agritourism (Winna et al., 2019), by allowing tourists to interact with rural settings, agricultural activities, and farm life. It allows individuals to bond with nature, participate in hands-on farming activities, and develop a better understanding of where their food comes from. This authenticity and immersion in agriculture establish a distinct selling feature for agritourism firms, distinguishing them from other types of tourism. Lindstrom and Olofsson (2016) claimed that agritourism offers a variety of experiences and options, giving flexibility in the kinds of services and activities that can be rendered. Farmers are free to be inventive when it comes to the activities they provide: farm stays, tours, fruit and vegetable picking, animal interactions, cooking classes, and training on sustainable farming methods. Because of this flexibility, farmers can customize their products to the tastes and interests of their intended market, which distinguishes and appeals to customers for their agritourism enterprise.

Moving to the next good of agritourism entrepreneurship, it gives farmers new sources of income and assists neighborhood businesses. Agritourism can boost the local economy (Escalante et al., 2020). Spending money on lodging, dining, shopping, and other services by visitors to agritourism sites supports nearby companies and generates jobs. This economic stimulus promotes cooperation between farmers and other business owners and fortifies the local community. Moreover, agritourism draws travelers from many areas and even nations, promoting cross-cultural understanding and respect. Farmers and the neighborhood have the chance to engage with guests, and exchange customs, tales, and delectable food while learning about other cultures. The community's tolerance, admiration, and knowledge of other cultures are improved by this exchange of ideas and experiences. It also frequently highlights the distinctive qualities and history of a nearby community. Agritourism supports the preservation and development of the local identity by showcasing historical sites, handicrafts, cuisine, and customs. This promotes a feeling of affiliation among locals and increases communal pride.

By fostering economic growth, cross-cultural interaction, educational opportunities, and a sense of community among the local population, agritourism can serve as a bridge between farmers and the community. Agritourism allows farmers to stay flexible and adapt to changing market trends (Fernandes and McElwee, 2018). They can monitor consumer preferences and emerging tourism trends to modify their offerings accordingly. For instance, if there is an increased demand for organic farming experiences or eco-friendly accommodations, farmers can adapt their practices and infrastructure to meet these preferences. By staying attuned to market trends, agritourism businesses can maintain their competitiveness and appeal to evolving consumer demands. Farms often have abundant land and natural resources, such as scenic landscapes, forests, rivers, or lakes (Williams and Quinn, 2018). Agritourism allows farmers to showcase and leverage these natural assets by providing opportunities for outdoor activities like hiking, fishing, wildlife observation, or nature trails. By utilizing the existing natural resources, farmers can attract visitors who appreciate the beauty and tranquility of rural environments.

When it comes to the aspect of education and awareness, agritourism provides a platform for educating visitors and the local community about agriculture, food production, and rural lifestyles. Farmers can offer guided tours, workshops, or demonstrations that highlight sustainable farming practices, environmental stewardship, or the importance of local food systems. By sharing their knowledge and expertise, farmers raise awareness among visitors and the community about the value of agriculture and the challenges faced by rural communities. More interestingly, food shortages are reduced as a result of hard work and target farming by smallholder farmers (Dzingirai, 2021). Agritourism activities often involve visitors directly engaging with the farm's produce, such as picking fruits or vegetables. This can help reduce food waste by allowing visitors to harvest only what they need and appreciate the effort involved in food production. Additionally, farmers can educate visitors about reducing food waste at home through proper storage, meal planning, and utilization of leftovers. Agritourism can strengthen the local community by providing economic opportunities for farmers and supporting local businesses. By choosing to support agritourism ventures, community members contribute to the sustainability of rural economies and help maintain the social fabric of farming communities.

Also, agritourism often incorporates elements of local culture, traditions, and heritage (Chase, 2019; Escalante et al., 2020). This helps preserve and promote cultural diversity and encourages responsible tourism practices that respect and celebrate the local community's way of life. By engaging with local communities, visitors gain a deeper understanding and appreciation of the region's cultural heritage. It provides opportunities for learning about agriculture, sustainable practices, and rural life. Visitors can gain knowledge about farming techniques, food production, and environmental stewardship (Chase, 2019). Engaging in educational activities and workshops can foster personal growth, expand one's skill set, and deepen understanding of the agricultural industry. Interacting with farmers and other community members allows individuals to form connections, learn from different perspectives, and develop a sense of belonging and shared humanity.

For individuals who value sustainable living, supporting local communities, or connecting with the source of their food, agritourism provides an opportunity to align their values with their travel experiences. Engaging in agritourism allows individuals to witness and support sustainable farming practices, learn about the origins of their food, and contribute to the local economy. This alignment between personal values and travel experiences can lead to a sense of fulfillment and purpose. Farms and rural areas often provide a peaceful and serene environment, allowing visitors to escape the noise and stress of city life. The slower pace, natural surroundings, and simple pleasures of agritourism can promote relaxation, rejuvenation, and a sense of inner peace. Agritourism can be a family-friendly activity, providing an opportunity for individuals to spend quality time with loved ones in a natural setting. Families can engage in farm activities together, share meals made from farm-fresh ingredients, and create lasting memories. These shared experiences and connections can contribute to personal fulfillment and strengthen family bonds.

Challenges Faced in Adopting Agritourism Entrepreneurship

Despite the existence of the benefits of agritourism entrepreneurship, there is a myriad of formidable challenges that can be encountered by farmers when adopting agritourism. Lack of knowledge related to accounting of operational costs of agritourism entrepreneurship. Running an agritourism venture involves ongoing expenses such as maintenance and repairs, utilities, insurance, marketing, staff wages, and supplies (Vincenzo and Costa, 2016). These costs can add up quickly and need to be factored into the business's financial planning. Seasonal fluctuations in revenue can make it challenging to cover these costs during slower periods. In addition to operational costs, agritourism entrepreneurship is associated with high start-up costs. Establishing an agritourism business typically requires significant upfront investment. Entrepreneurs may need to invest in infrastructure development, such as accommodations, visitor facilities, recreational activities, signage, and marketing materials. Purchasing or upgrading equipment and machinery may also be necessary. These initial costs can be substantial and may strain the financial resources of the entrepreneur.

Another challenge associated with the adoption of agritourism revolves around marketing and promotion costs. It is important at this juncture to mention that effective marketing and promotion are crucial for attracting visitors to an agritourism business. However, marketing efforts can be costly, particularly in reaching a wide audience through various channels such as online advertising, social media, website development, and participation in trade shows or tourism fairs. Allocating sufficient funds for marketing activities can be a challenge, especially for entrepreneurs with limited budgets.

Revenue volatility is a challenge that can be faced by agritourism entrepreneurs (Purandare, 2023). In this respect, agritourism businesses often experience significant variations in revenue throughout the year. Peak seasons, such as harvest periods or tourist seasons, may generate high levels of income, while off-peak seasons may result in lower or even negligible revenue. This revenue volatility can make it challenging to maintain consistent cash flow and financial stability (Vincenzo and Costa, 2016). Moreover, seasonal fluctuations in demand require agritourism entrepreneurs to carefully manage their staffing levels. During peak seasons, additional staff may be required to handle increased visitor traffic and operational demands. However, during slower periods, it may be necessary to reduce staffing or adopt flexible employment arrangements to manage costs. Balancing staffing needs with seasonal variations can be a logistical and financial challenge. Agritourism businesses often invest in infrastructure, such as accommodations, visitor facilities, and recreational amenities, to cater to peak-season demand. However, during off-peak seasons, these facilities may remain underutilized or even sit idle. The underutilization of infrastructure can result in inefficient use of resources and impact the business's profitability.

Poor time management is another challenge that can be encountered by agritourism entrepreneurs in rural areas. Running a successful agritourism business requires significant time and effort, as it involves managing both farming operations and hospitality services. Farmers-turned-entrepreneurs often find themselves juggling multiple responsibilities, such as crop cultivation, animal care, maintenance of farm infrastructure, and customer service. Balancing these diverse tasks and allocating time effectively can be a challenge, especially during peak seasons when both farming and hospitality activities demand attention. Moreso, operation coordination challenges can creep in. Integrating farming and hospitality operations in an agritourism business necessitates seamless coordination. Farmers must ensure that farming activities do not interfere with guests' experiences and vice versa. For example, managing noise, odors, or safety hazards associated with farming operations can be a challenge while simultaneously providing a pleasant environment for visitors. Maintaining a well-coordinated and harmonious operation requires careful planning, clear communication, and efficient logistical arrangements.

Going forward, poor cash flow management can limit the adoption of agritourism entrepreneurship. Agritourism businesses often experience seasonality in demand. Peak tourist seasons or specific harvest periods may generate higher revenue, while other times of the year may be slower. Managing cash flow becomes crucial to cover expenses during leaner periods and ensure the business's sustainability (Vincenzo and Costa, 2016). Entrepreneurs need to plan and budget accordingly to avoid cash flow shortages. Furthermore, securing financing for agritourism ventures can be challenging for agritourism entrepreneurs. Traditional lenders may be hesitant to provide loans, especially to start-ups or businesses without a proven track record. Lack of collateral or insufficient credit history may further complicate the financing process. Entrepreneurs may need to explore alternative financing options, such as grants, government programs, partnerships, or crowdfunding, to access the necessary capital.

Notably, determining the right pricing strategy for agritourism experiences can be challenging for agritourism entrepreneurs. Entrepreneurs need to strike a balance between attracting visitors with competitive prices and ensuring profitability. Setting prices too low may compromise the business's financial viability, while setting them too high may deter potential customers. Careful market research, cost analysis, and competitive benchmarking are essential to establish optimal pricing structures. Moreover, lengthy approval processes are another formidable challenge that limits the success of agritourism entrepreneurs. Obtaining the necessary permits and licenses for agritourism ventures can involve lengthy approval processes. This delay can hinder the timely launch or expansion of the business, impacting revenue generation and growth opportunities. Entrepreneurs may face delays due to bureaucratic procedures, interagency coordination, public consultations, and other factors.

Methodology

This chapter adopted a structured literature review methodological approach in an attempt to collect, organize, and synthesize the existing body of knowledge related to the critical success factors for agritourism in Africa.

This study aimed to address this research question: what are the critical success factors for agritourism entrepreneurship in Africa? Notably, secondary data was gathered from the Scopus database, Google Scholar, and other organizational sources. Structured literature review refers to a systematic analysis of secondary data on a particular topic of interest (Arcese et al., 2021; Kitcharoen, 2004; Paul et al., 2021). Specifically, Paul et al. (2021) document that structured literature review refers to thoroughness in the assessment of published material related to a specific topic. Structured literature review is the most scientific, informative, and rigorous review of literature which allows the researcher to spot and poke holes in the literature (Acampora et al., 2022; Palmatier et al., 2018; Paul and Criado, 2020). The following keywords were used as search strategy: "agritourism", "agritourism entrepreneurship", "critical success factors", and "agritourism sustainability. Published works that were not written in the English language were excluded. The collected data was then subjected to thematic analysis.

Results and Discussion

This chapter focused on the critical success factors for agritourism in Africa. With the adoption of thematic analysis, the results revealed six themes, namely, central government support, adult entrepreneurship education, conducive legislation framework and robust policy, financial resources, rural infrastructure development, and agritourism entrepreneurship awareness campaigns. These results are reported and discussed as follows:

Theme 1: Central Government Support

Thematic analysis shows that central government support is one of the critical success factors for agritourism entrepreneurship. It is within this context that governments in Africa are much expected to support agritourism in rural areas. For instance, the South African government supported agritourism entrepreneurship through its land reform program as indicated below:

> Post-apartheid in South Africa, agricultural support by government has been largely focused on the disadvantaged and formerly neglected small-scale farmers … . Government support, … .manifested through an array of initiatives, with the largest expenditure being in Land Reform and the Comprehensive Agricultural Support program. The Land and Agricultural Development Bank of South Africa operates in the primary agriculture and agribusiness sectors (Litheko, 2022, p. 1057).

In light of the above quotes, it is evident that the South African government is stimulating agritourism by supporting small-scale farmers across the country. The government's commitment towards agritourism entrepreneurship has a multiplier effect when it comes to economic development and growth, as well as agritourism sustainability.

Theme 2: Financial Resources

It has been evident that financial resources are a prerequisite for the stimulation of agritourism entrepreneurship in Africa. This means that agritourism activities require funding in the form of loans and equity. In the context of Ghana, it has been documented that "the knock-on effects on small-scale producers can be substantial: one year without tourists may mean the collapse of a small tourism enterprise without an adequate financial safety net to get through the tough times" (Holland et al., 2003, p. 31). Moreover, in the context of South Africa, it has been observed that:

> the Land and Agricultural Development Bank of South Africa was expected to focus on growing its loan offerings and, in doing so, create appropriate equity investment structures and opportunities, enable existing farmers to undertake transformational projects to grow their enterprises, increase funding for agricultural activities, and continue to assist small-scale farmers access supplier and enterprise development programs to expand their operations (Litheko, 2022, p. 1057).

Theme 3: Adult Entrepreneurship Education

Adult entrepreneurship also emerged as one of the critical success factors for agritourism entrepreneurship. Notably, the conscientization of farmers about entrepreneurship in the context of agritourism emerged as a strategic move. Given that most of the farmers in rural areas do not have a high level of formal education like tertiary education, it appears to be prudent to offer adult entrepreneurship education in the unique context of agritourism. In the South African context, it emerged that:

> The owners at the bottom of the pyramid of entrepreneurship development, characterized by no formal education, the business being the primary source of income and operating at the barest level of survival, require one-on-one mentorship programs which are tailor-made to fit their diverse needs. This could vary from basic, intermediate and advanced tourism business management and skills development programs which are tailor-made to suit their business needs (Lebambo, 2019, p.16).

In light of the above quote related to South Africa, it is observable that the provision of high-quality entrepreneurship education to farmers in the form of mentorship programs opens the door for the advancement of agritourism entrepreneurship in rural areas. In the same vein, this same pattern emerged in the case of Uganda. In this respect, "the formation of the community associations, their business planning, product development and marketing training took much longer than anticipated to reach a reasonable standard for foreign and domestic tourism markets" (Holland et al., 2003, p. 31). This implies that entrepreneurship education can go a long way in assisting agritourism entrepreneurship in terms of marketing, product and market development, and business planning. As such, the ultimate goal is to foster agritourism sustainability in neglected rural areas.

Theme 4: Infrastructure Development

It is worth highlighting that infrastructure emerged as the fourth theme related to critical success factors for agritourism entrepreneurship. It is particularly necessary to note that the success of agritourism entrepreneurship mainly depends on the quality and availability as well as affordability of the supporting infrastructure in rural areas. With high-quality supporting infrastructure, agritourism sustainability can be guaranteed in marginalized rural areas. In the context of Ghana, it emerged that "accessibility to agrotourism destinations has been recognized as problematic for many destinations since most of the farms are located in rural areas with poor road and transportation networks from urban centers" (Eshun and Mensah, 2020, p. 330). In addition, in the case of Lesotho, it emerged that "the lack of infrastructure such as network connectivity, coverage, broadband and absence of electricity in rural communities hinders the growth and development of agritourism industry" (Mpiti and De-la-Harpe, 2015, p. 5). With the above African cases related to challenges faced by entrepreneurs in rural areas, it is not surprising why most African countries are failing to effectively sustain agritourism. As such, it can be deduced that infrastructure development is a critical success factor for agritourism entrepreneurship.

Theme 5: Conducive Legislative Framework and Robust Policy

The fifth critical success factor for agritourism entrepreneurship that emerged from data analysis is the conducive legislative framework and robust policy. For agritourism entrepreneurship to be a success story in African countries, it was observed that there is a need for drafting and execution of legal framework and policy that articulate the challenges related to agritourism entrepreneurs.

Theme 6: Agritourism Entrepreneurship Awareness Campaigns

The last theme that emerged from the thematic analysis is agritourism entrepreneurship awareness campaigns. As a critical success factor for agritourism entrepreneurship, agritourism entrepreneurship awareness campaigns are necessary in an effort to unleash the potential of farmers in rural areas. The ensure related to the diversification of income streams of farmers is at the heart of such campaigns. In the Zimbabwean case, "the farmers highlighted that they required awareness programs as well as workshops that would teach them about agritourism, how to start it up as well as how to manage it successfully so that they get income" (Baipai et al., 2022, p. 623).

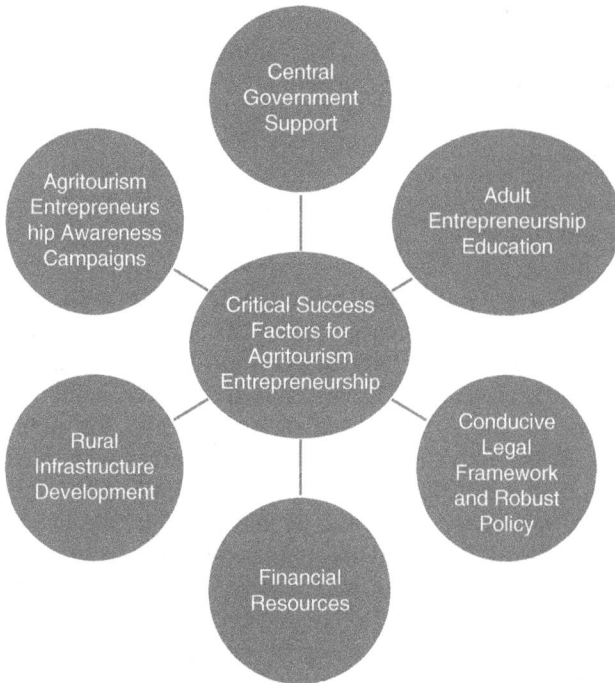

Figure 3.1 Critical success factors for agritourism entrepreneurship.
Source: Authors' compilation.

Based on this narrative, it is clear that most of the farmers in rural areas are not aware of the venture creation process and business registration process. More interestingly, within the context of Kenya, it emerged that "local stakeholders' participation has been enhanced through incentives and education campaigns to minimize the tribulations associated with agritourism and any other form that could emerge in the locality" (Kipkorir et al., 2022, p. 51). To this end, agritourism entrepreneurship campaigns are a necessity for ensuring agritourism sustainability in rural areas.

In light of the above critical success factors for agritourism entrepreneurship, Figure 3.1 shows the pictorial presentation of the critical success factors for agritourism entrepreneurship.

Implications for Theory, Policy, and Practice

This study opened avenues for theoretical, policy, and practical implications. In terms of the theoretical perspective, this study extends the application of the opportunity-based theory by applying it in the context of agritourism entrepreneurship. This implies that farmers can seize the opportunities that are embedded in agritourism by creating new ventures with the aim of maximizing profit. More importantly, this study enriches the existing body of knowledge related to agritourism entrepreneurship in the sense that it captures a framework of critical success factors as reported in Figure 3.1. Moreover, it aggregates the scattered extant literature related to agritourism entrepreneurship in Africa.

Equally important, there are some policy implications whereby the policymakers from the Ministry of Tourism and Agriculture can craft policies and frameworks to foster agritourism entrepreneurship in African rural areas. This may be in the form of tax holidays for agritourism enterprises in rural areas. This can go a long way in fostering agritourism entrepreneurship in rural areas. Finally, practitioners like bankers and managers of microfinance institutions can offer loan facilities to agritourism entrepreneurs in Africa. Specifically, the financial institutions should have a desk dedicated to agritourism entrepreneurs in rural areas in an attempt to provide the required financial products and services. This could arrest the challenges related to limited financial resources in the eyes of agritourism entrepreneurs in Africa. Consequently, SDG 8 can be achieved.

Study Limitations

There are some limitations associated with this study, which can act as avenues for further research in agritourism entrepreneurship in Africa. Firstly, this study did not focus on primary data. Therefore, future research work can focus on the collection of primary data on the critical success factors for agritourism in Africa. This could enhance the empirical evidence related to the critical success factors for agritourism entrepreneurship in Africa. Secondly, this study focused on Africa only. As such, this opened the avenue for further research on a comparative study on critical success factors between Africa and the United States of America. The comparative research can provide the contextual factors that can be linked to critical success factors for agritourism entrepreneurship in Africa.

Conclusion

This chapter focused on the critical success factors for agritourism in Africa. The agritourism-driven entrepreneurship is an emerging concept in Africa that is gaining traction in the mainstream literature. Although agritourism entrepreneurship is not a new concept, especially in developed countries like Italy and the United States of America, it is unfortunate to observe that not much attention has been paid to agritourism in Africa. More worryingly, the extant limited and sparse literature concerning agritourism in Africa is fragmented and scattered. Therefore, this study enriches the current stock of knowledge by capturing six critical success factors for agritourism entrepreneurship, namely, central government support, adult entrepreneurship education, conducive legal framework and robust policy, financial resources, rural infrastructure development, and agritourism entrepreneurship awareness campaigns. The results from this study will benefit policymakers, tourism professionals, agricultural practitioners, entrepreneurs, and researchers.

References

Acampora, A., Lucchetti, M. C., Merli, R., and Ali, F. (2022). The theoretical development and research methodology in green hotels research: A systematic literature review. *Journal of Hospitality and Tourism Management*, 51, 512–528.

Arcese, G., Valeri, M., Poponi, S., and Elmo, G. C. (2021). Innovative drivers for family business models in tourism. *Journal of Family Business Management*, 11(4), 402–422.

Baipai, R., Chikuta, O., Gandiwa, E., and Mutanga, C. N. (2022). Critical Success Factors for Sustainable Agritourism Development in Zimbabwe: A Multi-Stakeholder Perspective. *African Journal of Hospitality, Tourism and Leisure*, 11 (SE1), 617–632.

Baipai, R., Chikuta, O., Gandiwa, E., and Mutanga, C. N. (2023). A framework for sustainable agritourism development in Zimbabwe. *Cogent Social Sciences*, 9(1), 2201025.

Boone, H., and Duim, V. R. (2017). Agritourism entrepreneurship in the Netherlands: Exploring the multifunctional farm as sources of innovation. *Geoforum*, 84, 25–35.

Cheteni, P., and Umejesi, I. (2023). Evaluating the sustainability of agritourism in the wild coast region of South Africa. *Cogent Economics and Finance*, 11(1), 2163542.

Chiodo, E., Fantini, A., Dickes, L., Arogundade, T., Lamie, R. D., Assing, L., ... and Salvatore, R. (2019). Agritourism in mountainous regions—Insights from an international perspective. *Sustainability*, 11(13), 3715.

Ciolac, R., Adamov, T., Iancu, T., Popescu, G., Lile, R., Rujescu, C., and Marin, D. (2019). Agritourism—A sustainable development factor for improving the 'health' of rural settlements. case study Apuseni mountains area. *Sustainibility*, 11(5), 1467.

Dzingirai, M. (2021). The role of entrepreneurship in reducing poverty in agricultural communities. *Journal of Enterprising Communities: People and Places in the Global Economy*, 15(5), 665–683.

Escalante, A., Hurtado-Torres, N. E., Fernández, J. A., and Tovar, B. (2020). Prospects of Agri-Tourism Development in Mexico. *In Revitalising Rural Economies through Agri-Tourism* (pp. 111–126). Springer.

Eshun, G., and Mensah, K. (2020). Agrotourism niche-market in Ghana: A multi-stakeholder approach. *African Journal of Hospitality, Tourism and Leisure*, 9(3), 319–334.

Fernandes, L., and McElwee, G. (2018). Agritourism entrepreneurship and sustainable rural development: A mixed-methods study. *International Journal of Tourism Research*, 20(3), 337–346.

Giaccio, V., Mastronardi, L., Marino, D., Giannelli, A., and Scardera, A. (2018). Do rural policies impact on tourism development in Italy? A case study of agritourism. *Sustainability*, 10, 2938.

Grillini, G., Sacchi, G., Chase, L., Taylor, J., Van Zyl, C. C., Van Der Merwe, P., ... and Fischer, C. (2022). Qualitative assessment of agritourism development support schemes in Italy, the USA and South Africa. *Sustainability*, 14(13), 7903.

Holland, J., Burian, M., and Dixey, L. (2003). *Tourism in poor rural areas: Diversifying the product and expanding the benefits in rural Uganda and the Czech Republic*. PPT Working Paper No. 12. http://www.tanzaniagateway.org/docs/PPT_ Pro_Poor_Tourism_working_papers.pdf

Jean, B. P., Stael, K. P. R., Strasser-King, D. E. U., Cephas, T. K. S. L., Brown, F., and Enyan, M. (2023). The socio-economic impact of rural and agrotourism on the economy of West African rural folks. *EPRA International Journal of Socio-Economic and Environmental Outlook*, 10(4), 32–42.

Khazami, N., and Lakner, Z. (2022). The development of social capital during the process of starting an agritourism business. *Tourism and Hospitality*, 3(1), 210–224.

Kipkorir, N., Twili, N. S., and Gogo, A. (2022). Effects of agritourism development on the local community in Kericho County, Kenya. *Journal of Tourism, Culinary and Entrepreneurship*, 2(1), 34–53.

Kitcharoen, K. (2004). The importance-performance analysis of service quality in administrative departments of private universities in Thailand. *ABAC Journal*, 24(3).

Koens, K., Gössling, S., and Papp, B. (2018). Agritourism entrepreneurship: A review of the literature and direction for future research. *Journal of Sustainable Tourism*, 26(5), 724–743.

Lebambo, M. (2019). The role of entrepreneurial policies in developing rural tourism entrepreneurship in South Africa. *African Journal of Hospitality, Tourism and Leisure*, 8(3), 1–21.

Leonelli, S., Iaia, L., Masciarelli, F., and Vrontis, D. (2022). Keep dreaming: How personality traits affects the recognition and exploitation of entrepreneurial opportunities in the agritourism industry. *British Food Journal*, 124(7), 2299–2320.

Lindstrom, K., and Olofsson, P. (2016). Farm tourism entrepreneurship in developed and industrialized regions: Conceptualizing agritourism as a mainstream business activity. *International Journal of Tourism Research*, 18(4), 353–362.

Litheko, A. (2022). Development and management of small agro-tourism enterprises: A rural development strategy. *Development*, 11(3), 1053–1069.

Lovelock, B., Degarege, G., and Adeloye, D. (2023). Growing agri-heritage tourism in Africa: Challenges and opportunities. *Cultural Heritage and Tourism in Africa*, 137–153.

Mpiti, K., and De la Harpe, A. (2015). ICT factors affecting agritourism growth in rural communities of Lesotho. *African Journal of Hospitality, Tourism and Leisure*, 4(2), 1–11.

Palmatier, R. W., Houston, M. B., and Hulland, J. (2018). Review articles: Purpose, process, and structure. *Journal of the Academy of Marketing Science*, 46, 1–5.

Paul, J., and Criado, A. R. (2020). The art of writing literature review: What do we know and what do we need to know? *International Business Review*, 29(4), 101717.

Paul, J., Lim, W. M., O'Cass, A., Hao, A. W., and Bresciani, S. (2021). Scientific procedures and rationales for systematic literature reviews (SPAR-4-SLR). *International Journal of Consumer Studies*, 45(4), O1–O16.

Pérez-Olmos, K. N., Aguilar-Rivera, N., and Serna-Lagunes, R. (2023). Spatial analysis of Fortín, Veracruz, Mexico: Agritourism entrepreneurship public policy. In *SDGs in the Americas and Caribbean Region* (pp. 1453–1478). Cham: Springer International Publishing.

Purandare, S. (2023). Krishivan agri tourism: Challenges for sustainability. *The CASE Journal*, 19(6), 896–919.

van Zyl, C. C., and van der Merwe, P. (2021). The motives of South African farmers for offering agri-tourism. *Open Agriculture*, 6(1), 537–548.

Vincenzo, G., and Costa, R. (2016). Exploring the business potential of agritourism: Case studies from Italy. *Agricultural and Food Economics*, 4(1), 1–12.

Williams, G., and Quinn, B. (2018). Agritourism entrepreneurship and on-farm experiences: Insights from two Australian case studies. *Journal of Rural Studies*, 63, 134–143.

Winna, S., Djajasaputra, A., and Nugraha, F. P. (2019). The development of agri-tourism as ecotourism in Indonesia. In *International Conference on Science, Technology and Interdisciplinary Research (IC-STAR)* (pp. 235–240). Springer.

Zvavahera, P., and Chigora, F. (2023). *Agritourism: A Source for Socio-economic Transformation in Developing Economies*. Qeios. https://www.qeios.com/read/DXTYIG

4 Economic Potentials of Precision Agritourism in Alleviating Income Poverty among Farming Communities in Tanzania

The Cottage Farming Approach

Proscovia Paschal Kamugisha

Introduction

Agriculture and tourism sectors employ more than two-thirds and one-tenth of the national labor force, respectively (URT, 2021). These sectors also account for 25% and 17% of the national GDP, respectively (Busungu, 2022; URT, 2021). The agriculture sector's annual growth maintained a relatively low rate of (4%) for the last decade while the tourism sector grew by 18.9% for the period between 2016 to 2019 (URT, 2022) before being hit by COVID-19. The tourism sector is rejuvenating and starting to stabilize (URT, 2023a) with determined government efforts. Despite the promising growth of the tourism sector in many developing countries, macroeconomists argue that the revenue receipts from the tourism sector do not effectively reach the marginalized poor people, who constitute a significant proportion of the population in developing nations. (Lwoga, 2013). In light of this, any pro-poor efforts to link the two sectors are a prerequisite step to alleviate poverty among the masses. Pro-poor tourism in Tanzania should actively engage rural farming communities, which constitute the majority (80%) of the impoverished population in the country.

More than three-quarters of Tanzanian households engage in agriculture and a significant proportion (one-third) derive at least half of their income from agricultural activities (Finscope, 2017, 2023). The agricultural sector employs more than two-thirds of the national labor force (URT, 2021a, 2021b) and contributes a quarter of the national GDP (Embassy of Switzerland in Tanzania, 2023). Intertwining of large proportion of national labor force employed in agriculture sector and small contribution to GDP signifies existence of high poverty levels among farming communities. Leyaro & Morrissey (2013) noted an average income of 131,000Tsh/acre (USD 78.6/acre) is accrued from crop production equivalent to a monthly income of USD 13.1 for farming households residing in regions experiencing unimodal rainfall patterns and USD 26.2 for households located in bi-modal rainfall pattern regions. This is based

DOI: 10.4324/9781032696188-4

on the fact that smallholder farmers cultivate up to 2 acres of land size per season (URT, 2021b). This income is insufficient to meet the basic needs of an average household, which typically consists of five members. (Kilombele et al., 2023) and it is far below the international poverty line of USD 1.9 and national basic poverty line of 36,482 Tsh/month. Actually, only two-fifths of the farmers have adequate income to buy food and clothes for the family requirements all year long; while the rest (three-fifths) face difficulties in meeting household basic needs throughout the year (Anderson et al., 2016).

The poverty level is aggravated by seasonal nature of the incomes accumulated, price volatility of agricultural produce and agricultural risks involved in the production process (Borhara et al., 2020; Hamisi, 2013; Leyaro & Morrissey, 2013; Tanzania Meteorological Authority (TMA), 2023; Tumaini, 2009). Farmers fetch low prices during the peak season (URT, 2015) because they grow similar crops, which leads to market saturation during harvest (Bank of Tanzania (BOT), 2022; Huka et al., 2014; Mgale et al., 2022). Aggravated with cultivating small land areas of up to 2 acres per household (URT, 2021b), low productivity levels (Nyaligwa et al., 2022; URT, 2021b; World Bank, 2021) and large risks associated with agricultural production (Anderson et al., 2016); farmers end up earning low incomes to meet their basic needs. In this case, any pro-poor efforts to link agriculture and tourism sectors have been perceived as an important tool to alleviate poverty among the rural farming communities. Agricultural food accounts for more than one-fifth of tourism earnings in Tanzania whereby horticultural and beverage produce form a vital role in the country's tourist hotels and restaurants (UNCTAD, 2022). Pro-poor tourism in Tanzania has been implemented to improve rural livelihoods that constitute the majority (80%) of the poor people in Tanzania.

A number of pro-poor tourism approaches have been implemented in Tanzania. Improvement of tourism sector to accommodate the poor include launching of Community-Based Tourism (CBT) approach during the early 1990s. The CBT encompasses all community engagement tourism activities aiming at improving income welfare of the poor. Apart from wild life tourism, the CBT in Tanzania involves cultural tourism and agricultural tourism. While cultural tourism entices visitors with local culture and norms practiced by the community (ACT, 2018), agritourism entails supply of the agricultural produce to tourist destinations including tourist hotels, campsites and supermarkets where tourist source their needs (ACT, 2018; Sirima, 2023). Other mode of agritourism is practiced by devising farming activities that will not only be sold along the tourist destinations but also entails welcoming tourists to either observe the practiced and/or get involved in the farming activities for their leisure at a price (Barbieri & Mshenga, 2008). To date, agritourism in Tanzania exists in eight regions with 12 agritourism groups averaging 70 annual tourist visits per group, and nine success cases (ACT, 2018).

Numerous studies have highlighted the significant potential of agri-tourism as a means of diversifying farmers' sources of income, particularly in the rural settings where the majority of the population lives in poverty (Busungu, 2022; Mahaliyanaarachchi, 2016; Sirima, 2023). However, these studies have focused on normal farming activities which are prone to climate-related catastrophes. This book chapter complements this body of knowledge by exploring economic potentials of deploying precision agritourism to reap advantages of increased farm yield and meeting tourist attractions throughout the year.

Literature Reviews and Theoretical Foundation

Rogers' technology diffusion theory entails an innovation that seeks to reduce uncertainties in causal-effect relationship deemed to achieve the anticipated outcome. Rogers used the terms innovation and technologies as synonyms as will be used in this chapter. The theory considers technological adoption among users as a decision to use innovation which seems to be the best option available. Rogers' innovation diffusion theory assumes technology diffusion as the process through which the innovation is communicated via social system among members of the targeted population. Four pillars of technology are innovation, time assumed to elapse during which diffusion is taking place, communication routes through which the innovation is passing and ecological social networks governing the adoption.

Although agritourism is aged globally, its deployment in Tanzania is relatively a new concept. Given the newness of agritourism investment in the country, clarification of gains and drawbacks needs to be executed through personal communication and/or mass media such as radio broadcasting, newspapers and television advertisements. The ecological networks involve interrelated units aiming to solve common challenges and achieve a common goal. While agriculture sector stakeholders seek to diversify farmers' income sources and reduce poverty levels in the country's farming communities, the tourism industry stakeholders seek to diversify tourism products and improve the economic well-being of communities surrounding tourist sites. Henceforth, stakeholders from both industries would join hands on how to improve the economic welfare of farming communities located near tourist sites.

It can be concluded that the application of Rogers' innovation diffusion theory to agritourism in the Tanzanian context involves knowledge creation about the innovation, the decision-making process to adopt the technology, the implementation of the innovation, and the confirmation of its effectiveness. Rogers' theory reasoned innovation diffusion to be attributed with simplicity of the technology, observation on its applica-bility and the relative advantage the innovation offers compared to others. Accordingly, 13.5% of the targeted population is postulated to be early adopters while more than one-third (34%) as early majority who are assumed to be part of early adopters of agritourism innovation. Given the empirical literature in Tanzania (Neligwa, 2018), a 10% early adoption of precision agritourism was assumed in estimating of economic potentials of agritourism innovation to upgrade the incomes of farming communities in the country. Rogers' diffusion theory has been employed in studying innovations diffusion process in multidisciplinary sectors including marketing, agriculture, public health and information technology since its development in the early 1960s (Garc, 2020). The successful application of the theory in different disciplines makes a foundation of its application in this study.

Climate smart agricultural investments like precision agritourism seek to increase farmers' income since they supplement the incomes accrued from agricultural activities within the same production unit. Precision agriculture entails the optimal use of inputs in agricultural fields by applying the actual input requirements on the site and/or crop specific with the aid of digital technologies (Mueller et al., 2012); precision agritourism embarks on the use of precision agriculture principles to ensure increased yield and meet tourist demand requirements throughout the year. Indeed, precision agritourism offers diversified farmers' income sources, reduced agricultural loss risks and offers more job employment opportunities in the national farming communities. By enabling employment opportunities and increasing agricultural productivity levels, precision agritourism contributes to the Sustainable Development Goals (SDGs), African Development Agenda 2063 and other national initiatives geared to improve economic welfare and human livelihood of vulnerable groups.

Specifically, precision agritourism contributes to the achievement of SDG 1, which aims to halve income poverty by 2030 by creating increased income opportunities. Simultaneously, it plays a role in the attainment of SDG 2, addressing food and nutrition insecurity concerns. The achievement of SDG 2 is facilitated by the elevation of incomes for households engaged in agritourism investments, thus improving access to food and nutrition. In addition, precision agritourism aligns with SDG 13.3, which aims to establish robust adaptive climate change agricultural investments to protect vulnerable groups from climate change-related catastrophes. Equally important, precision agritourism investments will lead to the achievement of SDG 5, which seeks to narrow gender gap in vulnerable groups like farmers.

It is acknowledged that investment in agritourism and other hospitality industries are useful tools to empower women and youths in different localities of the globe such as France (Wright & Annes, 2014), Indonesia (Komariah et al., 2019) and Africa at large (Apaza-Panca et al., 2020). Halim et al. (2016) noted pertinent challenges facing women involved in agritourism, including limited access to economic resources, gender norms perspectives and balancing of gender traditional roles tied to women to make the opportunity a burden due to double roles at home and at employment. The challenges also face Tanzanian women in other employment positions. However, these challenges can be overcome by soliciting housemaids as it is done in other employment opportunities for women. Tanzanian households whose women are formally employed solicit the maids to take care of children, elders and other domestic chores (International Labour Office (ILO), 2016).

However, the precision agritourism investment is not a panacea for all economic downturns facing farmers. Thus, precision agritourism should not be considered as a single economic activity in the society. Solely depending on this economic activity could have negative consequences with pandemic incidences and economic recession that limits operation of tourism industry. A good example is the recent incidence of COVID-19 pandemic that led to a decreased household income by 12% in 2020, and reduced labor demand by 26%, 8.3% and 5.2% in the transport sector, construction subsector and retail industry, respectively (Henseler et al., 2022). Equally important, the occurrence of cyclical world economic recessions typically has a negative impact on the tourism industry. For instance, during the recession in 2008/09, the Tanzanian tourism sub-sector witnessed a 9% decline in tourist arrivals, a 50% drop in porters' incomes, and a 25% reduction in incomes among stall owners and operators in the Northern circuit of the country. (World Tourism Organisations (UNWTO) & International Labour Organisation (ILO), 2013). Similar negative effects were reported by Ahmada (2010) who recorded a 17% decline of tourist arrivals and 50% dropout of airline flights between 2008 and 2009 in Zanzibar. The recorded negative effects in tourism industry are associated with Precision agritourism since the diversified income for farmers is to be accrued from tourism industry. Henceforth, while investing in precision agritourism is attractive, risk mitigation mechanisms need to be developed to safeguard investors in the sector.

Other complications may arise due to the substantial capital investments involved in the farming cottage approach. The cottage farming approach entails constructing small houses within farm attractions to accommodate visitors. These cottages could be strategically placed near waterfalls, amidst specific crops, fish farming units, or any other attraction. Additionally, they would be facilitated with cooking facilities to allow tourists to prepare meals according to their preferences. Although the approach provides more privacy to both visitors and households accommodating the visitors; the cottage approach requires relatively huge investment costs to be invested apart from other agricultural activities. The huge investment cost outlay is acknowledged to be a barrier for agritourism development in South Africa (Niekerk, 2013) Namibia (Gaeb, 2023) and India (Goyal et al., 2023). The constraints can be resolved by regulating the financial sector to offer credit facilities for those who wish to invest in the industry, or collective action to a group of individuals who pull together resources and invest collectively.

Methodology

Tanzania's tourism sector annual growth rate accounted for almost one-fifth and ranks as the second most vibrant sector accounting for 17% of the national GDP before COVID-19; yet the sector has started to rejuvenate since mid-2021 due to determined strategic policy actions undertaken by the government. Meanwhile, the agricultural sector employs two-thirds of the national labor force whose major portion is poor practicing rain-fed agriculture. Precision agritourism is thought to contribute to improving economic welfare of the farming communities' proxy to tourist attraction sites through increased and diversified incomes. This book chapter seeks to explore potentials of precision agritourism to alleviate income poverty among Tanzanian farmers.

Five search queries, namely Agritourism, Precision agriculture, poverty levels among farming communities, Tanzania, and tourism regulatory framework, were employed to gather information from various search engines, including Google Scholar, *African Journals Online* (AJOL), IEEE Xplore, and Emerald. Additional search engines utilized included Research Gate, Taylor and Francis, MDPI, Springer, EBSCO, Elsevier, and Routledge. Furthermore, publications from authoritative bodies such as the World Bank Group, Country Authority Bodies, and UNCTAD were consulted to obtain statistics on the national, regional, and global status of agritourism, among other relevant data (refer to Appendix 4.1). These bibliographic databases are recognized as reliable and reputable within scientific communities, establishing them as valid sources of knowledge.

Additionally, the span of time (1997–2023) i.e. 26 years under review is relatively long to reflect the historical development of agritourism concept in Tanzania. However, the critical review has been based on literature published mainly within a seven-year time period from 2017 to 2023 which is a relatively long enough time to examine the national agritourism pattern. The inclusion criteria in searching published literature are depicted in Table 4.1 below. Such findings on published literature have been discussed throughout the chapter to reveal the needed information at hand.

Table 4.1 Literature review selection criteria

1	Year of publication	1997–2024
2	Language	English
3	Keywords	Agritourism, Precision agriculture
4	Types of publications	Peer-reviewed journals, international conferences, renowned authoritative reports & working papers, university thesis, policies and other regulatory frameworks governing agritourism industry

Limitation of the Study

The narrative review approach is known to lack systematic review procedures, raising concerns about potential biases if not handled carefully. Nevertheless, this approach remains valuable for generating evidence-based advocacy to inform investors, policymakers, regulators, Non-Governmental Organizations, and development partners to intervene in the agritourism industry. Furthermore, deployment of Rogers' theory of diffusion lacks a time specification under which innovation could be diffused in a particular society. To address this weakness, the author employed a combination of theory and empirical evidence in the Tanzanian context. Henceforth, it is expected that upon successful implementation of a promotion campaign through different mass media coupled with physical visits in smart farming agricultural fields and agritourism enterprises, precision agritourism is most likely to be adopted within a five-year period.

Poverty Levels of Agricultural Communities in Tanzania

More than a quarter of Tanzanians, amounting to 15 million people, were considered poor, living with less than the national poverty line of a monthly per capita spending of 49,320 shillings (The World Bank, 2019); almost half of the residents spend less than $ 1.9 per capita a day (UNCTAD, 2022). It is estimated that a national increase of GDP by 1% reduces proportion of the poor by only 0.45% (Worldbank, 2020). Indeed, 51% of the national income is owned by only 10% of the population; while a majority of Tanzanians (50%) own less than one-fifth of the national income (UNCTAD, 2022). This reflects national non-inclusive nature of economic development. Despite impressive annual economic growth of 7% for the last two decades before it fell to 4.9% since 2020 after COVID-19 (Embassy of Switzerland in Tanzania, 2023), a massive proportion of the population is still poor. Actually, economic progress emanated in sectors including industry, service and mining, each employing a small proportion of the national labor force accounting for only 3% (UNCTAD, 2022; Worldbank, 2020).

Poverty is more pronounced in rural areas where 81% of the poor reside (URT, 2023b) among farming communities compared to other sectors of the economy. In spite of the sector employing more than two-thirds of the national labor force (URT, 2021a, 2021b), it contributes only 25% of the national GDP (Embassy of Switzerland in Tanzania, 2023). Finscope (2023) noted that three out of four farmers struggled to meet basic regular expenses. In addition to failing to meet educational expenses (4%), a significant proportion of farmers (21%, 13% and 3%) could not afford to buy farm, livestock and farming equipment, respectively (*ibid.*). Given this background, it is necessary to devise a sustainable dynamic pro-poor income-generating activities to enhance the inclusiveness of massive farming communities in the country. Anderson et al. (2016) noted that smallholder farmers acknowledge the risks in agricultural production and are ready to diversify income-generating options even at a cost. Although it is acknowledged that a one-shoe-fits-all kind of solution does not exist, this book chapter advocates the deployment of precision agritourism to alleviate poverty. Precision agritourism suits more diversifying income sources as it mitigates climate-related vagaries for the produce, and ensures increased incomes from charges made to tourists.

Evolution of Agritourism in Tanzania

Barbieri & Mshenga (2008) defined agritourism as practices developed in agricultural fields to attract visitors. The concept encompasses two sectors, agriculture and tourism, aiming to enhance the profitability of farming undertakings. This book chapter adopts the Community-Based Tourism (CBT) and farming cottage Approaches, wherein farming communities cultivate crops and/or raise livestock to generate income directly from the farms. Simultaneously, they attract tourists to their farms, offering leisure experiences at a price. In scientific communities, CBT is recognized as an approach wherein communities are actively involved in the execution of tourism within their context, incorporating both tangible and intangible aspects of their cultural lives.

The existence of agritourism in human life can be traced back to the late 19th century in America, when urban residents started visiting rural relatives' farms during the spring season. Over time, rural visits gained popularity among urban dwellers, and this trend further expanded after the First World War in 1929, encompassing various parts of the globe (Mckenzie & Wysocki, 2002; Wicks and Merrett, 2003). Since then, urban occupants enjoyed different farm activities during vacation that included riding ponies, petting animals and encountering rural life that formed the basis of commencing commercial farm visits in America (Chase et al., 2018; Walke et al., 2017). In the 1990s, agritourism boomed into commercial aspects of the nations during vacations in the country. Agritourism became more famous and was practiced in European countries such as France, German, England and Italy during the late 1960s (Ilyukhina et al., 2021). Since then, agritourism has been spreading in numerous developing countries and Africa at large. South Africa's agritourism is relatively advanced compared to other African economies which has existed only within the last 1.5 decades (Grillini et al., 2022).

Tanzania's agritourism dates back to the late 2000s during which the tourism sector seemed to perform well economically due to structural reforms made during late 1980s and early 1990s, though the sector's earnings to local communities were limited. The minimal benefits to the communities were associated with the incapacity of local tourism entrepreneurs to compete with foreign investors. During that period, it is estimated that Tanzania experienced an annual loss of two-thirds of its foreign currency from tourism earnings (Salazar, 2009). Limited earnings to local communities were aggravated by National Tourist Policy 1999 and Tourist Master Plan 2002 which promoted the tourist industry but were silent about socio-economic development of surrounding communities (Lwoga, 2013). Limited access to economic progression of tourism industry called for tourism product diversification and local community inclusiveness. To spearhead socio-economic inclusiveness of the community, a number of Community Based Tourism (CBT) initiatives were launched including Cultural Tourism Programmes (CPT), Eco-tourism and agritourism. SNV (Netherland Development Organizations) for example supported the CBT initiatives as part of her Corporate Social Responsibility.

Agritourism is comprehensively articulated in the national policies, laws, and guidelines governing the agriculture and tourism sectors in Tanzania. However, these legal and regulatory tools are separate for each sector, despite both embracing agritourism. Tanzania's Integrated Tourism Master Plan actively promotes the diversification of tourism products to attract more visitors and foster a vibrant tourism sector. In Part IV, under Caption 14.3, the plan explicitly states the government's commitment to expanding tourist products, including agritourism. Similarly, the Tanzania Wildlife Policy advocates for the diversification of tourist products beyond protected areas to meet tourist expectations and enhance shared values with communities surrounding tourist sites. Similarly, the Ministry of Natural Resources and Tourism Strategic Plan 2021–2026 admit inadequate diversification of tourism products as a barrier to fast track the tourism sector growth, and stipulate commitment of the government to promote agritourism for enhancing the sector's growth and inclusive economic development at large. Moreover, tourism product diversification is in line with Tanzania's third Five Year Development Plan of 2021/22–2025/26 that seeks to increase tourist visits up to 5,000,000 annually with promotion of agritourism as part of the development plan.

Equally important, agritourism is embraced in agricultural-related policies, strategies and regulations as a means of diversifying income-generating activities to mitigate poverty among farming communities. Section 3.17 of National Agricultural Policy 2013 admits multiple risks associated with the agricultural sector, and embraces non-farm (agriculturally based) income-generating activities to mitigate the risks. One of the objectives of the second phase of Agricultural Sector Development Programme (ASDP II 2015/16–2025/26) is to enhance value addition and commercialize rural settings by promoting diversification of agricultural production. Agritourism being one of the means of diversifying agricultural products is embraced in this framework.

Institutionally, agritourism is governed by Agricultural Sector Development Lead Ministries (ASLMs) and Ministry of Natural Resources and Tourism (MNRT). While ASLMs promote development issues in their specific disciplines, MNRT manages issues related to tourism and natural resource governance. ASLMs include Ministry of Agriculture, Ministry of Livestock and Fisheries, Ministry of Land, Housing and Human Settlement Development, Ministry of Water and Ministry of Industries and Trade. The policies, laws and guidelines devised at ministerial levels are implemented in collaboration with the President's Office, Regional Administration and Local Governments (PO-RALG). Moreover, there are different ministerial boards and agencies that are extensions of government boards including crops boards, Research Institutes, Tanzania Investment Center (TIC), National Environment Management Council (NEMC), Tanzania Tourist Board (TTB), Tanzania National Parks Authority (TANAPA), Ngorongoro Conservation Area Authority (NCAA) and Tanzania Wild Life Research Institute (TAWIRI), to mention a few. Generally, the government plays a regulatory role while the private sector takes the operational role in agritourism sub-sector.

Concurrently, the private sector is represented by different associations that influence undertakings of the agricultural sector. Apex of cooperatives in both sectors of agriculture and tourism industry include but not limited to Tanzania Association of Cultural Tourism Organizers (TACTO), Tanzania Horticultural Association (TAHA), Community Wildlife Management Areas Consortium (CWMAC), Tourism Confederation of United Republic of Tanzania (TCT) and Responsible Tourism United Republic of Tanzania (RTTZ).

Current Status of Agritourism in Tanzania

A number of agritourism initiatives exist in Tanzania (Table 4.2). However, most of the initiatives are concentrated in the northern zone (Arusha, Dar es Salaaam, Kilimanjaro and Tanga), probably due to improved infrastructure and proximity to tourist attractions. Fewer agritourism activities are pronounced in other regions including Morogoro, Zanzibar, Mwanza, Mbeya, Iringa and Songwe regions. (Agricultural Council of Tanzania (ACT), (2018) mapped eight regions of Morogoro, Arusha, Dodoma, Kilimanjaro, Iringa, Coast, Mbeya, and Zanzibar as successful examples for agritourism in the country. Likewise, Neligwa (2018) presented a presence of 'Rural Tourism Project' in 12 villages of rural Tanzania Coordinated by *Mtandao wa Vikundi vya Wakulima Tanzania* (MVIWATA) and TAMADI. The coordination has facilitated farmers' groups in Morogoro, Dodoma, Manyara, Tanga, Kilimanjaro, Arusha and Zanzibar regions to operate agritourism activities accommodating guests in their residential areas. The project modes were done by visitors sleeping in farmers' homes and share each and every thing with the family. Although this project stands as a success case, it limits the privacy of both farmers' households and tourist visitors, and unbiased selection of households to upkeep the guests on the other hand. To date, 12 agritourism groups in the aforementioned regions enjoy tourist visitation averaging 70 visits per annum (ACT, 2018).

Moreover, there are a number of private agricultural companies engaged in the supply of agricultural produce to tourist hotels and camping sites

Table 4.2 Status of agritourism industry in Tanzania

S/N	Author	Location	Title/objective	Findings/status of agritourism	Recommendations and way forward
1	ACT (2018)	Morogoro, Arusha, Dodoma, Kilimanjaro, Iringa, Coast, Mbeya, and Zanzibar	Evaluating the linkage of tourism on Tanzanian' sustainable agriculture	Agritourism exhibit potential to improve farmers' economic status. Yet, it is currently inhibited by lack of knowledge on the opportunity, insufficient promotion of pro-poor tourism and poor deployment of Good Agricultural Practices (GAP) in their agricultural fields.	Advocation of pro-poor tourism, facilitating farmers to implement GAP in their undertakings and creation of awareness on the opportunity
2	Mkwizu et al. (2020)	Lindi, Shinyanga, Morogoro	Perception of residents on resources and promotion of rural tourism in Tanzania	Lack of awareness on agritourism among rural residences regarding the prevailing farm resources.	Promotion awareness campaigns on rural tourism
3	Sirima (2023)	Morogoro (Uluguru mountains)	Assessing prospects of agritourism (strawberry picking) to improve rural economy.	The strawberry farming is at its infancy stage in the area characterized by lack of awareness, poor infrastructure to farm site limited marketing instruments of the agritourism products.	Combination of strawberry farming & agritourism portrays huge potential to diversify income-generating activities to farmers
4	Neligwa (2018)	Morogoro, Dodoma, Manyara, Arusha, Kilimanjaro, Tanga, Zanzibar	Evaluation of rural tourism on smallholder farmers livelihood	The agritourism improves rural dwellers' livelihood sustainably. However, agritourism concept is relatively new.	Trainings of villagers and other potential stakeholders in the rural settings is important to stimulate investment in the localities.

5	Busungu (2022)	Tanzania	To review impact of pro-poor tourism approaches on poverty reduction in Tanzania	Agritourism activities with high potential to reduce poverty include coffee farming, rice farming, floriculture and supply of agricultural products to tourist destinations.	Agritourism possesses huge potentials of poverty eradication and can be enhanced through provision of training services to potential participants in agritourism industry, facilitate regulatory guidelines and advocate the tourism.
6	Mgonja (2020)	Development of Tourism in Tanzania: Strengthening Agriculture–Tourism Linkages	Evaluation of linkage between Tourism and Agricultural sectors of the economy	Weak linkages (both forward and backward) between agriculture and tourism sectors.	Strategic solution to link the two sectors
7	(Anderson, 2018)	Lindi, Shinyanga, Morogoro	To examine means of integrating local communities along tourism value chain	Existence of linkage between agrarian with tourists along the chain though the linkage is hampered by relatively low quality attributes of the products supplied.	Infrastructure development and skilled labor are prerequisite steps to integrate the local community along tourism supply chains.

Source: Study result.

(ACT, 2018; Bengesi & Abdalla, 2018). These companies secure premium prices to the supplied produce, though quality attributes required in tourist hotels require adherence to stringent safety standards. Despite the notable progress in agritourism initiatives, there is still ample room to deploy precision agritourism in the country. Mkwizu et al. (2020) noticed that the use of abundant available resources to be engaged in agritourism in rural settings is limited by insufficient knowledge of the agritourism opportunity and poor infrastructure to access the farms all year long.

Success Cases on Deploying Cottage Agritourism Approach in Alleviating Poverty

Agritourism has proved to be a means to improve the livelihood of residents in Tanzanian rural settings. Nelson (2003) revealed that the CBT has increased village earnings by 24 folds from 25 in 1995 to 600 in 2000 in Longido amounting to $10,000 per annum. Moreover, agricultural food accounts for more than one-fifth of tourism earnings in Tanzania whereby horticultural and beverage produce form a vital role in the country's tourist hotels and restaurants (UNCTAD, 2022).

Agritourism is acknowledged to be a facilitative gear in enhancing the economic development of rural dwellers' incomes in developing countries like Tanzania (Lak & Khairabadi, 2022). This is because it enables farming communities to have increased and diversified incomes in the same land unit of production (Sirima, 2023). Tanzania's agritourism industry portrays vast investment ventures due to its richness in agricultural terrains that offer tourists local farming experiences and cultural norms in the rural settings (Zvavahera & Chigora, 2023). The coffee farms in Arusha, Kilimanjaro, Kagera and Mbeya regions, and spice farming in Zanzibar, floriculture farming in mountain and hilly areas present main potential agritourism investment opportunities in Tanzania. However, the prevailing agritourism practices in the country is characterized by low productivity levels that are aggravated by climate change catastrophes (Kilungu, 2023).

Specifically, the cottage agritourism approach is exercised in different developing countries, and has proved to be a solution to lessen income poverty among farming households. The Afe Babalola University (ABUAD) farm located in Ado Ekiti State Nigeria is a successful example of agritourism enterprises that employ the cottage agritourism farming approach. The farm owns 3,000 ha and practices mixed farming approaches with a combination of crops, animal husbandry and aquaculture sub-projects (Balogun, 2020). The animals raised include poultry, pigs and snail; while crops grown include plantains, mushrooms, ground nuts and oil palm to mention a few. The farm is complemented with fish farming. The cottages are built in some areas of the farm for the tourist to enjoy their vacations on the farm.

Likewise, the Al-Baqura Village in Jordan is another success case where cottage farming agritourism is practiced. With this case, farm houses in the village are offered to tourists as lodges (Obeidat & Hamadneh, 2022). Moreover, Keisies Cottage (Montagu) and Tierhoek Cottages (Robertson) in South Africa depict other success cases. Both farm cottages provide agritourism services that increase employment opportunities and diversify income sources to the residents. A number of horticultural crops are grown in both cottages such as pecan nuts, gooseberries, clementines, quinces, lemons, mangos and apricots. The fruits raised in the farm are used to make a wide range of juices and jams. Tourists stay on the farm facilitated with self-catering services whereby different activities are arranged for them including swimming, mountain biking and hiking (Niekerk, 2013). It can be concluded that a number of agritourism cottage approach cases exist that can be up scaled to other premises to diversify the tourism products and farmers' income sources in other parts of the globe like Tanzania to improve the livelihood of marginalized groups near tourist sites.

However, agricultural practices in the cases described above entail utilization of rain-fed agricultural systems that limit agricultural production and even tourist visits during the dry season. Dry season in most of Tanzania's agroecological zones amounts to six months equivalent to 50% of the whole year (Kamugisha, 2023; Mbwambo et al., 2016). It is postulated that the six-month dry season would lessen agritourism incomes by 50% that would improve the livelihood and economic welfare of these vulnerable groups of people. To address the climate-associated challenges confronting the agritourism sub-sector, the adoption of climate-smart technologies is deemed a necessary intervention. Consequently, it is assumed that precision agritourism will not only promote increased income through sustained tourist arrivals on the farm but will also boost incomes through cost reduction and enhanced farm yield. Precision agritourism's necessity is evidenced by associated climate change prevalence of declining rainfall patterns, increased temperature levels and incidences of floods and persistent drought (Borhara et al., 2020; Luzi-Kihupi et al., 2015). Adaptation of climate change mitigation is advocated by tourism stakeholders in COP26 that was declared as part of the urgent strategy to decarbonize the planet (WTTC, 2021) at the global level, and the Tanzania National Determined Contribution (NDC) of 2021 at national level (URT, 2021b).

Evidence suggests increased efficiency in agricultural production through precision agriculture. Employment of Weed Seeker 2 has sped up the herbicide spraying period by 30% and reduced the costs of herbicides by 90% equivalent to $70/ha in a Brazilian corn farm (Luccio, 2022). Sarri et al. (2020) noted a significant result for Smart Machine for Agricultural Solutions High-tech (SMASH) project for the four modules agrobot in weed control. Likewise, Soylu and Carman found that application of AI agricultural robots reduced both slippage and fuel costs by more than two-fifths compared to conventional tractors. Moreover, Lowenberg-DeBoer et al. (2019) noted that employment of agricultural robots saved time for weeding by 100 h/ha in carrot and sugar beet farms whereby weeding costs were reduced by 50%.

Regarding water use efficiency, the utilization of tensiometer irrigation facilities with a scheduled time frame and predetermined moisture requirements has been shown to result in increased crop yield (Kumar & Ashok, 2020). Moreover, alternative wetting and drying (AWD) techniques led to the improvement of rice yield by 15% and saved water for irrigation by 42% (Yang et al., 2018). It can be concluded that precision agritourism will increase farmers' income through increased productivity and diversified revenue accrued from tourist visits to the farm.

Potentials of Precision Agritourism in Tanzania

Precision agritourism can improve the prevailing agritourism initiatives. To date, there are 18 groups of farmers engaged in agritourism activities (Agricultural Council of Tanzania (ACT), 2018; Neligwa, 2018). This can be a starting point to enhance sustainable agritourism in the farming communities throughout the year. Moreover, to date there are 780 tourist attraction sites in the country including 22 national parks, 3 RAMSAR sites, 22 game reserves, 38 wildlife conservation areas, 28 Controlled areas and Ngorongoro Conservation Area Authority (NCAA) (URT, 2021b). Other tourist resources are 465 forest reserves, 23 forest plantations, 19 nature forest reserves, 133 cultural heritage sites and 7 Museums (*ibid*).

With 780 tourist attraction resources in the country, the promotion of precision agritourism has the potential to uplift at least one village adjacent to each tourist site. This implies that the nation could support precision agritourism in at least 780 villages. Assuming the Rogers' diffusion theory of adoption and empirical studies in Tanzania (Neligwa, 2018), 10% of households in each village is postulated to join precision agritourism initiatives (Table 4.3).

Based on the 2022 Census report, a normal Tanzanian village is composed of an average of 500 households with the size of five members (Hante, 2023). Henceforth, at least 39,000 members have potential to be employed in the agritourism initiatives (Table 4.2). A monthly pay of 28 billion Tanzanian shillings equivalent to USD 136,800,000 per annum would be earned by farming communities located near tourist attractions.

Conclusion and Recommendations

Despite the achievements made so far in Tanzanian agritourism, the venture currently relies on rain-fed agricultural mechanisms, making it vulnerable to climate-related uncertainties. Additionally, the practice involves tourists staying with host households, limiting the privacy of both tourists and the accommodating households. Both the agriculture and tourism sectors still have the potential to grow, diversify products and services offered, and improve the economic welfare of farming communities surrounding tourist sites. The existing strong foundations of agritourism networking activities

Table 4.3 Potential economics of adopting precision agritourism in Tanzania

S/N	Tourist attraction	Number of tourist site attractions	Average number of households in the village	10% of households adopting agritourism	Number of household members engaged in agritourism	Average monthly per capita gains (Tsh)	Average household monthly gains (Tsh) (Percapita income* number of household members engaged in agritourism)	Total income accrued from agritourism (Household income* number of households* number of tourist attractions) (Tsh)
1	National parks	22	500	50	3	250,000	750,000	825,000,000
2	NCAA	1	500	50	3	250,000	750,000	37,500,000
3	Game reserves	22	500	50	3	250,000	750,000	825,000,000
4	Controlled areas	28	500	50	3	250,000	750,000	1,050,000,000
5	Wild life conservation areas	38	500	50	3	250,000	750,000	1,425,000,000
6	Ramsar site	3	500	50	3	250,000	750,000	112,500,000
7	Forest reserve	465	500	50	3	250,000	750,000	17,437,500,000
8	Nature forest reserves	19	500	50	3	250,000	750,000	712,500,000
9	Forest plantations	23	500	50	3	250,000	750,000	862,500,000
10	Cultural heritage sites	133	500	50	3	250,000	750,000	4,987,500,000
11	Museums	7	500	50	3	250,000	750,000	262,500,000
12	**Grand total**							**28,537,500,000**

Source: Author's computation.

along major national tourist sites enable the deployment of precision agritourism in a relatively straightforward manner. In this regard, it is proposed to implement precision agritourism using a cottage farming approach, wherein small houses are constructed on the farm to accommodate guests. These cottages can be strategically placed near waterfalls, fish farming units, or other attractions. Furthermore, the cottages can be equipped with cooking facilities, allowing tourists to prepare meals according to their preferences. Upon the request of tourists, catering services on farm sites can be arranged to serve those who prefer to have meals in restaurants, providing them with a unique lifestyle experience. The introduction of catering services adds more value to the farming communities, as individuals can be employed in the respective restaurants.

Building of farmhouses and offering of catering services to tourists necessitate pooled investment resources (capital) that warranty joint effort investments among smallholder farmers. A number of models can be adopted to operate the cottage farming agritourism approach. (1) A group of dedicated farmers can invest in this agritourism through farmers' organizations whereby the accrued profits are distributed to group members. These groups can be operated through the *Mtandao wa Vikundi vya Wakulima Tanzania* (MVIWATA) networks. Actually, a success story exists as described by Neligwa (2018). This model can be improved with precision agricultural technologies and upscaled to other villages near tourist attractions. The government and financial institutions can embrace precision agritourism by arranging a special window to facilitate farmers who want to invest in this juncture. A Build a Better Tomorrow (BBT) program model that is currently running nationwide for the youth could be employed to upgrade precision agritourism as a strategic solution.

Alternatively, Agricultural Marketing Cooperative Unions (AMCOs) can be a good entry point upon creating awareness campaigns on benefits of the investment opportunity. The AMCOs are almost found in every district for specific crops. Therefore, apart from MVIWATA, AMCO leaders can be sensitized to organize groups of farmers who are interested in precision agritourism to make an investment. Likewise, livestock and fish farmers could also be organized in groups of their associations to access the credit to finance precision agritourism farming. Otherwise, combination of crops, livestock and fish farming could offer the best experience to visitors in the combining farms. Indeed, farmers' associations would be the entry point for interested farmers to access loans from specified banks such as TIB development bank with single-digit interest rates. Otherwise, banks like CRDB bank that has a special window to finance agricultural-related investments with an interest rate of only 9% can be consulted for such an interesting endeavor.

Large-scale and/or medium-scale investors could invest in precision agritourism and involve smallholder farmers through out-grower arrangements or employ them as workers in farming activities. This model could serve as another approach to facilitate precision agritourism. Additionally, farmers can be engaged in agribusiness activities that provide commodities and/or services, adding value to the agritourism ecosystem. Government agencies responsible for roads, including the Tanzania National Road Agency (TANROAD) and Tanzania Rural and Urban Road Agency (TARURA), should invest in the construction and rehabilitation of infrastructure to ensure access to cottage farming for communities surrounding tourist sites throughout the year. It is hypothesized that the rehabilitation of 100 km of roads surrounding tourist sites in rural settings would create an enabling environment to expedite precision agritourism investments. The rehabilitation of road infrastructures would not only facilitate precision agritourism but also contribute to overall rural development.

Areas for Further Research

Based on the available literature, there is limited information on the economic feasibility analysis of agritourism in Tanzania and developing countries as a whole. Many existing studies perceive agritourism as an alternative source of income-generating activity, yet this perspective lacks substantiation through quantitative data. Considering that agritourism necessitates additional investment costs for farmers, it becomes imperative to conduct an economic feasibility analysis of this investment.

References

Agricultural Council of Tanzania (ACT). (2018). *Agritourism Regions for Enhancing Linkages between Tourism and Sustainable Agriculture in the United Republic of Tanzania Responsible Tourism Tanzania (RTTZ), the United Repub-lic of Tanzania Organic Agriculture Movement (TOAM) and UNCTAD. Additional cont.*

Ahmada, M. A. (2010). *Paraninfo Digital* [KDI School of Public Policy and Management]. 10.1016/j.earlhumdev.2006.05.022

Anderson, J., Marita, C., & Musiime, D. (2016). *National Survey and Segmentation of Smallholder Households in Tanzania Understanding Their Demand for Financial, Agricultural and Digital Solutions* (Issue May).

Anderson, W. (2018). Linkages between tourism and agriculture for inclusive development in Tanzania: A value chain perspective. *Journal of Hospitality and Tourism Insights, 1*(2), 168–184. 10.1108/JHTI-11-2017-0021

Apaza-Panca, C. M., Arévalo, J. E. S., & Apaza-Apaza, S. (2020). Agritourism: Alternative for sustainable rural development. *Dom. Cien., 6*(4), 207–227.

Balogun, K. (2020). Agritourism development and communal socio-economic sustainability in Nigeria. *Afro Asian Journal of Social Sciences, 11*(11), 0–17. https://www.researchgate.net/publication/343810070

Bank of Tanzania (BOT). (2022). *Monthly Market Bulletin.*

Barbieri, C., & Mshenga, P. M. (2008). The role of the firm and owner characteristics on the performance of agritourism farms. *Sociologia Ruralis, 48*(2), 166–183. 10.1111/j.1467-9523.2008.00450.x

Bengesi, K. M. K., & Abdalla, J. O. (2018). Forces driving purchasing behaviour of tourists hotels along tourist-agricultural supply chain in Zanzibar. *International Journal of Marketing Studies, 10*(2), 36. 10.5539/ijms.v10n2p36

Borhara, K., Pokharel, B., Bean, B., Deng, L., & Wang, S. S. (2020). On Tanzania's precipitation climatology, variability, and future projection. *Climate, 18.*

Busungu, C. (2022). Enhancing urban tourism: The role of Urban Agro-forestry and landscaping to enhancing city tourism in Mwanza City. *Eastern African Journal of Hospitality Tourism, Leisure and Tourism, 7*(1).

Chase, L., Stewart, M., Schilling, B., Smith, B., & Walk, M. (2018). Agritourism: Toward a conceptual framework for industry analysis. *Journal of Agriculture, Food Systems, and Community Development, October 2019,* 1–7. 10.5304/jafscd.2018.081.016

Embassy of Switzerland in Tanzania. (2023). *Economic Report 2023 Tanzania* (Issue August).

Finscope. (2017). *Insights that Drive Innovation* (Issue 60705).

Finscope. (2023). *Insights That Drive Innovation.* www.peakon.com

Gaeb, M. D. (2023). *Assessing the Potential of Agritourism at Neudamm, Namibia* (Issue April). The University of Namibia.

Garc, J. A. (2020). Diffusion of Innovation. In *The International Encyclopedia of Media Psychology* (Issue May, pp. 1–9). 2020 John Wiley & Sons, Inc. Published 2020 by JohnWiley & Sons, Inc. 10.1002/9781119011071.iemp0137

Goyal, P., Chadha, S., & Singh, P. S. R. (2023). *A Study on the Problems and Prospects of Agro-tourism in Rajasthan State Chaudhary Charan Singh National Institute of Agricultural Marketing, Jaipur* (Issue March).

Grillini, G., Sacchi, G., Chase, L., Taylor, J., Van Zyl, C. C., Van Der Merwe, P., Streifeneder, T., & Fischer, C. (2022). Qualitative assessment of agritourism development support schemes in Italy, the USA and South Africa. *Sustainability (Switzerland), 14*(13), 1–23. 10.3390/su14137903

Halim, M. F., Morais, D. B., Barbieri, C., Jakes, S., & Zering, K. (2016). Challenges faced by women entrepreneurs involved in agritourism. *Travel and Tourism Research Association: Advancing Tourism Research Globally, 6*(3), 213. 10.595 8/2321-5828.2015.00027.3

Hamisi, J. (2013). *Study of Rainfall Trends and Variability Over Tanzania Supervisors;* (Issue August). University of Nairobi.

Hante, M. (2023). *Voluntary Sub-National Localization of SDGs in Tanzania, 2023 – Review Report 2023.*

Henseler, M., Maisonnave, H., & Maskaeva, A. (2022). Economic impacts of COVID-19 on the tourism sector in Tanzania. *Annals of Tourism Research Empirical Insights, 3*(1), 100042. 10.1016/j.annale.2022.100042

Huka, H., Ruoja, C., & Mchopa, A. (2014). Price Fluctuation of Agricultural Products and its Impact on Small Scale Farmers Development: Case Analysis from Kilimanjaro Tanzania. *European Journal of Business and Management, 6*(36), 155–161.

Ilyukhina, N. A., Parushina, N. V., Chekulina, T. A., Gubina, O. V., Suchkova, N. A., & Maslova, O. L. (2021). Global trends and regional policy in agricultural tourism. *IOP Conference Series: Earth and Environmental Science, 839*(2). 10.1088/1755-1315/839/2/022054

International Labour Office (ILO). (2016). *A Situational Analysis of Domestic Workers in the United Republic of Tanzania ILO Country Office Dar es Salaam* (1st ed.).

Kamugisha, P. P. (2023). Fodder Commercialization as strategic Solution for Increased Ruminants' Productivity and Reduced Lad Use Conflicts in Tanzania. In A. Z. Sangeda, G. M. Msalya, I. S. Seleman, & E. J. Mtengeti (Eds.), *Health Rangelands for Sustainable Natural Resource Productivity* (pp. 147–175).

Kilombele, H., Feleke, S., Abdoulaye, T., Cole, S., Sekabira, H., & Manyong, V. (2023). Maize productivity and household welfare impacts of mobile money usage in Tanzania. *International Journal of Financial Studies, 11*(1). 10.3390/ijfs11010027

Kilungu, H. (2023). Decade of climate change and tourism research in Tanzania: Where are we? *Tanzania Journal of Forestry and Nature Conservation, 92*(1), 185–201.

Komariah, N., Padjajaran, U., Saepudin, E., Padjajaran, U., Rodiah, S., & Padjajaran, U. (2019). Women empowerment in the development of agro tourism village. *Advances in Social Science, Education and Humanities Research, 203*(Iclick 2018), 69–72.

Kumar, P., & Ashok, G. (2020). Design and fabrication of smart seed sowing robot. *Materials Today: Proceedings, November.* 10.1016/j.matpr.2020.07.432

Lak, A., & Khairabadi, O. (2022). Leveraging agritourism in rural areas in developing countries: The case of Iran. *Frontiers in Sustainable Cities, 4*(July). 10.3389/frsc. 2022.863385

Leyaro, V., & Morrissey, O. (2013). *Expanding Agricultural Production in Tanzania Scoping Study for Surveys* (Issue April).

Lowenberg-DeBoer, J., Huang, I., Grigoriadis, V., & Blackmore, S. (2019). Economics of robots and automation. *Precision Agriculture.*

Luccio, M. (2022). Satellite driven Precision Agriculture. *GPS World, 33*(1), 1–44.

Luzi-Kihupi, A., Killenga, K., & Bonsi, C. (2015). A review of maize, rice, tomato and banana research in Tanzania. *Tanzania Journal of Agricultural Sciences, 14*(1), 1–20.

Mahaliyanaarachchi, R. (2016). Role of agri tourism as a moderated rural business. *Tourism, Leisure and Global Change, 2*(March).

Mbwambo, N., Nandonde, S., Ndomba, C., & Desta, S. (2016). *Assessment of Animal Feed Resources in Tanzania* (Issue May).

Mckenzie, N., & Wysocki, A. (2002). Agritainment: A Viable Option for Florida Producers 1. *University of Florida Extension,* 1–3. http://citeseerx.ist.psu.edu/ viewdoc/download?doi=10.1.1.527.4872&rep=rep1&type=pdf

Mgale, Y. J., Timothy, S., & Dimoso, P. (2022). Measuring rice price volatility and its determinants in Tanzania: An implication for price stabilization policies. *Theoretical Economic Letters, 12,* 546–563. 10.4236/tel.2022.122031

Mkwizu, K. H., Ngaruko, D. D., & Mtae, H. G. (2020). Resources and promotion of rural tourism in Tanzania: Residents' perspective. *ACM International Conference Proceeding Series.* 10.1145/3440094.3440397

Mueller, N. D., Gerber, J. S., Johnston, M., Ray, D. K., Ramankutty, N., & Foley, J. A. (2012). Closing yield gaps through nutrient and water management. *Nature, 490*(7419), 254–257. 10.1038/nature11420

Neligwa, M. (2018). Rural tourism enhances smallholder farmer livelihoods. In *Experience Capitalization: Learning from Farmer Organisation* (pp. 54–58).

Nelson, F. (2003). Community-based tourism in Northern Tanzania: Increasing opportunities, escalating conflicts and an uncertain future. In *Association for Tourism and Leisure Education Africa Conference, Community Tourism: Options for the Future* (TNRF Occasional Paper Number 2; Issue 2).

Niekerk, C. van. (2013). *The Benefits of Agritourism: Two Case Studies in the Western Cape (Issue March).* Stellenbosch University.

Noel Biseko Lwoga. (2013). Tourism development in Tanzania before and after independence: Sustainability perspectives. *The Eastern African Journal of Hospitality, Leisure and Tourism, 1*(April), 23.

Nyaligwa, Masuki, L., & Karwani, G. (2022). Variation in nutrient use efficiency under varying soil conditions in maize growing areas in Northern Zone, Tanzania – East Africa. *International Journal of Current Research, 1.*

Obeidat, B., & Hamadneh, A. (2022). Agritourism – A sustainable approach to the development of rural settlements in Jordan: Al-Baqura village as a case study. *International Journal of Sustainable Development and Planning, 17*(2), 669–676. 10.18280/ijsdp.170232

Salazar, N. B. (2009). A troubled past, a challenging present, and a promising future? Tanzania's tourism development in perspective. *Tourism Review International, 12*(3–4), 259–273. https://lirias.kuleuven.be/handle/123456789/200400

Sarri, D., Lombardo, S., Lisci, R., Pascale, V., & Vieri, M. (2020). AgroBot smash a robotic platform for the sustainable precision agriculture. *Elettronico, 67*(May), 793–801. 10.1007/978-3-030-39299-4

Sirima, A. A. (2023). Strawberry picking as an agritourism activity at Uluguru Mountains, Tanzania. *Open Journal of Social Sciences, 11*, 332–339. 10.4236/jss.2023.117023

Tanzania Meteorological Authority (TMA). (2023). *Statement on the Status of Tanzania Climate in 2022.*

The World Bank. (2019). Tanzania Mainland Poverty Assessment. In *Tanzania Mainland Poverty Assessment.*

Tumaini, E. (2009). *Analysis of Rainfall Characteristics in Tanzania for Climate Change Signals (Issue September).* University of Nairobi.

United Nations Conference on Trade and Development (UNCTAD). (2022). *Enhancing productive capacities in the United Republic of Tanzania: A coherent and operational strategy.* United Nations.

United Republic of Tanzania (URT). (2015). *Agricultural Sector Development Strategy – II 2015/2016–2024/2025.*

United Republic of Tanzania (URT). (2021a). 2020 Tanzania in figures. In *National Bureau of Statistics, United Republic of Tanzania (Issue June).* http://www.nbs.go.tz/nbs/takwimu/references/Tanzania_in_Figures_2015.pdf%0Ahttps://www.nbs.go.tz/nbs/takwimu/references/Tanzania_in_Figures_2019.pdf

United Republic of Tanzania (URT). (2021b). Ministry of Natural Resources and Tourism Strategic Plan 2021/22–2025/26. In *The United Republic of Tanzania Tanzania Forest Service Agency (TFS).*

United Republic of Tanzania (URT). (2022). *Investment Update: A Look Into the Tourism Sector in Tanzania: Policy, Law, Incentives (Issue October 2022).*

United Republic of Tanzania (URT). (2023a). *Ministry of Finance and Planning Budget Speech 2023/2024 (Issue July).*

United Republic of Tanzania (URT). (2023b). *Tanzania's Voluntary National Review Report on the Implementation of the 2030 Agenda for Sustainable Development (Issue July).* https://ghana.un.org/en/195640-ghana-2022-voluntary-national-review-report-implementation-2030-agenda-sustainable

URT. (2021). *Integrated Labour Force Survey 2020/21, Analytical Report.*

Walke, S. G., Kumar, A., & Shetiya, M. M. (2017). *Study of Global, National and Regional Evolution of Agritourism. 5*(12), 30–37.

Wicks, B. E., & Merrett, C. D. (2003). *Agritourism: An Economic Opportunity for Illinois Defining Agritourism: The Agricultural Perspective Defining Agritourism: The Tourism Perspective. 14*(9), 1–8.

World Bank. (2021). *Implementation Completion and Result Report on a Grant from the Global Agriculture and Food Security Program to the United Republic of Tanzania for the Expand Rice Production Project.*

World Tourism Organisations (UNWTO) and International Labour Organisation (ILO). (2013). *Economic Crisis, International and its Impact on the Poor.* The World Tourism Organization. https://www.ilo.org/wcmsp5/groups/public/@ed_dialogue/@sector/documents/publication/wcms_214576.pdf

World Travel & Tourism Council (WTTC). (2021). *A Net Zero Roadmap for Travel & Tourism (Issue November)*.

Worldbank. (2020). *Overview Tanzania Mainland Poverty Assessment Executive Summary*. http://www.copyright.com/.

Wright, W., & Annes, A. (2014). Farm women and agritourism: Representing a new rurality. *Sociologia Ruralis, 54*(4), 477–499. 10.1111/soru.12051

Yang, L., Gao, D., Hoshino, Y., Suzuki, S., Cao, Y., & Yang, S. (2018). Evaluation of the accuracy of an auto-navigation system for a tractor in mountain areas. *SII 2017 – 2017 IEEE/SICE International Symposium on System Integration, 2018-January*, 133–138. 10.1109/SII.2017.8279201

Zvavahera, P., & Chigora, F. (2023). Agritourism: A source for socio-economic transformation in developing economies. *Qeios*, 1–16. 10.32388/dxtyig.2

Appendices

Appendix 4.1 Agritourism book chapter Matrix of reference publishing databases

Author	Title/area	Publishing database
Busungu, 2022	Agritourism	Research gate
Lwoga, 2013	Agritourism	Taylor and Francis
Leyaro & Morrissey, 2013	Agricultural production	International Growth Center (IGC)
Kilombele et al., 2023	Poverty levels among Tanzanian farmers	MDPI
Anderson et al., 2016	Poverty levels among Tanzanian farmers	Google Scholar
Huka et al., 2014	Farmers agricultural risks	journals.iau.ir
Nyaligwa et al., 2022	Low agricultural productivity	Google Scholar
Sirima, 2023	Agrotourism	Google Scholar
Barbieri & Mshenga, 2008	Agritourism	Repository of Iași University of Life Sciences, ROMANIA
Mahaliyanaarachchi, 2016	agritourism	EBSCO
Jose, 2020	Roger's diffusion theory	Research gate
Wright & Annes, 2014	Agritourism	Springer
Komariah & Rodiah, 2019	agritourism	Atlantis Press
Apaza-Panca et al., 2020	Agritourism	Revista Cientifica
Hanseler et al., 2022	Agritourism and COVID-19	Elsevier
Niekerk, 2013	Agritourism	Stellenbosch University http://scholar.sun.ac.za
Mckenzie & Wysocki, 2002	Agritourism	SAGE
Wicks & Merrett, 2003	Agritourism	Research gate
Chase et al., 2018	Agritourism	Research gate
Ilyukhina et al., 2021	Agritourism	IOP Publishing Ltd
Grillini et al., 2022	Agritourism	MDPI
Mgonja, 2020	Agritourism	Routledge
Salazar, 2009	Agritourism	Elsevier
Mkwizu et al., 2020	Agritourism	DOAJ
Anderson et al., 2018	Agritourism	Emerald
Zvavahera & Chigora, 2023	Agritourism	Qeios
Kilungu, 2023	Climate change	AJOL
Borhara et al., 2020	Climate change	MDPI
Luzi-Kihupi et al., 2015	Climate change	AJOL
Sarri et al., 2020	Precision agriculture and productivity	MDPI
Soylu and Carman, 2021	Precision agriculture and productivity	MDPI
Lowenberg-DeBoer et al., 2019	Precision agriculture and productivity	Springer
Kumar & Ashok, 2020	Precision agriculture and productivity	Research gate
Yang et al., 2018	Precision agriculture and productivity	IEEE Xplore

5 Artificial Intelligence and Sustainable Agritourism for Human Development in Africa

Leslie Wellington Sirora and Martin Muduva

Introduction

Sustainable agritourism is a thriving industry that has gained significant attention worldwide due to its potential not only to contribute to economic development, sustainable agriculture practices, and rural revitalization, but also to human development (Ammirato et al., 2020). It offers visitors the opportunity to experience and engage with agricultural activities, local food production, and rural culture. Sustainable agritourism can create jobs in rural areas, provide additional income for farmers, and promote environmentally friendly agriculture practices. It can also help preserve cultural traditions and promote cultural exchange between urban and rural communities. In addition, it can provide opportunities for education and skill building, particularly for young people interested in agriculture or tourism. Artificial Intelligence (AI), on the other hand, is capable of promoting sustainable agritourism practices by helping farmers optimize crop yields, reduce waste, and improve soil health. AI can also help agritourism businesses better understand their customers' preferences and tailor their offerings accordingly. By leveraging AI's capabilities, agritourism businesses can improve their operations and offer experiences that are more engaging to visitors. This can lead to increased revenue for farmers and rural communities while preserving cultural heritage and promoting sustainable agriculture practices (Hillsberg, 2021).

Agritourism, defined as the intersection of agriculture and tourism, has become a thriving industry, boasting a global market share projected to reach 117.37 billion by 2027 (Agritourism, 2022). One of the leading factors in propelling agritourism is the focus on sustainable development that has become topical in recent years. The concept of sustainable agritourism emphasizes the integration of environmentally friendly practices, community involvement, and economic viability. By promoting sustainable farming techniques, preserving cultural heritage, and fostering local economic development, agritourism has the potential to drive positive change in rural communities. In essence, it promotes human development. Human development focuses on improving the well-being and capabilities of individuals,

DOI: 10.4324/9781032696188-5

aiming for sustainable progress that benefits both present and future generations. It encompasses various aspects, including access to education, healthcare, economic opportunities, and participation (UNDP, 2015).

Agritourism has been successful in various parts of the world (Bosmann et al., 2021) (Bosmann, Hospers, & Reiser, 2021) (Baipai, 2021). However, Africa, despite having the world's second-largest arable land, comprising about 23% of the global total, has immense untapped potential for the industry. (Statista, 2023) Only a few countries, such as South Africa, Kenya, Morocco, and Tanzania, have achieved competitive agritourism goods and experiences (Ammirato et al., 2020). This is a missed opportunity as agritourism has the potential to play a significant role in promoting sustainable development and human well-being in Africa. The African agritourism sector faces unique challenges that require tailored solutions to unlock its full potential. These hurdles include market inefficiencies, limited access to information, and concerns regarding environmental sustainability. These challenges hinder the industry's ability to contribute effectively to economic growth and human development (Agritourism, 2022). To address these challenges, the integration of AI has emerged as a potential solution. AI, with its ability to process vast amounts of data, learn from patterns, and make informed decisions, holds promise for overcoming market inefficiencies, improving environmental sustainability practices, and enhancing information dissemination in the agritourism sector (Bhagat et al., 2022).

In the context of Africa, the use of AI in agritourism remains largely unexplored. Thus, there is a pressing need to investigate how AI can be effectively utilized to promote sustainable agritourism and facilitate human development in the African continent. To provide a comprehensive under-standing of the potential impact of AI on sustainable agritourism in Africa, this study will focus on specific countries in Africa, namely Kenya, Tanzania, South Africa and Zimbabwe. Kenya, Tanzania and South Africa have performed notably well in this regard. Zimbabwe has also seen great improvement over the years (Ammirato et al., 2020; Baipai, 2021). By examining agritourism practices and challenges in these countries, the study will shed light on the broader African context while providing valuable insights into the specific opportunities and barriers faced by the country.

Research Problem

Despite the potential of agritourism to promote economic growth and human development in Africa, the industry faces various challenges, including market inefficiencies, environmental sustainability concerns, and limited access to information. The application of AI in the context of sustainable agritourism remains largely unexplored in Africa. Therefore, there is a need to investigate how AI can be effectively utilized to address these challenges and promote sustainable agritourism, thereby facilitating human develop-ment in the African continent.

Objectives

1 To analyse the specific market inefficiencies in the agritourism industry in Africa with a view to how AI can be implemented.
2 To explore the potential of AI to address the identified market inefficiencies in the agritourism industry in Africa.
3 To assess the potential applications of AI in improving the overall efficiency and effectiveness of agritourism operations in Africa, with a focus on enhancing human development.
4 To recommend AI-enabled sustainable agritourism practices that contribute to the empowerment and socio-economic development of local communities in Africa.

Research Questions

The study sought to find answers to the following questions:

1 What are the specific market inefficiencies in the agritourism industry in Africa, and how can AI be leveraged to address these inefficiencies effectively?
2 How can AI technologies and data-driven approaches contribute to enhancing environmental sustainability in agritourism practices in Africa?
3 What are the potential applications of AI in improving the overall efficiency and effectiveness of agritourism operations in Africa, with a focus on enhancing human development?
4 How does the adoption of AI-enabled sustainable agritourism practices contribute to the empowerment and socio-economic development of local communities in Africa?

Theoretical Development

The theoretical framework underpinning this study was based on the triple bottom line (TBL) framework, which emphasizes the need to balance economic, social, and environmental considerations in decision-making (Hillsberg, 2021). This framework was key in guiding the data collection, comprehensive literature review and analysis process by identifying key variables and relationships that needed to be explored.

Literature Review

The literature review presented herein provided a critical analysis of the existing literature and studies on the topic of artificial intelligence and sustainable agritourism in Africa. It first examined the interconnected ecosystems of AI and sustainable agritourism to illuminate their intersection. Subsequently, it delved into the current state of agritourism in Africa and outlined the associated opportunities and challenges. Finally, it emphasized the potential of AI to address specific sustainable development goals through agritourism.

Bridging the ecosystems

Artificial Intelligence

AI has been defined as the study of "intelligent agents," entities that perceive their environment and act to maximize their chances of achieving goals. This definition encompasses various subfields of AI such as machine learning and reasoning systems which allow computing systems to learn from data without being explicitly programmed. However, a comprehensive definition of AI is not easy to provide, as different authors and disciplines may have different perspectives and criteria for what it constitutes. Over the years, AI's historical trajectory showcases its constant redefinition and expansion of capabilities; some have argued that AI's essence goes beyond technical prowess. In his book, "Artificial Intelligence: A Very Short Introduction", Wallach avoids a singular, rigid definition of AI and adopts a nuanced approach. He acknowledges the ongoing debate and complexities surrounding the concept and explores different historical and contemporary definitions of AI, highlighting the contestation within the field. Wallach argues that no single criterion can fully capture the breadth of AI, emphasizing the importance of considering various approaches and techniques used in AI. While not providing a specific definition, Wallach emphasizes that AI aims to exhibit intelligent behaviour, encompassing aspects such as perception, learning, planning, and decision-making.

Boden, in her thought-provoking work entitled "AI: Its Nature and Origins", challenges the very notion that intelligence is the exclusive domain of biological brains. She champions alternative frameworks like function-alism, which focuses on mental states and functions, and embodied cognition, where the body shapes the very fabric of intelligence. While she agrees that the goal of AI is to simulate human intelligence like problem-solving, learning and reasoning, she stresses that this doesn't necessarily imply replicating the human brain's architecture or processes. Boden further highlights that AI isn't limited to one specific approach and an AI system is to be judged by its ability to produce intelligent outputs, regardless of the underlying mechanisms. In essence, she suggests that if a system can solve problems, make decisions, and adapt to new situations in a way that can be deemed intelligent, it qualifies as AI, even if it operates differently from a human mind. Although different authors diverge on the nature of AI, what has been a point of intersection is that it concerns making machines do things that humans would ordinarily do.

AI systems can be classified into different types, such as weak or strong, narrow or general, symbolic or sub-symbolic, reactive or cognitive, and so on, depending on their capabilities, goals, and methods. They typically learn how to perform tasks by processing massive amounts of data, looking for patterns to model in their own decision-making. In many cases, humans will supervise an AI's learning process, reinforcing good decisions and discouraging bad ones. Some AI systems, however, are designed to learn without supervision – for example, by controlling a greenhouse over and over until they eventually find the optimal settings for irrigation and temperature. Researchers have concurred that AI needs to be regulated in order to address potential ethical issues such as bias, discrimination and privacy.

Sustainable Agritourism

Sustainable agritourism is a multifaceted concept that combines agriculture, tourism, and sustainability to create economic opportunities, foster cultural exchange, and preserve the environment. It requires a holistic approach that considers economic viability, social equity, cultural integrity, and environmental preservation. By leveraging agricultural heritage and resources, sustainable agritourism attracts conscious travellers interested in rural experiences, local food, and traditional farming practices.

However, sustainable agritourism faces complexities and challenges. Unequal access to resources and infrastructure in rural communities can hinder its development. It is crucial to ensure that agritourism benefits are distributed equitably, avoiding the pitfalls of enclave development that only benefit a select few and exacerbate existing inequalities. Additionally, the risk of cultural commodification must be addressed to preserve local traditions and cultures from being exploited for commercial gain. Sustainable agritourism should foster cultural exchange while respecting and preserving the authenticity and integrity of local cultures.

Despite these challenges, sustainable agritourism holds immense potential. It can revitalize rural economies, empower local communities, and promote sustainable development. By involving local stakeholders in decision-making processes and emphasizing community engagement and capacity building, sustainable agritourism can be a collaborative endeavour that benefits both tourists and the communities they visit. By carefully balancing economic gain, environmental stewardship, and cultural integrity, sustainable agritourism can create a harmonious and inclusive tapestry that nourishes communities and ecosystems alike.

Agritourism in Africa

One example of agritourism in Africa is the Maasai Mara Serena Safari Lodge in Kenya. Nestled within the Maasai Mara, this eco-lodge blends luxury with conservation. Guests enjoy wildlife viewing while contributing to community development projects and wildlife protection initiatives. The lodge sources local food, minimizes waste, and uses solar power, showcasing responsible tourism practices. Another innovative agritourism partnership can be seen in Rwanda with Zipline drones. By delivering fresh produce directly from farms to hotels and restaurants, Zipline minimizes food waste and bolsters the livelihoods of small-scale farmers. This initiative enhances the tourism experience with local, high-quality ingredients while empowering rural communities. In Uganda, tourists seeking to glimpse mountain gorillas can opt for homestays with local families near the Bwindi Impenetrable Forest. This generates income for communities, reduces pressure on the park, and fosters cultural exchange. Additionally, sustainable farming practices are encouraged among homestay families, promoting environmental responsibility. Zanzibar offers eco-tours and spice farms as part of its agritourism offerings Visitors can embark on guided tours of organic spice farms, learning about traditional cultivation methods and enjoying farm-to-table meals. This educates tourists about sustainable agriculture and supports local farmers who prioritize responsible resource management. Ethiopia, the birthplace of coffee, offers fair trade coffee tours as a form of agritourism. Tourists can learn about coffee cultivation and processing while ensuring fair compensation for farmers. This promotes sustainable farming practices and empowers local communities through economic participation.

Research on agritourism in Africa is generally low despite the fact that there is so much potential on the continent. Only a few countries have made significant achievements in agritourism, and these include South Africa, Tanzania, Kenya and Morrocco (Ammirato et al., 2020). Lombard (2020) conducted a study that shed light on the considerable growth of the tourism industry in South Africa, which accounted for 4.4% of jobs at the time. The research highlighted the potential of agritourism in diversifying farmers' operations, increasing resilience, generating additional income, and promoting sustainable farming practices. By integrating tourism activities into their farms, farmers can tap into the growing demand for authentic and immersive experiences in rural areas. Van Zyl's (2019) work delved into the growing prevalence of agritourism in South Africa, revealing that approximately 24% of South African farmers already host some form of agritourism on their farms. This finding indicates a growing interest among farmers to engage in agritourism activities, recognizing the potential benefits it offers in terms of income diversification and community engagement. The study also highlighted the various types of agritourism activities being offered, such as farm stays, farm tours, agricultural workshops, and on-farm dining experiences.

Opportunities

Agritourism has emerged as a means of sustainable development for rural areas, combining agricultural activities with tourism to attract visitors and generate income for local communities. This approach emphasizes the rational and planned exploitation of agricultural resources and the valorisation of cultural and naturalistic heritage as levers for sustainable growth. Legislation in countries like Italy has supported agritourism as a sustainable strategy to diversify and complement the economic activities of individual farms, with key objectives including the conservation of the rural environment and the socio-economic development of the local farming population.

The integration of AI into sustainable agritourism in Africa holds tremendous potential for human development. AI technologies, such as machine learning and data analytics, can enhance decision-making processes, improve resource management, and optimize operational efficiency in agritourism enterprises. By analysing large datasets related to visitor preferences, market trends, and environmental factors, AI can develop personalized experiences for tourists while minimizing the impact on natural resources. Moreover, AI-powered systems can facilitate communication and collaboration among different stakeholders in the agritourism system, leading to mutually beneficial relationships and the successful implementation of agritourism initiatives.

Studies conducted in New Jersey and the UK have highlighted the economic benefits of agritourism and its impact on farm profitability. The research reveals that agritourism is an increasingly popular form of alternative agriculture enterprise development aimed at expanding farm income through the fuller employment of existing farm resources. Moreover, the impacts of agritourism on farm profitability have been found to be statistically significant, with positive effects observed, particularly among small farms operated by individuals primarily engaged in farming.

Agritourism has also been recognized as a "smart chance" for ensuring the sustainability of the rural environment, particularly in mountainous and arid regions. It offers a multiplier effect on important aspects of economic and social life in the community, such as diversifying alternative income sources for farms, capitalizing on local products, and addressing depopulation and abandonment issues. Furthermore, the linkages between tourism and agriculture through agritourism have been emphasized as a means to enhance smallholders' livelihoods and rural community development, fostering sustainability principles.

The integration of artificial intelligence (AI) in agriculture has been recognized for its potential to enhance crop production, automate tasks, and minimize environmental damage caused by the excessive use of fertilizers (Gast, 2022). This technological advancement has the capacity to bolster food production, mitigate environmental impact, and foster sustainable agritourism across Africa. Furthermore, the potential of AI in addressing the challenges of sustainable agritourism in Africa has been highlighted, emphasizing the need for African companies to take a proactive role in developing AI applications and creating an AI ecosystem (Candelon, 2021). Collaboration with national governments, addressing challenges related to data privacy and connectivity, and investing in talent development are essential steps to enhance the competitiveness of African companies in the AI landscape.

Moreover, Suanpang and Pothipasa (2021) conducted a study on the development of AI-recommended agrotourism to support community-based tourism in the post-COVID-19 era. Their decision support system utilized a mobile application to collect traveller's information and generate appropriate tourism programs using the Extreme Learning Machine. The proposed system showed high satisfaction among users and could be implemented to support and promote community-based tourism, revitalizing the travel industry in the post-COVID-19 period. Additionally, Bhagat et al. (2022) conducted a systematic bibliometric analysis of articles on AI's role in sustainable agriculture, highlighting the need for more research in this field, particularly in Africa. Their study provides a framework for future researchers to identify key areas of interest in AI and sustainable agriculture.

Several case studies further highlight the potential of AI in addressing specific challenges in agritourism. Adewumi et al. explore the use of AI technologies to enhance food security and income generation for smallholder farmers in Nigeria. Moyo et al. examine the impact of AI technologies in improving water management and conservation for agriculture businesses in Zimbabwe. Ncube et al. investigate the effect of AI technologies on increasing customer satisfaction and loyalty for agritourism businesses in South Africa. These studies collectively underscore the transformative potential of AI in shaping the future of sustainable agritourism in Africa, offering insights into the diverse applications and benefits of AI technologies in the agricultural and agritourism sectors.

Challenges

While AI presents significant opportunities for sustainable agritourism in Africa, there are also challenges that need to be addressed. One of these challenges is the digital divide and limited access to AI technologies in rural areas. Infrastructure development and capacity building initiatives are required to bridge this gap and ensure that rural communities can fully benefit from AI-powered solutions. It is crucial to invest in the necessary technological infrastructure and provide training and resources to empower local communities in utilizing AI technologies effectively.

Government support and policies play a vital role in driving the adoption and integration of AI in agritourism enterprises. By providing the necessary resources and incentives, governments can encourage the use of AI technologies in the agricultural sector. This support can include funding research and development, offering tax incentives, and creating regulatory frameworks that promote responsible and sustainable use of AI.

Furthermore, the ethical and social implications of AI in agritourism must be carefully considered. It is essential to ensure that AI technologies are used responsibly and ethically, without compromising the cultural and natural heritage of rural areas. Transparency and privacy protection should be prioritized to address concerns about data collection and usage. Inclusive decision-making processes that involve local communities and stakeholders are crucial to ensure that AI benefits all parties involved and respects their rights and interests.

AI Stimulated Agritourism towards the Achievement of Sustainable Development Goals

The Sustainable Development Goals (SDGs), outlined by the United Nations, provide a comprehensive framework for addressing global challenges and achieving sustainable development. These goals encompass a wide range of interconnected issues, including poverty, hunger, health, education, gender equality, economic growth, infrastructure, climate action, and more. The SDGs aim to create a more equitable, inclusive, and sustainable world by 2030 (Ghidouche and Ghidouche, 2021). Agritourism, the combination of agriculture and tourism, holds significant potential for contributing to the achievement of specific SDGs. When coupled with the transformative power of AI, agritourism can become an even more powerful catalyst for sustainable development in Africa (Bhagat et al., 2022).

Karampela and Kizos (2018) discuss the potential of agritourism to revitalize rural areas, enhance income diversification, and reduce poverty. By leveraging AI-powered precision agriculture, resource management can be optimized, leading to improved yields and income opportunities for rural communities. This in turn alleviates poverty and thus addresses sustainable development goal number one.

Community involvement and government support are critical success factors for sustainable agritourism development. By incorporating AI technologies in education and training programs, agritourism can enhance access to quality education and promote learning opportunities related to sustainable agriculture and technological advancements, contributing to SDG 4. Gender disparities can also be bridged through community involvement (Baipai, 2021). Several other SDGs such as good health, zero hunger, good health and economic growth can also be addressed through sustainable agritourism, which in turn can be improved by the adoption of AI technologies.

These studies show that AI has potential in addressing the challenges facing sustainable agritourism in Africa. It emphasizes the need for African companies to actively champion AI applications, collaborate with national governments, address challenges related to data privacy and connectivity, and invest in talent development. The development of AI-recommended agrotourism systems can support and promote community-based tourism. However, further research is required to explore specific AI applications in agritourism, their impact on sustainability, and the current state of AI adoption in African agritourism. The limited literature on AI's role in sustainable agriculture indicates the need for more research in this field, particularly in Africa.

Research Method

The study employed a mixed-methods approach grounded in the Triple Bottom Line (TBL) framework to analyse the potential of AI in promoting sustainable agritourism for human development in Africa.

Conceptual Framework

The conceptual framework for this chapter posits that AI can be used to promote sustainable agritourism in Africa by addressing the following challenges:

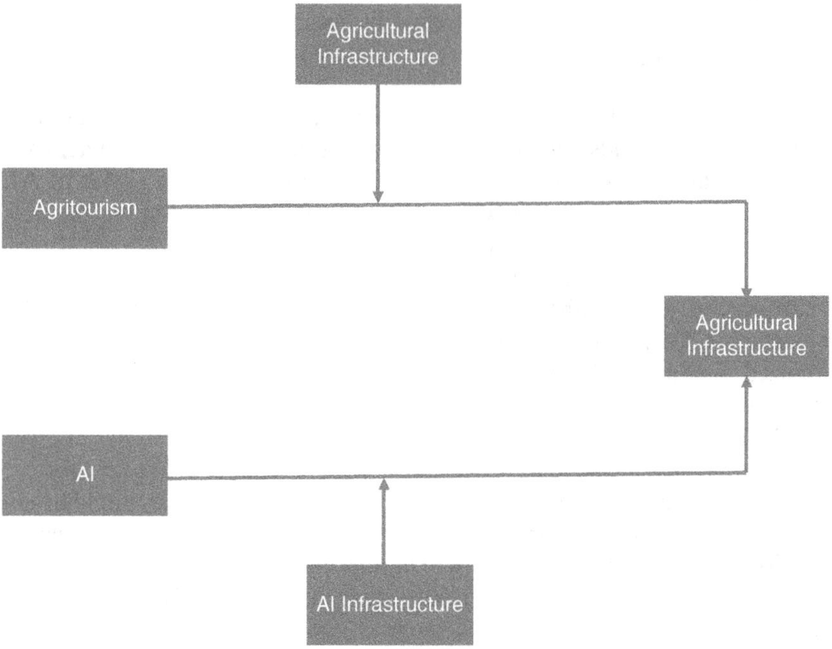

Figure 5.1 A conceptual framework for the study.

• Market inefficiencies
• Environmental sustainability
• Limited access to information

Figure 5.1 shows the conceptual framework for this study.

Primary Data Collection

The data sources of this study included primary and secondary data. Primary data was collected directly from the participants through interviews and surveys from stakeholders in the agritourism industry in Africa. These surveys and interviews were administered online. Secondary data comprised of existing literature. The primary and secondary data sources were chosen based on the research objectives and the availability and accessibility of the data.

Sample Selection

To ensure a representative sample for the primary data collection, purposive sampling approach was used to select key stakeholders from various regions in Africa, including 15 farmers, 10 tour operators, and 5 government representatives, and 5 AI experts. The sample size for the qualitative interviews was determined based on the principle of data saturation, where new information ceased to emerge from the interviews. The sample size for the survey was determined using a sample size calculator, ensuring a sufficient number of respondents to achieve statistical significance.

Primary Data Analysis

The qualitative data collected from the interviews were transcribed, coded, and analysed using thematic analysis. A total of five key themes emerged, including market inefficiencies, environmental sustainability, information dissemination, economic growth, and community empowerment. These themes provided a rich description of the research phenomenon. Word clouds were also used for exploratory analyses, visualization and comparison.

For the quantitative data, the survey responses were computed using Python programming language for analysis. Numpy, Matplotlib, ScikitLearn, Seaborn and Statsmode were the libraries used to perform the analyses in Python. Descriptive statistics were conducted to summarize the data. The survey had a response rate of 91%, with a total of 32 participants. The demographic characteristics of the survey participants were as follows: 60% were male, 40% were female, and 100% were between the ages of 25 and 65. The participants represented 4 countries in Africa, namely, South Africa, Zimbabwe, Tanzania and Kenya.

Systematic Literature Review

To analyse the secondary data, that is the existing literature, the systematic literature review (SLR) methodology was adopted. The methodology involved a five-step method, namely data collection evaluation of data, searching and locating information sources, developing a conceptual framework, and collating research and summarizing. The research data for the study included variables such as keywords, authors, journals, citations, and collaboration between authors, year of publication, and the theme of research. The data on these variables was obtained from two databases namely Web of Science and Scopus. The two databases were chosen based on the fact that they curate most of the research done in the field of agritourism.

The keywords "Africa" and "agritourism", together with the synonyms of the latter, "agrotourism", "rural tourism", and "farm tourism" were used to search through the databases. The following prompt was used:

("agritourism" OR "agrotourism" "rural tourism" OR "farm tourism") AND Africa

Using the Python environment, specifically the Matplotlib and WordCloud libraries, word cloud analysis was then carried out on these research using the titles, abstracts and keywords. Figure 5.2 shows the word cloud of the major

Figure 5.2 Word cloud of themes identified in the study.

Source: Anaconda working environment.

recurring keywords in the reviewed research. The size of the word in the depiction is directly proportional to the frequency of occurrence in the corpus.

This process helped in the identification of the primary keywords used in the literature associated with agritourism in Africa, which were then used to further identify additional literature from the databases on the same subject. In line with the TBL theory, research questions corresponding to the objectives of the study were formulated and studies relevant to those questions were sought among those collated from the databases. The methodologies, findings and recommendations of these studies were then compared and contrasted.

Rigor

To ensure the rigor of the study, the following measures were taken:

• The conceptual framework was developed based on a review of the relevant literature.
• The data collection methods were appropriate for the research questions.
• The data analysis methods were transparent and replicable.
• The findings were triangulated across multiple data sources.

Results

Market Inefficiencies in the Agritourism Industry in Africa and the Role of AI

The analysis of the primary data revealed several market inefficiencies in the agritourism industry in Africa. These inefficiencies include inadequate access to market information, lack of transparency in pricing, limited collaboration between stakeholders, and inadequate marketing strategies. These findings indicate the need for innovative solutions to address these challenges (Figure 5.3 and Table 5.1).

Market Inefficiencies for Agritourism in Africa

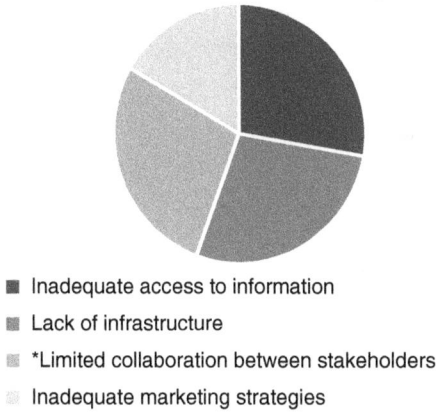

- Inadequate access to information
- Lack of infrastructure
- *Limited collaboration between stakeholders
- Inadequate marketing strategies

Figure 5.3 Market inefficiencies for agritourism in Africa.

Table 5.1 Market inefficiencies in the agritourism industry in Africa

Inefficiency	Frequency
Inadequate access to information	High
Lack of infrastructure	High
Limited collaboration between stakeholders	High
Inadequate marketing strategies	Moderate

Contributing to Environmental Sustainability through AI in Agritourism Practices

The study also investigated how AI technologies and data-driven approaches can contribute to enhancing environmental sustainability in agritourism practices in Africa. The findings revealed that AI can play a crucial role in optimizing resource allocation, reducing waste, and promoting sustainable farming practices. By utilizing AI-powered systems, agritourism operators can monitor environmental conditions, optimize water and energy usage, and minimize the ecological footprint of their operations. Additionally, AI can assist in predicting and mitigating environmental risks, facilitating the adoption of sustainable practices that safeguard natural resources.

Addressing Limited Access to Information in Agritourism through AI-Based Solutions

The study identified key challenges related to limited access to information in the context of agritourism in Africa. These challenges include insufficient dissemination of market information, lack of awareness about agritourism

Distribution of major challenges associated with access to
information on agritourism in Africa

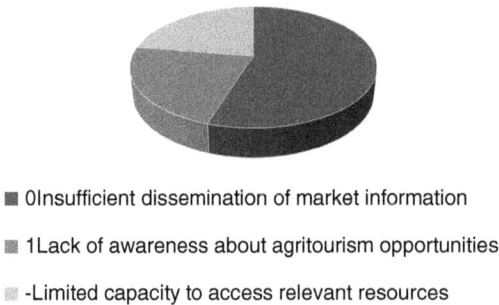

■ 0Insufficient dissemination of market information

■ 1Lack of awareness about agritourism opportunities

▨ -Limited capacity to access relevant resources

Figure 5.4 Distribution of major challenges associated with access to information on
agritourism in Africa.

Table 5.2 Challenges related to limited access to information in agritourism in Africa

Challenge	Frequency
Insufficient dissemination of market information	High
Lack of awareness about agritourism opportunities	Moderate
Limited capacity to access relevant resources	Moderate

Source: Developed by the authors.

opportunities, and limited capacity to access relevant resources (Figure 5.4
and Table 5.2).

Potential Applications of AI in Improving Efficiency and Effectiveness in Agritourism Operations

The study explored the potential applications of AI in improving the overall
efficiency and effectiveness of agritourism operations in Africa, with a specific
focus on enhancing human development. The results from the primary data
were consistent with findings of the literature reviewed that have shown how AI
can enhance various aspects of agritourism operations, such as product
development, marketing, management, and networking. Suanpang and
Pothipasa (2021) developed an AI-recommended system for agrotourism that
can provide personalized and customized tourism programs for travellers based
on their preferences and needs. Makworo and Mireri explored how ICTs, such
as mobile phones, the internet, and social media, can facilitate the integration
of community-based tourism enterprises with other sectors, such as agriculture,
education, and health. It was noted that AI technologies can automate routine

Impact of AI-enabled Sustainable Agritourism Practices on Local
Communities in Africa

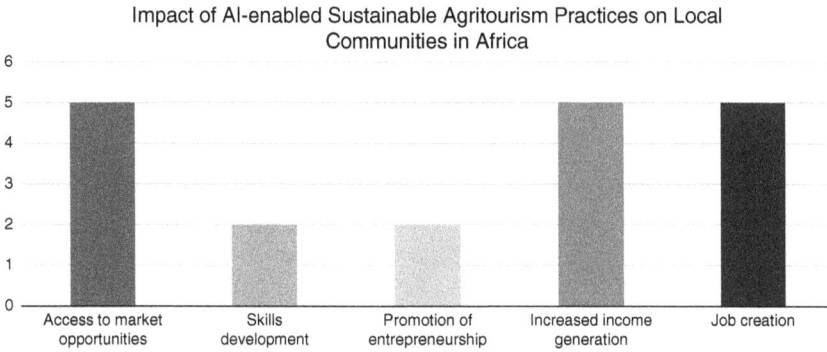

Figure 5.5 Impact of AI-enabled sustainable agritourism practices on local communities in Africa.

tasks, such as inventory management and scheduling, freeing up time for agritourism operators to focus on higher-value activities. Additionally, AI can enhance decision-making by analysing vast amounts of data, providing insights on market trends, customer preferences, and resource optimization. This, in turn, can lead to improved profitability, increased productivity, and enhanced customer satisfaction (Suanpang and Pothipasa, 2021).

The Empowerment and Socio-Economic Development of Local Communities through AI-Enabled Sustainable Agritourism Practices

The study investigated how the adoption of AI-enabled sustainable agritourism practices contributes to the empowerment and socio-economic development of local communities in Africa. The findings highlighted several positive impacts of AI in this context.

AI technologies can empower local communities by providing access to market opportunities, enhancing skills development, and promoting entrepreneurship. By leveraging AI-powered platforms, local farmers and tour operators can connect with a wider customer base, access valuable resources, and expand their business networks. This, in turn, can lead to increased income generation, job creation, and overall socio-economic development within these communities. It was noted that from the literature reviewed, there is not much data on the implementation of AI in agritourism in Africa. The primary data collected from the surveys, however, revealed the data in Figure 5.5.

Discussion

The study aimed to investigate the specific market inefficiencies in the agritourism industry in Africa and explore the potential of AI in addressing these inefficiencies effectively. Additionally, the study aimed to examine how AI technologies and data-driven approaches can contribute to enhancing environmental sustainability in agritourism practices in Africa. The research also aimed to identify the key challenges related to limited access to information in the context of agritourism in Africa and propose AI-based solutions to improve information dissemination and accessibility. Furthermore, the study aimed to explore the potential applications of AI in improving the overall efficiency and effectiveness of agritourism operations in Africa, with a focus on enhancing human development. Lastly, the research aimed to examine how the adoption of AI-enabled sustainable agritourism practices contributes to the empowerment and socio-economic development of local communities in Africa. While the research confirms previous studies, it highlights gaps for future introspection.

The research findings presented in this study align with existing studies that have identified market inefficiencies in African agritourism. These inefficiencies include inadequate market information, pricing issues, limited collaboration, and insufficient marketing strategies. The proposed use of artificial intelligence (AI) to address these inefficiencies is supported by previous research that highlights AI's potential for enhancing market intelligence, optimizing pricing, fostering collaboration, and improving marketing reach in African agritourism.

Furthermore, the identified challenges regarding limited information access, such as market data dissemination, awareness gaps, and resource access limitations, are in line with previous research. The proposed AI-based solutions for improving information dissemination and accessibility through platforms and applications align with research that suggests how AI can bridge the information gap in African agritourism.

The study's findings on AI's potential for automating tasks, enhancing decision-making through data analysis, and improving profitability and customer satisfaction in agritourism are supported by existing research on AI's applications in tourism and hospitality. Moreover, the study highlights the role of AI in empowering local communities through market access, skills development, and entrepreneurship support, which resonates with previous research on AI's potential for rural development and inclusive growth in Africa.

However, it is important to note that the study does not consider the possible negative impacts of AI on the social and cultural aspects of agritourism in Africa. Agritourism is not only a form of economic activity but also a way of preserving and promoting the local heritage, traditions, and values of rural communities (citation needed). The use of AI may disrupt the authentic and human interactions between agritourism operators and visitors, and reduce the opportunities for learning and exchange (citation needed). Therefore, future studies could zoom in on the importance of balancing the technological and human dimensions of agritourism in Africa.

Implications

The study on sustainable agritourism in Africa has several implications for the development and promotion of agritourism on the continent. The findings highlight the potential benefits of agritourism, such as income diversification, community engagement, cultural preservation, and sustainable farming practices. However, the study also identified challenges that hinder the growth and development of agritourism in Africa, including limited infrastructure, lack of marketing and promotion, and the need for training and capacity building among farmers. Moreover, the study showed that the optimization of management of natural resources and expansion of capabilities contribute towards human development. On the other hand, the study both supported and challenged the TBL theory by demonstrating the potential of AI technologies and data-driven approaches to enhance the profitability, environmental sustainability, and human development in the agritourism industry in Africa while also highlighting limitations and risks associated with their application. It showed that AI can help agritourism operators monitor and manage natural resources, reduce waste and emissions, and optimize energy efficiency, thereby enhancing their environmental sustainability. On the other hand, the study acknowledges that AI technologies are not readily available and accessible to agritourism operators in Africa, and that there are many barriers and challenges to their adoption and deployment, such as lack of infrastructure, skills, data, funding, regulation, and awareness. The study also recognizes that AI technologies may disrupt the authentic and human interactions between agritourism operators and visitors, and reduce the opportunities for learning and exchange, thereby affecting the social and cultural aspects of agritourism. Furthermore, the study admits that AI technologies may pose ethical, social, and environmental issues, such as privacy, security, accountability, bias, and inequality, thereby undermining the trust and legitimacy of agritourism.

The findings also suggest that policymakers and practitioners should support and facilitate the adoption and deployment of AI technologies and data-driven approaches in the agritourism industry in Africa. This can be done by providing adequate infrastructure, skills, data, funding, regulation, and awareness to agritourism operators and stakeholders. Policymakers and practitioners should also foster collaboration and coordination among different actors and sectors involved in agritourism, such as farmers, tourism operators, researchers, educators, and civil society organizations. Moreover, policymakers and practitioners should promote ethical, inclusive, and participatory AI governance and innovation, ensuring that the benefits and risks of AI are shared equitably and transparently among all stakeholders.

Additionally, the research implies that policymakers and practitioners should monitor and evaluate the impacts and outcomes of AI technologies and data-driven approaches on the livelihoods, capabilities, and well-being of rural communities involved in agritourism in Africa. This can be done by developing and applying appropriate indicators, metrics, and tools to measure the social, economic, and environmental effects of AI on agritourism. Policymakers and practitioners should also ensure that the voices and perspectives of rural communities are heard and considered in the design, implementation, and assessment of AI technologies and data-driven approaches. Furthermore, policymakers and practitioners should encourage and enable learning and exchange among rural communities and other stakeholders on the best practices and lessons learned from AI-enabled sustainable agritourism.

It also indicates that policymakers and practitioners should leverage and integrate AI technologies and data-driven approaches with other existing and emerging technologies and innovations that can enhance the sustainability and human development of agritourism in Africa. This can be done by exploring and exploiting the synergies and complementarities between AI and other technologies and innovations, such as renewable energy, biotechnology, blockchain, and mobile applications. Policymakers and practitioners should also foster a culture of creativity and experimentation among agritourism operators and stakeholders, encouraging them to adopt and adapt AI technologies and data-driven approaches to their specific contexts and needs. Additionally, policymakers and practitioners should stimulate and support the development and dissemination of local and indigenous knowledge and innovations that can enrich and improve the quality and diversity of agritourism in Africa.

Study Limitations

This chapter presents limitations that should be considered when interpreting the findings of the study. These limitations include the small sample size, selection and recruitment bias, and time constraints. The study's sample size was relatively small, consisting of only 32 participants. This limited sample size may have restricted the generalizability of the findings to the larger population of Africa. It is important to note that Africa is a diverse continent with 195 countries, and the participants in this study were predominantly from only four countries. This potential lack of representation may impact the generalizability of the results to the entire African population. The reason for this was that the study relied on convenience sampling and snowballing for participant recruitment. While the field of agritourism is trending, it has been sparsely adopted around Africa, hence snowballing proved to be a reasonable approach. Additionally, the study was conducted over a relatively short period, imposing time constraints on data collection and analysis. This limitation may affect the generalizability of the findings as social, economic, or environmental factors can fluctuate over time and influence the observed outcomes. A longer study duration would have allowed for a more comprehensive understanding of potential changes or variations that may occur. It is essential to acknowledge these limitations when interpreting the results of the study. Despite these limitations, the findings still provide valuable insights within the context and scope of the study. Future research with larger sample sizes, diverse participant representation, and longer study durations may help to address these limitations and provide a more comprehensive understanding of the topic.

References

Agritourism. (2022, May). Retrieved from AGMRC: https://www.agmrc.org/commodities-products/agritourism

Ammirato, S., Felicetti, A. M., and Raso, C. (2020). Agritourism and sustainability: What we can learn from a systematic literature review. *MDPI, 12*(22), 9575. doi: 10.3390/su12229575

Baipai, R. (2021). A Critical Review of Success Factors for Sustainable Agritourism Development. *African Journal of Hospitality, Tourism and Leisure*.

Bhagat, P. R., Naz, F., and Magda, R. (2022). Artificial intelligence solutions enabling sustainable agriculture: A bibliometric analysis. *Plos One*. doi:10.1371/journal.pone.0268989

Bosmann, M., Hospers, G., and Reiser, D. (2021). Searching for Success Factors of Agritourism: The Case of Kleve County (Germany). *European Countryside, 13*(3).

Candelon, F. (2021). *Developing an Artificial Intelligence for Africa strategy*. Retrieved from OECD Development Matters: https://oecd-development-matters.org/2021/02/09/developing-an-artificial-intelligence-for-africa-strategy/

Gast, A. (2022). *Why artificial intelligence is vital in the race to meet the SDGs*. Retrieved October 30, 2023, from World Economic Forum: https://www.weforum.org/agenda/2022/05/artificial-intelligence-sustainable-development-goals/

Ghidouche, A.-Y. K., and Ghidouche, N. L. (2021). Achieving sustainable development goals through agritourism in Algeria. *Worldwide Hospitality and Tourism Themes, 13*(1), 63–80. doi:10.1108/WHATT-08-2020-0092

Hillsberg, A. (2021, April). *Agritourism: A Sustainable Way of Tourism and Maintain Socio-Economical Balance*. Retrieved from Allied Market Research: https://blog.alliedmarketresearch.com/agritourism-a-sustainable-way-of-tourism-and-maintain-socio-economical-balance-918

Karampela, S., and Kizos, T. (2018). Agritourism as a tool for rural development: An analysis of stakeholders' perspectives in Greece. *Journal of Rural Studies, 57*, 49–60.

Lombard. (2020). *The Role of Agritourism in the Agricultural Economy*. Retrieved from Agriorbit: https://agriorbit.com/the-role-of-agritourism-in-the-agricultural-economy/

Statista. (2023). *Agricultural land in Africa from 2010 to 2019*. Retrieved from Statista: https://www.statista.com/statistics/1287280/agricultural-land-in-africa/#:~:text=Africa%20had%20around%201%2C119%20million,minimal%20fluctuations%20in%20the%20continent.

Suanpang, P., and Pothipasa, P. (2021). AI recommended agrotourism supporting community-based. *Journal of Management Information and Decision Sciences, 24*(1), 1–10.

UNDP. (2015). *What Is Human Development?* Retrieved from UNDP Human Development Reports: https://hdr.undp.org/content/what-human-development

van Zyl, C. C. (2019). The size and scope of agritourism in South Africa. *Development Southern Africa*, 385–399.

6 Exploring the Ethical Dimensions of Sustainable Agritourism in Central Angola

Unveiling the Role of Local Actors in Fostering Economic Growth and Environmental Conservation

Jabulani Garwi

Introduction and Background to the Study

This chapter explores the ethical dimensions of sustainable agritourism in the Caála Municipality, located in Huambo Province, Central Angola. The primary objective of this research is to examine the crucial roles played by various local actors, including farmers, community leaders, government authorities, and other stakeholders, in promoting economic growth and environmental conservation in the agritourism sector. It sheds light on the challenges and opportunities in balancing economic aspirations with environmental sustainability and cultural preservation. Sustainable agritourism is a global phenomenon that combines agriculture and tourism to drive economic prosperity, protect the environment, and foster rural progress (Budeanu et al., 2016; Silvestri et al., 2022). This approach recognizes the interconnectedness of economic development, environmental stewardship, cultural heritage preservation, and social inclusivity, emphasizing the need for a delicate balance (Wang et al., 2022; Suriyankietkaew et al., 2022). Ethical considerations play a crucial role in achieving this balance by promoting sustainable livelihoods, equitable benefit distribution, and economic resilience within local communities, while also focusing on sustainable farming practices, environmental mitigation, resource conservation, and biodiversity protection (Ciolac et al., 2021; Torabi et al., 2023). Sub-Saharan Africa, particularly Southern Africa, is witnessing the growth of agritourism as a means to diversify economies, reduce poverty, and promote rural development, leveraging the region's agricultural landscapes, cultural heritage, and biodiversity (Baiocco and Paniccia, 2023). This form of tourism offers small-scale farmers additional income sources, improved livelihoods, and opportunities for rural economic growth (Baipai et al., 2021; van Zyl and van der Merwe, 2021).

Angola has recognized agritourism as a vital component of its ongoing economic diversification efforts. Historically reliant on oil as its primary source of revenue, the nation has been prompted to explore alternative

DOI: 10.4324/9781032696188-6

strategies for economic growth due to the decline in oil income. In pursuit of this objective, Angola is actively implementing policies and projects to broaden its economic sectors, with tourism playing a central role in this transformation (MINAMB, 2021). The overarching aim is to foster regional and provincial development, alleviate the challenges faced by marginalized populations, and promote a diverse range of economic activities. As Angola seeks to reduce its dependence on oil and drive sustainable development, agritourism emerges as a promising avenue for catalyzing rural economic expansion (MINAMB, 2021). The Caála Municipality has been selected for this study due to its agricultural significance and untapped tourism potential, offering fertile ground for agritourism initiatives that can alleviate poverty, create jobs, and revitalize rural areas in Angola's agriculture-dependent economy (MINAMB, 2018). However, ethical considerations in sustainable agritourism practices in Angola are largely unexplored. Integrating agricultural production with immersive tourism experiences in Caála can showcase farming traditions, promote local products, and engage visitors in rural life, leading to revenue generation, cultural preservation, resource conservation, and community empowerment (MINAMB, 2018).

This study employed a qualitative research design rooted in a social constructivist framework to investigate the ethical dimensions of sustainable agritourism in Caála. The research approach involved qualitative interviews and policy analysis, offering a robust foundation for gaining valuable insights into the topic. An interpretivist lens was adopted to comprehensively understand the perspectives of various stakeholders involved in agritourism development. Data collection utilized a non-probability sampling technique, resulting in the selection of 21 key informants from relevant institutions directly engaged in agritourism development in Caála Municipality. These informants represented diverse entities, including the Ministry of Agriculture and Rural Development, the Ministry of Tourism, the Institute for Agricultural Development (IDA), the Forum of Angolan Non-Governmental Organizations (FONGA), and the Confederation of Farmer Associations and Agricultural Cooperatives of Angola (UNACA). UNACA specifically advocates for the interests of small-holder farmers. The research findings will enhance understanding of the complex interplay between economic development, environmental conservation, and cultural heritage preservation in agritourism. These outcomes will guide policymakers, entrepreneurs, and local communities in promoting sustainable and ethical agritourism practices, as well as foster a harmonious future for all stakeholders involved.

Literature Review

Sustainable agritourism – a fusion of agriculture and tourism, has gained significant scholarly attention due to its potential for driving economic growth, environmental conservation, and rural development (Chiodo et al., 2019). Extensive research has greatly enhanced the understanding of sustainable agritourism by exploring various aspects such as accessible and affordable sustainability (Ait-Yahia et al., 2021), farmer perceptions of sustainability (Mgonja et al., 2017), empowered actions for sustainable progress (Moswete et al., 2016; Nicolaides, 2020), recognition of sustainable practices (Ramaano, 2023), support for sustainability through ecological citizenship (Sambajee et al., 2022; Suwannasat et al., 2022), and rejection of unsustainable practices (Satta et al., 2019; Tang and Xu, 2023). The prevailing concept of sustainable agritourism emphasizes the crucial need to ensure long-term economic, environmental, and social well-being (Wiltshier, 2017). However, a significant gap in many sustainability studies lies in the oversight of ethical considerations when devising and implementing sustainability strategies and actions (Park et al., 2023). While sustainable agritourism aims to enhance the well-being of all stakeholders, as well as economic, environmental, and social aspects, there is a potential for unethical practices that can lead to undesirable outcomes. For instance, some producers and tourists may exploit the sustainability trend without genuinely improving business practices or minimizing negative impacts on the socio-environmental aspects. This raises concerns about sustainability myopia, a situation where short-term gains overshadow the long-term sustainability goals (Satta et al., 2019). To gain a comprehensive understanding of sustainable agritourism, it is crucial to incorporate ethical considerations alongside economic, environmental, social, and technological factors, thereby fostering collective well-being in the ecosystem. Furthermore, it is important to note that the majority of studies focus on sustainability in urban areas of developed countries, leaving a significant gap in research on agritourism. To address the existing research gap, this chapter seeks to provide valuable insights into the underexplored aspects of agritourism in a developing African country. By delving into the interplay of ethics and sustainability practices in agritourism, the study sheds light on previously unexplored dimensions of this critical intersection.

Extensive research has emphasized the paramount importance of ethics in the realm of sustainability initiatives. These initiatives, including sustainable agritourism, heavily depend on the ethical conduct of all stakeholders involved, such as farmers, rural businesses, tourism operators, local communities, and tourists (Mnisi and Ramoroka, 2020). The significance of ethical behaviour lies in its ability to prevent negative impacts that could hinder sustainability efforts. However, ethics are a complex and culturally nuanced subject. Different communities may hold diverse views on what is considered morally acceptable or unacceptable. For example, consequentialist communities prioritize the greater good for the majority, while deontologist societies emphasize conformity to local customs and the protection of the environment (Bhatta and Ohe, 2020). In the field of agritourism studies, ethics are often explored through policies and regulations established by formal institutions, such as rural councils and governments. These codes of ethics serve as guiding principles for decision-making processes that encompass various sustainability aspects. However, it is crucial to further explore the subject, especially in contexts where the ethical landscape is shaped by both formal institutions and the practices of local communities. Therefore, this chapter directs its attention to the ethical dimensions of sustainable agritourism in the African continent. It investigates the interplay of ethics within a framework that encompasses the influence of both formal institutions and local community practices, as highlighted by Baiocco and Paniccia (2023). By delving into this context, a more comprehensive understanding of the ethical challenges and opportunities associated with sustainable agritourism can be attained.

In a study on the ethical dimensions of agritourism in Italy, Fanelli and Romagnoli (2020) emphasized the need to balance economic imperatives with the preservation of environmental and cultural assets. The establishment and operation of sustainable agritourism ventures heavily rely on ethics. Similarly, Broccardo et al. (2017) stressed the significance of ethical considerations when tourists interact with local communities. Their research highlights the importance of responsible tourism practices that respect cultural values, preserve natural resources, and ensure equitable distribution of benefits among stakeholders. Likewise, Madanaguli et al. (2022) explored the ethical challenges of achieving a delicate balance between economic advancement and environmental sustainability in agritourism destinations. They argued for ethical decision-making processes that prioritize long-term sustainability over short-term economic gains. Additionally, Ngo et al. (2018) investigated the ethical dilemmas faced by Mexican agritourism entrepreneurs, highlighting the tension between economic prosperity and environmental preservation. While these studies provide global insights, there is a need to examine these complex ethical issues within the context of African agritourism. Thus, this study presents an opportunity to explore the ethical dimensions of agritourism in Africa, focusing on its unique challenges and potential solutions.

The critical ethical challenge faced by Africa's agritourism sector involves striking a delicate balance between advancing economic development and safeguarding indigenous cultures and ecosystems. Compared to other regions, agritourism ecosystems in Africa are characterized by unique complexities stemming from diverse cultural traditions, the value of biodiversity, and community-centric approaches to land management. Baipai et al. (2023) emphasize the importance of prioritizing the well-being and interests of local communities over profit-driven motives in agritourism ventures. This entails respecting indigenous land rights and ensuring the equitable distribution of economic benefits derived from tourism. Ait-Yahia et al. (2021) highlight the significance of responsible interactions between agritourists and local communities, as demonstrated in their study in Algeria. They emphasize the need to uphold the dignity of local communities and to avoid commodifying their cultural heritage. From an environmental ethics perspective, Nicolaides (2020) emphasizes the need for agritourism to promote biodiversity conservation through the adoption of sustainable land-use practices aligned with community values. Manalo et al. (2019) discuss how ethical decision-making paradigms can be propagated through well-crafted policy frameworks and robust collaboration among multiple stakeholders. They argue that Africa's rich cultural diversity, wildlife populations, and breathtaking natural landscapes present a remarkable opportunity for cultivating ethically managed agritourism enterprises that foster sustainable development. This aligns with the Sustainable Development Goals (SDGs), specifically goals 1 (poverty alleviation), 15 (environmental sustainability), and 11 (community well-being). These insights are of utmost importance as they form the bedrock for comprehending the ethical aspects associated with agritourism. This chapter will further expand upon the acquired knowledge and serve as a guide for analysing the distinctive challenges encountered in the agritourism sector in Angola. One of the challenges is in striking a delicate balance between economic progress and the conservation of indigenous cultures and ecosystems.

Moreover, agritourism in Africa offers significant opportunities for the incorporation of ethical principles, thereby contributing to various aspects of SDG 8. A few examples serve as compelling illustrations of this concept. For instance, Gitau et al. (2022) conducted a study in Kenya, focusing on the ethical dimensions of agritourism in the Maasai Mara conservancies. These conservancies prioritize the preservation of wildlife habitats, promote responsible tourism practices, and ensure the involvement and benefits of local Maasai communities. By doing so, they align with the overarching objective of achieving sustainable economic growth, job creation, and community empowerment. Similarly, in South Africa, Haywood et al. (2020) demonstrated the positive impact of the Fair Trade Tourism (FTT) certification program on ethical agritourism practices. FTT-certified establishments, such as sustainable vineyards and eco-lodges, prioritize fair wages, community development, and environmental sustainability. These initiatives contribute to fostering inclusive economic growth, creating decent work opportunities, and promoting responsible consumption. In the Gambia, ethical agritourism is observed through community-based eco-lodges like the Tumani Tenda Eco Lodge (Samdi and Nmadu, 2022). Through these lodges, local communities are actively involved in decision-making processes, supporting cultural preservation, and promoting sustainable agricultural practices. Consequently, the eco lodges contribute to sustainable tourism, economic diversification, and community well-being. Furthermore, Alphonse et al. (2023) shed light on Rwanda's ethical agritourism practices, specifically through the Kinigi Community Eco-Tourism Project near Volcanoes National Park. This project highlights community-led initiatives, revenue-sharing models, and the protection of endangered mountain gorillas. These efforts align with the objectives of sustainable economic growth, employment generation, and income equality. These examples vividly illustrate the potential of ethically managed agritourism enterprises in Africa to contribute to sustainable development, underlining the importance of integrating ethical dimensions. Thus, this study examines the ethical aspects associated with agritourism in Angola, considering the delicate balance between economic progress and the preservation of indigenous cultures and ecosystems. By doing so, the study contributes to the growing conversation on the role of ethical and sustainable agritourism in the attainment of the Sustainable Development Goals.

In fostering an ethically sound agritourism sector in Africa, ensuring inclusivity and accessibility is crucial. Janjua et al. (2021) highlight the importance of equitable participation of women, youth, and marginalized groups in agritourism entrepreneurship and employment opportunities. Concerns arise from infrastructure limitations that restrict access to agritourism sites for travellers with disabilities, as well as exclusionary practices based on ethnicity, gender, or income level, as highlighted by Dangi and Jamal (2016). Promoting inclusive agritourism practices that empower vulnerable groups is a socially responsible approach. Campbell and Kubickova (2020) propose well-designed policies and education programs for tourism operators across Africa to facilitate this. They argue that if managed ethically, agritourism in Africa can significantly contribute to key sustainable development goals. On the other hand, Van Zyl and van der Merwe (2022) suggest that inclusive agritourism models, providing equitable entrepreneurship and employment opportunities for youth, women, and marginalized groups, can catalyze poverty reduction, food security, and human development targets aligned with SDGs 1 and 2. Additionally, community-owned and managed agritourism enterprises, as highlighted by Torabi et al. (2023), can promote reduced inequalities within and between vulnerable rural communities (SDG 10) by directing tourism revenue into health, education, and social infrastructure for local residents. Wiltshier (2017) posits the potential of responsible interactions with wildlife and natural habitats within agritourism initiatives to advance conservation, rehabilitation, and biodiversity goals under SDG 15. Similarly, Xue and Shen (2022) argue that culturally sensitive and locally-driven agritourism approaches can strengthen sustainable communities (SDG 11) by preserving cultural heritage assets and generating sustainable livelihoods and income streams for residents. This study will demonstrate how Angola has involved diverse local players to promote ethical aspects associated with agritourism, thereby striking a balance between economic progress and the preservation of indigenous cultures and ecosystems.

Recent research shows the significance of ethical considerations in tourist interactions with local communities, both within the African continent and beyond. Park et al. (2023) highlight the importance of responsible tourism practices that uphold cultural values, preserve natural resources, and ensure a fair distribution of benefits among stakeholders. Their findings reveal the need for ethical decision-making processes that prioritize long-term sustainability over short-term economic gains. Similarly, Sahebalzamani and Bertella (2018) explore the ethical challenges faced in agritourism destinations, particularly in striking a balance between economic advancement and environmental sustainability. Their research advocates for ethical practices that take into account the preservation of ecosystems and biodiversity while supporting local communities. In Tanzania, Anderson (2015) conducted qualitative research examining the Kilimanjaro Indigenous Community Tourism Association (KICOTA) as an example of ethical agritourism. KICOTA promotes sustainable tourism practices that respect indigenous cultures, contribute to community development, and protect the natural environment. Through initiatives such as community-owned eco-lodges and guided tours led by local guides, KICOTA ensures equitable sharing of tourism benefits among community members, aligning with principles of inclusive economic growth and community empowerment. In addition, Melubo et al. (2019) conducted a study focusing on the ethical dimension of agritourism in specific sites in Tanzania, including Ndombo-Mfulony-Nkoarisambu. Their study highlights the importance of ethical considerations in tourist interactions with local communities, stressing the opportunity for visitors to experience the local culture, landscape, birdlife, and locally grown and roasted coffee. These studies provide valuable insights into the ethical dimensions of agritourism and offer guidance for sustainable and ethical practices. By drawing from these existing studies, conducting further research on the ethical considerations of agritourism in Angola will lead to a deeper understanding of the specific challenges and opportunities within the country. This study will benefit from incorporating insights, gaining valuable perspectives, benchmarks, and best practices, all of which will inform the development of sustainable and ethical agritourism practices in Angola.

While numerous studies have explored sustainable agritourism in various regions, there is still a critical need for further examination in the African continent, particularly in Angola. It is noteworthy that no research has specifically addressed the ethical challenges and opportunities faced by local actors in Central Angola, specifically within the Caála Municipality. Therefore, the primary objective of this study is to bridge this research gap by conducting an investigation into the ethical considerations within the agritourism industry in this specific area. By examining the role of local actors, the study aims to enhance our understanding of how agritourism can effectively promote economic growth while simultaneously ensuring the preservation of environmental integrity and cultural heritage. This research endeavour seeks to make a substantial contribution to the expanding body of knowledge on sustainable agritourism in Africa, with a specific focus on the unique context of Central Angola and the Caála Municipality.

Theoretical Framework

The investigation of the ethical dimensions of sustainable agritourism in Central Angola employs the Triple Bottom Line (TBL) approach as its theoretical framework (Figure 6.1). Developed by John Elkington in 1994, this sustainability framework encourages organizations and businesses to assess success by considering not only economic performance but also environmental and social impacts (Zaharia and Zaharia, 2021). The TBL

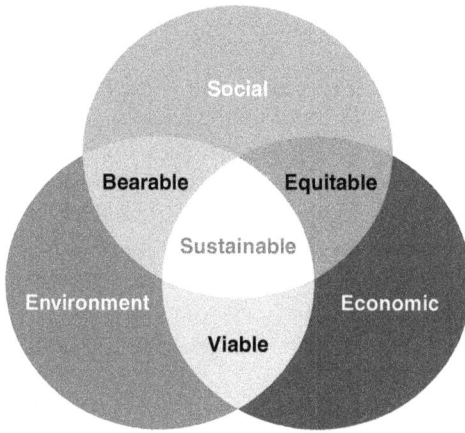

Figure 6.1 Triple bottom line (TBL) approach by John Elkington in 1994.

approach evaluates the performance and impact of an organization or initiative based on three dimensions: economic, environmental, and social. It provides a comprehensive assessment of success beyond financial profitability, promoting a balance between these dimensions to avoid environmental degradation and social inequality. The TBL framework supports a holistic view of sustainability, where economic prosperity aligns with environmental stewardship and social responsibility. Economically, the TBL approach focuses on financial aspects such as revenue, cost-effectiveness, and equitable distribution of economic benefits (Zaharia and Zaharia, 2021). Environmental sustainability considers ecological impact, including resource conservation, pollution reduction, and sustainable resource management. Social sustainability emphasizes community engagement, social equity, cultural preservation, and ethical practices (Dainienė and Dagilienė, 2015).

In this study, the TBL framework is used to analyze the ethical considerations in agritourism in Caala Municipality. It guides the examination of economic, environmental, and social sustainability dimensions simultaneously. The economic dimension explores agritourism's financial viability, revenue generation, and potential for local economic development, with a focus on assessing equitable benefit distribution and income generation. The environmental dimension delves into the ethical challenges and opportunities related to land use, resource management, and environmental preservation, aligning with principles of environmental stewardship. The social dimension addresses cultural preservation, community engagement, and social equity, examining the preservation of cultural heritage, respect for local traditions, and community benefits. The study also considers the role of local actors, including farmers, indigenous communities, and government authorities, in shaping the ethical aspects of sustainable agritourism.

Materials and Methods

Study Area

Caála, located in Huambo Province, Central Angola, is a municipality known for its thriving agricultural sector and emerging agritourism industry (MINAMB, 2021). The region offers picturesque landscapes, including fertile farmlands, charming villages, and captivating natural attractions, making it an attractive destination for tourists seeking an authentic rural experience. Agritourism activities in Caála encompass farm stays, agricultural tours, cultural exchanges, and opportunities for tourists to actively participate in farming activities (MINAMB, 2021). Huambo Province has successfully implemented agritourism initiatives, such as the "Farm-to-Table Experience" and the "Rural Homestay Programme" (MINAMB, 2018). The Farm-to-Table Experience is a program whereby tourists visit local farms, engage in agricultural activities, and gain insights into traditional farming practices, while supporting local farmers and promoting their agricultural products. In the Rural Homestay Programme tourists stay with local families in rural communities, participate in their daily routines, and learn traditional skills, thereby fostering cross-cultural understanding (MINAMB, 2018). These initiatives contribute to the economic growth of the region and provide unique cultural immersion opportunities for tourists.

Research Design

To delve into the ethical dimensions of sustainable agritourism in Caála Municipality, this study adopted a qualitative research design. Qualitative research was chosen for its ability to facilitate in-depth exploration, providing a nuanced understanding of participants' viewpoints, experiences, and behaviours. Through qualitative methods, the study sought to unravel the intricate relationship between economic advancement, environmental preservation, and the safeguarding of cultural heritage within the realm of agritourism. This approach allowed for a comprehensive examination of the ethical considerations inherent to this industry.

Sampling Technique

A purposive sampling technique was used in this study to select participants who possessed relevant knowledge and practical experiences in sustainable agritourism in Caála Municipality. The sample consisted of 21 key informants, made up of 14 males and 7 females. The participant selection process covered various stakeholder groups, ensuring an inclusive and diverse range of perspectives. The key informants included individuals in influential positions within the field, such as representatives from the Ministry of Agriculture and Rural Development, the Ministry of Tourism, the Institute for Agricultural Development (IDA), the Forum of Angolan Non-Governmental Organizations (FONGA), a conservation-focused NGO, and delegates from the Confederation of Farmer

Associations and Agricultural Cooperatives of Angola (UNACA). Participants from the local farming community, community leadership, and individuals involved in tourism operations were also included. For a comprehensive overview of the interview participants, please refer to Table 6.1 below.

The inclusion of 21 participants in this study adheres to the principles of qualitative research. Qualitative inquiries aim to gain a deep understanding of participants' experiences and perspectives rather than seeking broad generalizability to a larger population. The focus is on obtaining detailed and in-depth data from a smaller number of participants. By involving a diverse group of stakeholders, this study sought to comprehensively explore various perspectives, insights, and experiences related to the ethical dimensions of sustainable agritourism in Caála Municipality. The chosen sample size was

Table 6.1 Interview participants' profile

Participant	Gender	Age Range	Level of Education	Affiliation/Occupation
P1	Male	35–45	College	Ministry of Agriculture and Rural Development
P2	Female	30–40	University	Ministry of Tourism
P3	Male	40–50	College	Institute for Agricultural Development (IDA)
P4	Male	45–55	University	Forum of Angolan Non-Governmental Organizations (FONGA)
P5	Female	35–45	College	Conservation NGO
P6	Male	50–60	Secondary School	Confederation of Farmer Associations and Agricultural Cooperatives of Angola (UNACA)
P7	Male	45–50	Secondary School	Farmer
P8	Female	45–50	Secondary School	Farmer
P9	Male	45–55	Secondary School	Community Leader
P10	Female	40–50	Primary	Community Leader
P11	Male	30–40	College	Tourism Operator
P12	Female	30–40	College	Tourism Operator
P13	Male	35–45	Secondary School	Farmer
P14	Female	35–45	Secondary School	Farmer
P15	Male	45–50	Secondary	Farmer
P16	Male	35–40	Secondary	Farmer
P17	Female	30–40	University	Ministry of Agriculture and Rural Development
P18	Male	35–45	University	Ministry of Tourism
P19	Male	40–50	University	Institute for Agricultural Development (IDA)
P20	Male	45–55	University	Forum of Angolan Non-Governmental Organizations (FONGA)
P21	Male	50–60	College	Confederation of Farmer Associations and Agricultural Cooperatives of Angola (UNACA)

based on the expectation that it would yield rich and diverse data, facilitating a profound understanding of the subject matter.

Data Collection

The primary method of data collection in this study involved conducting semi-structured key informant interviews. These interviews were conducted face-to-face to encourage in-depth discussions and explore participants' perspectives on ethical challenges, practices, and future prospects in sustainable agritourism. The interviews included open-ended questions and probing inquiries to elicit insightful responses and personal experiences. Participants provided informed consent and agreed to have the interviews audio-recorded, ensuring accurate and authentic capture of their valuable insights. In addition to the interviews, detailed field notes were taken, capturing not only verbal exchanges but also non-verbal cues and contextual details, thereby enriching the collected data. Alongside the interviews, a thorough examination of local policies and practices was conducted. This involved carefully reviewing relevant government reports, which provided a deeper understanding of the implications of the regulatory framework for sustainable agritourism.

Data Analysis

For the analysis of the qualitative data collected from the key informant interviews, a thematic analysis approach was employed. This approach involved a systematic process of identifying patterns, themes, and categories within the data to gain insights into the ethical dimensions of sustainable agritourism. The first step in the analysis involved carefully reviewing the transcribed interviews and field notes. During this process, the data were coded to identify relevant concepts and themes related to the research topic. Codes were assigned to specific segments of the data that represented key ideas or topics discussed by the participants. Once the initial coding was complete, the codes were grouped into broader themes that captured commonalities and recurring patterns across the dataset. These themes were refined and organized to develop a comprehensive understanding of the ethical challenges, practices, and opportunities in sustainable agritourism. Throughout the analysis, an iterative approach was followed. This involved multiple readings of the data to ensure accuracy and reliability. The researchers revisited the coded data, reviewed the identified themes, and made adjustments as necessary.

Research Findings

Key Ethical Considerations

The research conducted on the ethical dimensions of sustainable agritourism in Caala Municipality yielded significant insights into the ethical considerations within sustainable tourism. The study specifically unveiled three pivotal dimensions: Economic Aspirations vs. Cultural Preservation, Environmental Conservation vs. Land Use, and Community Engagement vs. Resource Management.

Economic Aspirations vs. Cultural Preservation

The research findings revealed the relationship between economic aspirations and cultural preservation in agritourism. Stakeholders recognized the need to balance both aspects, with the understanding that preserving cultural heritage is crucial while pursuing economic growth. Cultural preservation not only enhances the tourist experience but also supports income generation and increased revenue. Participants stressed the importance of avoiding compromising cultural heritage for economic growth. They acknowledged that showcasing the region's cultural heritage within agritourism attracts tourists seeking authentic and immersive experiences. As one participant stated, "preserving our cultural heritage is essential for attracting tourists who want to fully immerse themselves in our local traditions and way of life." This viewpoint resonated with stakeholders who recognized the preference among tourists for genuine cultural encounters.

The research findings highlighted the socioeconomic benefits of cultural preservation in agritourism. Stakeholders identified opportunities for income generation and employment through cultural initiatives like traditional performances, culinary experiences, and craft workshops. One participant observed, "by incorporating cultural elements into agritourism, we can create new job opportunities and contribute to economic growth." Preserving cultural heritage attracts more tourists, leading to increased revenue. As one participant opined, "by safeguarding our cultural heritage, we can provide unique experiences that attract more tourists to our region." This strengthens the link between cultural preservation and the economic aspirations of agritourism entrepreneurs, benefiting the community's economic prosperity.

Environmental Conservation vs. Land Use

The research findings shed light on the ethical considerations of environmental conservation and land use in sustainable agritourism. Participants indicated the need for sustainable practices that minimize environmental harm while meeting tourist demands. One community leader expressed the challenge of balancing environmental protection and providing tourist experiences, saying, "the environment must be protected, but tourists want experiences. Sometimes this means adversely impacting our land." Stakeholders, including farmers acknowledged the significance of sustainable land-use practices in preserving biodiversity and enhancing agritourism experiences. As one participant stated, "sustainable land-use practices not only protect our environment but also enhance the quality and authenticity of agritourism experiences." This highlights the understanding that sustainable land use-practices contribute to the overall value and appeal of agritourism.

The adoption of organic farming emerged as a key sustainable land-use practice. Participants agreed that organic farming methods which prioritize the use of natural fertilizers and pest control measures, play a crucial role in protecting soil health, preserving water quality, and reducing the environmental impact of agricultural activities. One participant said, "we believe that organic farming is essential in preserving the integrity of our land and in ensuring the authenticity of the agritourism experiences we offer."

Conserving natural resources was also recognized as a fundamental aspect of sustainable land-use practices. Participants were of the view that responsible water management and preservation of natural habitats are crucial components. They suggested that there is a need to implement water conservation measures such as rainwater harvesting in order to optimize water usage and minimize waste.

Additionally, participants acknowledged the significance of preserving natural habitats in agritourism destinations to protect biodiversity and provide visitors with opportunities to engage with the region's natural environment. A participant elaborated, "we understand the importance of conserving our natural resources as they are key to sustaining our agritourism attractions and providing a unique experience for our visitors." By prioritizing sustainable land-use practices, participants believed that agritourism stakeholders in the Caala Municipality could establish a harmonious relationship between agricultural activities and environmental conservation. They posited that adopting such practices not only safeguards natural resources and ecosystems but also enhances the overall attractiveness and competitiveness of agritourism offerings.

Community Engagement vs. Resource Management

The research findings indicate the vital role of community engagement in ethical agritourism, particularly in resource management. Participants argued that it is important to involve local communities in decision-making processes and resource management as this has economic and socio-cultural benefits. As one environmental activist stated, "empowering local communities to actively participate in agritourism activities not only benefits them economically but also fosters a sense of ownership and pride". This perspective emphasizes the understanding that community involvement is crucial for the success and sustainability of agritourism. When communities are actively engaged, they are motivated to promote and support agritourism activities, contributing to their long-term viability. As one interviewee noted, "community involvement is crucial for the success of agritourism. We need to involve local communities in decision-making processes and capacity building to empower them to participate actively in agritourism activities". Recognizing the importance of community engagement reflects the understanding that when communities have a stake in agritourism ventures, they are more likely to contribute to their success.

The study revealed that in order to effectively promote and support agritourism there is a need for education and training programs that engage local communities. Farmers expressed a lack of knowledge and skills in developing and managing agritourism ventures, highlighting the importance of comprehensive awareness programs and workshops. One farmer expressed uncertainty, stating, "I am not really sure what agritourism is all about." Another farmer revealed their lack of knowledge about the concept and its profitability, saying "I do not really know about the concept. I do not have enough information about its profitability." Therefore, if provided with targeted information and skill-building initiatives, farmers can confidently participate in and contribute to the success of agritourism. Participants identified government agencies such as the Angola Institute of Agricultural Development as key stakeholders in education and training initiatives. They suggested that the Institute should actively involve local communities in decision-making processes as a way to foster a sense of ownership and pride in responsible resource management.

The participants further proposed that the Institute should provide the necessary facilities to complement the identified agritourism attractions. They also acknowledged the importance of tailored education, training programs, workshops, and technical assistance offered by the Institute. These initiatives would enable farmers to develop and manage successful agritourism ventures, ensuring that resource management practices align with sustainable principles. Through the collaborative efforts between the Institute and local communities, participants believed that the relationship between community engagement and resource management in agritourism could be comprehensively addressed and strengthened. These views reflect the perspectives shared by the participants, who recognize the potential of government agencies in supporting and promoting sustainable practices in agritourism through active community engagement and responsible resource management.

Ethical Dilemmas Faced by Local Actors

An examination of the ethical dilemmas of sustainable agritourism in Caala Municipality highlighted several significant challenges faced by local actors. The study identified three key ethical dilemmas that confront these actors: the tension between economic gains and environmental conservation, the preservation of cultural authenticity versus meeting tourist demands, and the imperative of equitable revenue distribution. These dilemmas represent the complex and multifaceted challenges that individuals and communities face in achieving a balance between economic growth, environmental preservation, and cultural integrity.

Balancing Economic Gains with Environmental Conservation

One prominent ethical dilemma identified in the research is the challenge of balancing economic gains with environmental conservation. Local actors, including government authorities and community members, expressed the desire for economic growth while recognizing the importance of preserving a healthy environment. They face the ethical tightrope of finding a sustainable middle ground that allows for economic development while minimizing negative environmental impacts. A government authority succinctly captured this dilemma thus, "we want economic growth, but we also want to pass down a healthy environment. It is a constant struggle." This statement reflects the ongoing ethical challenge faced by local actors who must navigate the need for economic prosperity while ensuring the long-term preservation of the natural environment.

Preserving Cultural Authenticity While Meeting Tourists' Demands

Another significant ethical dilemma identified in sustainable agritourism is the preservation of cultural authenticity while meeting the demands of tourists. Local actors, such as tourism operators and community members, grapple with providing authentic cultural experiences while catering to the expectations and desires of tourists, which include modern comforts and conveniences. Striking a balance between preserving the region's unique cultural identity and meeting the evolving demands of tourists is essential. A tourism operator expressed this dilemma. He said, "tourists come for the authentic experience, but they also expect modern comforts. Finding the middle ground is a challenge." This dilemma highlights the importance of developing strategies and offerings that provide an authentic experience while integrating appropriate amenities and services that enhance visitor satisfaction. The Ministry of Tourism also agreed that there is a need for careful consideration in creating offerings that maintain cultural authenticity while meeting tourists' desires.

Equitable Revenue Distribution

Equitable revenue distribution emerged as another pressing ethical concern in sustainable agritourism. Local actors, including community members and NGOs, indicated the necessity of ensuring that the benefits of agritourism are shared equitably among all stakeholders instead of accruing solely to a select few. This ethical dilemma is at the heart of promoting ethical agritourism. A member of an NGO spoke to the effect that efforts should be made towards ensuring that everyone instead of a select few benefits as this is at the heart of ethical agritourism. Thus, achieving equitable profit-sharing mechanisms and ensuring fair compensation for local communities, farmers, and other stakeholders is crucial for ethical and sustainable agritourism practices. Participants in the research felt that addressing this ethical dilemma requires the establishment of transparent and inclusive governance structures and mechanisms that promote equitable revenue distribution. Engaging all relevant stakeholders in decision-making processes and implementing policies that prioritize fair compensation and opportunities for local community members were cited as essential steps in addressing this dilemma.

Role of Local Actors in Shaping Ethical Agritourism

The study also explored the vital and active involvement of local actors in shaping the ethical dimensions of sustainable agritourism in Caala Municipality. The study found that the ethical framework of sustainable agritourism in the study area is not solely constructed by abstract principles; rather, it is forged and upheld through the active involvement of various local actors who play pivotal roles in its realization. Farmers, indigenous communities, government authorities, and NGOs each make unique contributions to the ethical fabric of agritourism in this region. Their collective efforts synergies to create a responsible and sustainable approach to tourism that is firmly grounded in ethical values.

Farmers as Stewards of the Land

The research indicates that farmers, as custodians of the land, occupy a central position in shaping the ethical landscape of agritourism. They are instrumental in promoting sustainable farming practices that minimize environmental impact while ensuring the production of high-quality agricultural products. Farmers actively engage in responsible tourism practices by adopting methods that protect soil health, preserve natural resources, and conserve biodiversity. Their dedication to sustainable land-use practices not only safeguards the environment but also enhances the authenticity and appeal of agritourism experiences. Speaking about the vital role of farmers, an environmental conservation group member said, "farmers are the custodians of the land. They play an instrumental role in maintaining its health."

Indigenous Communities as Guardians of Cultural Heritage

The study found that indigenous communities in Caala Municipality are at the forefront of preserving cultural authenticity within agritourism. Their traditions and heritage are not mere relics of the past; they are vibrant treasures actively shared with tourists by these communities. Through cultural showcases, workshops, and immersive experiences, indigenous communities enrich the agritourism landscape with their unique cultural identity. Their active participation in cultural preservation ensures that tourists have the opportunity to deeply engage with the local way of life. A community leader proudly elucidated on their role, saying, "our traditions are our treasures. We actively participate in cultural showcases and workshops to share our heritage with tourists."

Government Authorities as Guardians of Responsible Tourism

The study also reveals that government authorities wield significant influence in shaping the ethical contours of agritourism through policy and regulations. Their commitment to creating a conducive environment for responsible tourism is evident in their efforts to enact regulations that safeguard the environment and culture of Caala Municipality. By actively engaging in the development of responsible tourism policies, government authorities contribute to the ethical foundation upon which agritourism thrives. A government official reiterated this commitment, saying, "we are dedicated to creating a conducive environment for responsible tourism. This includes enacting regulations that safeguard our environment and culture."

NGOs as Advocates for Responsible Tourism and Community Engagement

The study further found that NGOs and community engagement initiatives play a pivotal role in advancing ethical agritourism in the region. These organizations work closely with local communities to raise awareness about responsible tourism practices and promote sustainable development. They focus on initiatives that support cultural preservation, environmental conservation, and community empowerment. Through their collaborative efforts with local actors, NGOs contribute to building capacity, fostering community engagement, and implementing sustainable practices in agritourism. In explaining their work, an NGO representative said, "we work closely with communities to raise awareness about responsible tourism. Our projects focus on sustainable development and cultural preservation."

Discussion of Findings

The study on the ethical dimensions of sustainable agritourism in Caala Municipality provides valuable insights into the complex and interconnected nature of ethical considerations within sustainable tourism. One important dimension explored in the study is the balance between economic aspirations and cultural preservation. Findings show that economic growth and the preservation of cultural heritage are intertwined in agritourism. Previous research has also indicated the economic benefits of cultural preservation in tourism (Bhatta and Ohe, 2020). Collaboration and cooperation among stakeholders are highlighted as significant factors in sustainable tourism development, aligning with the social aspect of the Triple Bottom Line (TBL) model (Aziz, 2016).

Another dimension addressed in the study is the tension between environmental conservation and land-use practices. Stakeholders acknowledge the need to minimize environmental harm while providing meaningful tourist experiences. This aligns with the principles of sustainable tourism, which emphasize responsible environmental practices (Kupika and Dube, 2023). The adoption of organic farming as a sustainable land-use practice exemplifies a commitment to environmental preservation, consistent with the planet-conscious practices reflected by the TBL model (Trinh, 2021).

The study also emphasizes the importance of community engagement and resource management as a dimension of ethical considerations. Active involvement of local communities in resource management reflects a commitment to social responsibility in agritourism. Community engagement has been highlighted in research on community-based tourism as crucial for sustainable development (Wang et al., 2022). Additionally, education and training programs are recognized as essential for community empowerment and align with the TBL's focus on people and social well-being (Aziz, 2016).

Furthermore, the examination of ethical dilemmas faced by local actors in agritourism aligns with the broader ethical considerations encompassed by the Triple Bottom Line model. The dilemmas include balancing economic gains with environmental conservation, preserving cultural authenticity while meeting tourist demands, and ensuring equitable revenue distribution. These dilemmas reflect the tensions between profit and planet, profit and people, and encourage fairness and social equity, respectively.

Study Implications and Limitations

Study Implications

The findings of the study on the ethical dimensions of sustainable agritourism in Caala Municipality have several important implications for theory, policy, and practice in the field of sustainable tourism. Theoretically, this study makes several important contributions to knowledge on sustainable agritourism. The application of the Triple Bottom Line model provides a nuanced understanding of how economic, environmental, and social dimensions intersect in complex ways in real-world sustainable agritourism initiatives. The findings reveal tensions between these dimensions, such as economic goals overriding environmental conservation and social welfare. This demonstrates the need for a holistic, ethical framework like the TBL model to highlight these complex dynamics. The research also enriches theoretical perspectives on stakeholder collaboration, indicating that meaningful community participation and multidirectional communication between all stakeholders are essential for avoiding unethical practices. Key theoretical concepts like the Circular Economy model and Alternative Food Networks are also shown to have salience in understanding challenges and pathways in ethical and sustainable agritourism.

On the practical level, the research points to multiple priorities for policy-makers, tourism promoters, and regional bodies overseeing tourism development. Firstly, incentives and training programs to assist agritourism farmers in transitioning to organic methods are vital as evidenced by environmental benefits. Secondly, regulations and voluntary certification schemes could promote sustainability and ethics by setting standards, though they must be co-developed with stakeholders rather than imposed in a top-down manner. Thirdly, facilitating collaborative governance through multi-stakeholder committees and partnerships is recommended to foster open dialogue and cooperation. Regional bodies could help establish these governance mechanisms. Fourthly, educational initiatives and promotional campaigns pitched to tourists, farmers, and the wider public can raise awareness of sustainability issues and influence behaviours. Regarding future research directions, studies should adopt longitudinal, ethnographic approaches to provide nuanced understanding of how sustainable agritourism initiatives evolve over time in specific locales. Investigating the perspective of farmers using qualitative methods could offer valuable insights. Comparative analyses between different contexts are needed to illuminate how socio-political, economic and cultural particularities shape sustainability outcomes. Research must also remain attentive to potential gaps between rhetoric and reality regarding the extent to which stated sustainable practices are implemented ethically. Overall, this study demonstrates the multifaceted nature of analyzing and promoting ethical, sustainable agritourism, underscoring the value of context-specific, community-engaged research.

Study Limitations

While the study provides valuable insights into the ethical dimensions of sustainable agritourism in Caala Municipality, it is important to acknowledge certain limitations that may impact the generalizability and robustness of the findings. Firstly, the research was confined to a single case study region, Caala Municipality, thereby limiting generalizability of findings to other locales. The specific social, economic, political and environmental context of Caala Municipality shapes the ethical dynamics of its agritourism sector in unique ways that may differ across regions. Additional comparative case studies in diverse locales would strengthen generalizability. Secondly, the qualitative methodology relying on interviews and document analysis has inherent biases such as response bias among interview participants. Incorporating quantitative data on impacts of agritourism could provide more objective measures to complement the qualitative insights. Thirdly, the study's findings are based on the perspectives and experiences of selected stakeholders in Caala Municipality. Other stakeholders, such as tourists and local residents, may have different perspectives and insights that are not captured in the study. Broadening data collection to wider groups of stakeholders could enrich understanding of the ethical dimensions involved. Fourthly, the study provides only a snapshot of the ethical dynamics at one point in time. Longitudinal research tracking changes over time could reveal crucial insights into how ethical considerations evolve as agritourism develops. Finally, the subjective nature of interpreting "ethical" practices means that conclusions drawn reflect researchers' perspectives that are shaped by their positioning. Critical reflexivity is required. Notwithstanding these limitations, this research remains valuable as it offers crucial preliminary insights into the intricate ethical dimensions of sustainable agritourism in Caala Municipality. To enhance the depth and applicability of future studies, it is recommended to employ mixed methods, conduct multi-case analyses, and foster inclusive stakeholder engagement. These approaches will contribute to the development of more robust and transferable knowledge, enabling a better understanding of how to effectively balance economic, environmental, and social objectives in the context of ethical and sustainable agritourism development.

Conclusion and Recommendations

Conclusion

This study sheds light on the ethical dimensions of sustainable agritourism in Caala Municipality. It highlights the intricate balance required between economic growth, cultural preservation, environmental conservation, and community engagement. Stakeholders, including farmers, indigenous communities, government authorities, and NGOs, each contribute to shaping the ethical fabric of agritourism in the region. The study's comparative analysis revealed diverse perspectives on economic aspirations and the intensity of ethical challenges. Tourism operators prioritized financial gains, while indigenous community members emphasized cultural preservation. Government authorities faced complex dilemmas, and NGOs focused on advocating for ethical practices. Overall, the study accentuates the multifaceted nature of ethical dilemmas in sustainable agritourism.

Recommendations

In order to promote ethical agritourism practices in the Caala Municipality, several recommendations are proposed. Firstly, the development of comprehensive ethical guidelines and standards that encompass economic, environmental, and social considerations is crucial. Additionally, capacity building programs should be implemented to educate and train farmers, agritourism operators, and local communities on sustainable farming practices, cultural preservation, and responsible tourism. Active community engagement, collaboration among stakeholders, and transparency in revenue distribution are essential to fostering ethical practices. Long-term planning, monitoring, and evaluation, along with government support, including facilities, technical assistance, and funding, are necessary. Lastly, promoting tourist education, research, and innovation will contribute to continuous improvement and adaptation to evolving ethical standards. By implementing these recommendations, the Caala Municipality can establish itself as a model for responsible and sustainable agritourism in Angola.

References

Ait-Yahia Ghidouche, K., Nechoud, L., and Ghidouche, F. (2021). Achieving sustainable development goals through agritourism in Algeria. *Worldwide Hospitality and Tourism Themes, 13*(1), 63–80.

Anderson, W. (2015). Cultural tourism and poverty alleviation in rural Kilimanjaro, Tanzania. *Journal of Tourism and Cultural Change, 13*(3), 208–224.

Alphonse, H., Ricardo, S., and Vince, S. (2023). Sustainable development strategies of domestic and international tourism in Rwanda. *International Journal of Trends in Scientific Research and Development, 7*(2), 71–96.

Aziz, R. C. (2016). Development of sustainable rural tourism at Sungai Sedim Amenity Forest, Kedah, Malaysia from a tourism stakeholders' perspective. PhD Thesis, Universiti Putra Malaysia.

Baipai, R., Chikuta, O., Gandiwa, E., and Mutanga, C. N. (2021). A critical review of success factors for sustainable agritourism development. *African Journal of Hospitality, Tourism and Leisure, 10*(6), 1778–1793.

Baipai, R., Chikuta, O., Gandiwa, E., and Mutanga, C. N. (2023). A framework for sustainable agritourism development in Zimbabwe. *Cogent Social Sciences, 9*(1), 2201025.

Baiocco, S., and Paniccia, P. (2023). Integrating the natural environment into tourism firms' business model for sustainability. *Environmental Science and Pollution Research*, 1–14.

Bhatta, K., and Ohe, Y. (2020). A review of quantitative studies in agritourism: The implications for developing countries. *Tourism and Hospitality, 1*(1), 23–40.

Broccardo, L., Culasso, F., and Truant, E. (2017). Unlocking value creation using an agritourism business model. *Sustainability, 9*(9), 1618.

Budeanu, A., Miller, G., Moscardo, G., and Ooi, C. S. (2016). Sustainable tourism, progress, challenges and opportunities: An introduction. *Journal of Cleaner Production, 111*, 285–294.

Campbell, J. M., and Kubickova, M. (2020). Agritourism microbusinesses within a developing country economy: A resource-based view. *Journal of Destination Marketing and Management, 17*, 100460.

Chiodo, E., Fantini, A., Dickes, L., Arogundade, T., Lamie, R. D., Assing, L., and Salvatore, R. (2019). Agritourism in mountainous regions — Insights from an international perspective. *Sustainability, 11*(13), 3715.

Ciolac, R., Iancu, T., Brad, I., Adamov, T., and Mateoc-Sîrb, N. (2021). Agritourism—A business reality of the moment for Romanian rural area's sustainability. *Sustainability, 13*(11), 6313.

Dainienė, R., and Dagilienė, L. (2015). A TBL approach based theoretical framework for measuring social innovations. *Procedia-Social and Behavioral Sciences, 213*, 275–280.

Dangi, T. B., and Jamal, T. (2016). An integrated approach to "sustainable community-based tourism". *Sustainability, 8*(5), 475.

Fanelli, R. M., and Romagnoli, L. (2020). Customer satisfaction with farmhouse facilities and its implications for the promotion of agritourism resources in Italian municipalities. *Sustainability, 12*(5), 1749.

Gitau, D. K., Njuguna, M., and Wahome, E. (2022). Rurality: Promotion of vernacular landscapes as a stimulus for sustainable rural tourism development in Kenya. *Sustainable Tourism Dialogues in Africa, 7*, 165.

Haywood, L. K., Nortje, K., Dafuleya, G., Nethengwe, T., and Sumbana, F. (2020). An assessment for enhancing sustainability in rural tourism products in South Africa. *Development Southern Africa, 37*(6), 1033–1050.

Janjua, Z. U. A., Krishnapillai, G., and Rahman, M. (2021). A systematic literature review of rural homestays and sustainability in tourism. *Sage Open, 11*(2), 21582440211007117.

Kupika, O. L., and Dube, K. (2023). A resilient tourism future for developing countries: Conclusions and recommendations. *In COVID-19, Tourist Destinations and Prospects for Recovery: Volume Three: A South African and Zimbabwean Perspective* (pp. 293–310). Cham: Springer International Publishing.

Madanaguli, A., Kaur, P., Mazzoleni, A., and Dhir, A. (2022). The innovation ecosystem in rural tourism and hospitality – A systematic review of innovation in rural tourism. *Journal of Knowledge Management, 26*(7), 1732–1762.

Manalo, C. C., Amboy, S. M., Gamil, R., Geroy, A., and Festijo, B. T. (2019). Benefits of agritourism in Batangas province. *Asia Pacific Journal of Education, Arts and Sciences, 6*(3), 8–16.

Melubo, K., Lovelock, B., and Filep, S. (2019). Motivations and barriers for corporate social responsibility engagement: Evidence from the Tanzanian tourism industry. *Tourism and Hospitality Research, 19*(3), 284–295.

Mgonja, J. T., Sirima, A., and Mkumbo, P. J. (2017). A review of ecotourism in Tanzania: Magnitude, challenges, and prospects for sustainability. *Ecotourism in Sub-Saharan Africa*, 170–183.

Ministry of Environment (MINAMB). (2018). *National Report for the Conference of the Parties to the Convention on Biological Diversity*. In Ministry of Urbanism and Environment of Angola; Government of Angola: Luanda, Angola. [Online] Available at: http://www.minamb.gov.ao/ [Accessed 16 February 2023].

Ministry of Environment (MINAMB). (2021). *National Action Plan for Adaptation in Angola*. Luanda, Angola: Ministry of Environment. [Online] Available at: http://www.minamb.gov.ao/ [Accessed 16 February 2023].

Mnisi, P., and Ramoroka, T. (2020). Sustainable community development: A review on the socio-economic status of communities practicing ecotourism in South Africa. *International Journal of Economics and Finance, 12*(2), 505–519.

Moswete, N., Saarinen, J., and Manwa, H. (2016). Cultural tourism in Southern Africa: Progress, opportunities and challenges. *Cultural Tourism in Southern Africa*, 180–188.

Ngo, T., Lohmann, G., and Hales, R. (2018). Collaborative marketing for the sustainable development of community-based tourism enterprises: Voices from the field. *Journal of Sustainable Tourism, 26*(8), 1325–1343.

Nicolaides, A. (2020). Sustainable ethical tourism (SET) and rural community involvement. *African Journal of Hospitality, Tourism and Leisure, 9*(1), 1–16.

Park, J., Zou, S., and Soulard, J. (2023). Transforming rural communities through tourism development: An examination of empowerment and disempowerment processes. *Journal of Sustainable Tourism*, 1–21.

Ramaano, A. I. (2023). Environmental change impacts and inclusive rural tourism development on the livelihoods of native societies: Evidence from Musina Municipality, South Africa. *International Journal of Ethics and Systems*.

Sahebalzamani, S., and Bertella, G. (2018). Business models and sustainability in nature tourism: A systematic review of the literature. *Sustainability, 10*(9), 3226.

Sambajee, P., Ndiuini, A., Mutinda, P., Masila, D. K., Baum, T., Wairimu, R., and Kiage, E. O. (2022). Undertaking research among marginalised tourism communities in Kenya: An important methodological lesson. *In Handbook of Innovation for Sustainable Tourism* (pp. 310–329). Edward Elgar Publishing.

Samdi, E. M., and Nmadu, T. M. (2022). Sustainable management and ecotourism destination governance in Africa. *Sustainable Tourism Dialogues in Africa, 7*, 49.

Satta, G., Spinelli, R., and Parola, F. (2019). Is tourism going green? A literature review on green innovation for sustainable tourism. *Tourism Analysis, 24*(3), 265–280.

Silvestri, C., Silvestri, L., Piccarozzi, M., and Ruggieri, A. (2022). Toward a framework for selecting indicators of measuring sustainability and circular economy in the agri-food sector: A systematic literature review. *The International Journal of Life Cycle Assessment*, 1–39.

Suriyankietkaew, S., Krittayaruangroj, K., and Iamsawan, N. (2022). Sustainable leadership practices and competencies of SMEs for sustainability and resilience: A community-based social enterprise study. *Sustainability*, *14*(10), 5762.

Suwannasat, J., Katawandee, P., Chandrachai, A., and Bhattarakosol, P. (2022). Site selection determinant factors: An empirical study from meeting and conference organizers' perspectives. *Journal of Convention and Event Tourism*, *23*(3), 209–239.

Tang, M., and Xu, H. (2023). Cultural integration and rural tourism development: A scoping literature review. *Tourism and Hospitality*, *4*(1), 75–90.

Torabi, Z. A., Pourtaheri, M., Hall, C. M., Sharifi, A., and Javidi, F. (2023). Smart tourism technologies, revisit intention, and word-of-mouth in emerging and smart rural destinations. *Sustainability 15*(14), 10911.

Trinh, T. T. T. (2021). *Agritourism as a tool for sustainable local development: The case of Thuy Bieu, Vietnam*. Doctoral Thesis, Auckland University of Technology.

Van Zyl, C. C., and van der Merwe, P. (2021). The motives of South African farmers for offering agri-tourism. *Open Agriculture*, *6*(1), 537–548.

Van Zyl, C. C., and Van der Merwe, P. (2022). Critical success factors for developing and managing agri-tourism: A South African approach. *International Conference on Tourism Research*, *15*(1), 536–545.

Wang, M., Rasoolimanesh, S. M., and Kunasekaran, P. (2022). A review of social entrepreneurship research in tourism: Knowledge map, operational experiences, and roadmaps. *Journal of Sustainable Tourism*, *30*(8), 1777–1798.

Wiltshier, P. (2017). Community engagement and rural tourism enterprise. In *Rural Tourism and Enterprise: Management, Marketing and Sustainability* (pp. 53–67). Wallingford UK: CABI.

Xue, L. L., and Shen, C. C. (2022). The sustainable development of organic agriculture: The role of wellness tourism and environmental restorative perception. *Agriculture*, *12*(2), 197.

Zaharia, R. M., and Zaharia, R. (2021). Triple bottom line. *The Palgrave Handbook of Corporate Social Responsibility*, 75–101.

7 African Tourism and Hospitality Experience

The Zimbabwean Perspective

*Julius Tapera, Purity Hamunakwadi,
Rahabi Mashapure, and Enet Mukurazita*

Introduction

Globally, the significance of tourism is recognised for its myriad associated benefits which capture wealth creation initiatives from its vast sectors to curb poverty and unemployment. To understand the significance of tourism, the United Nations World Trade Organisation (UNWTO) (2018), Musavengane et al. (2019) and Zhou (2022) affirm that national economies benefit from revenue generation and employment creation provided by the sector. To this end, more than one billion tourists travel every year to international destinations which shows valuable evidence that tourism is playing a leading role in the economic sector. The Global Market Report in 2020 asserts that the global hospitality industry accounts for about USD4.5 trillion in consumer spending from hotels (accommodation), casinos, restaurants, cruise ships, bars, tour operators, and travel agencies (Aksoy et al., 2022). Across the world, the World Travel and Tourism Council (WTTC) reports that during the period between 2014–2019, prior to the COVID-19 pandemic, travel and tourism segments created 1 to 5 new jobs (WTTC, 2020). More precisely, in 2019 the global GDP from the travel and tourism sector accounted for 10.4%, equating to USD10 trillion and 10.3% of all jobs or USD334 million (WTTC, 2023). More recently, in 2022, tourism has had a global average increase in contribution to GDP of 7.6% to 22% from 2021, though remaining below the 2019 levels of 23%. To buttress the contribution of the GDP, 22 million jobs were created in 2022 representing 7.9% which was an increase from 2021 and 11.4% only below 2019 levels (WTTC, 2023). To add, in 2022 domestic visitor spending rose by 20.4% which was only 14.1% below 2019 (WTTC, 2023). Likewise, in 2022 international visitor spending rose by 81.9% but it remained behind by 14.1% from 2019 percentages (WTTC, 2023). To add, the Hospitality Global Report confirms that in 2022 the global hospitality market increased from USD4,390.59 billion to USD4,699.57 billion in 2023 at a compound annual growth rate (CAGR) of 7.0%. Likewise, the global hospitality market in 2023 reached approximately USD4.7 trillion with a projection to escalate to USD5.8 trillion in 2027 at a compound annual growth rate (CAGR) of 5.5% (Statista Research

DOI: 10.4324/9781032696188-7

Development, 2023). In the same vein, in the United Kingdom (UK) the third largest employer is the hospitality industry with a contribution of 3.2 million jobs which generated an enormous 72 billion pounds that directly added value to her economy (UK Hospitality, 2018). These remarkable global trends are similar in other developed countries in job creation and revenue generation though they differ in geographical areas.

Although the impact of COVID-19 has been catastrophic to developmental tools in place for resuscitating the tourism industry and hospitality sector, in Africa the industry has been growing exponentially in contemporary times (Muresherwa and Tichaawa, 2022). Over the past few years, tourism has been recognised as undoubtedly a key feature in the development of African economies (Monnier, 2021). The key role played by the tourism and hospitality sectors shows that several opportunities can be utilised by communities, policy makers, NGOs, and all relevant stakeholders to enhance economic growth and local economic development for the benefit of local people. In support of this, Muresherwa and Tichaawa (2022) predict economic complementarity and community consolidation, especially with community-based tourism, as socio-ecological resilience is generated from the significant economic activities. Dube (2021) confirms that tourism has the potential to contribute to all 17 sustainable development goals (SDGs) and 169 targets directly or indirectly through local and regional connections. Hence, governments have been instituting policies that aim for the sustainability of the tourism industry as the lapse of agenda 2030 SDGs draws closer, though inequality, hunger, and poverty are worsening.

The African Union Agenda 2063 remains in line with the global stance that recognises the importance of socio-economic development and structural transformation from jobs created by the tourism and hospitality sector (Zhou, 2022). In developing countries, the BRICS (Brazil, Russia, India, China, and South Africa) countries represent a potential bloc that caters to major tourists from developed countries. These countries showed an increase to 30%, a jump from 11%, of the world's GDP in the period between 1990 to 2014. According to Bama et al. (2022), in 2019, the top-ranked African countries for travel and tourism were Mauritius, South Africa, Seychelles, Morocco, Namibia, Kenya, Tunisia, Cape Verde, Botswana, and Tanzania. From international tourist arrivals (visits) Morocco welcomed 13.93 million visitors and South Africa received 10.23 million visitors (Bama et al., 2022:558). However, about 12.4 million employment opportunities were lost in Africa from the COVID-19 pandemic (Bama et al., 2022). In contextualising, the Zimbabwean tourism and hospitality sectors are distinguished by their major attractions, uniqueness of certain iconic destinations, service quality and other distinctive features and experiences.

Muresherwa et al. (2022) assert that Zimbabwe, being home to one of the Seven Wonders of the World, presents unique and exquisite tourist attractions. These range from Victoria Falls, a renowned world heritage site (Great Zimbabwe), a myriad of cultural heritage sites, various national game parks, and wildlife conservancies with diverse wildlife species including the 'big five.' These highly attractive destinations on the African continent have been significantly contributing to the continuous growth and development of the African tourism and hospitality industry. In 2018, endowed with many natural resources that draw international, regional, and local tourists, the country received 957 752 visitors in the rainforest of Victoria Falls and national parks (Zimbabwe Tourism, 2020). The uniqueness of the Zimbabwean tourism sector is that it is also steeped in some unique protected areas which are normally visited by domestic visitors (individuals, church groups, and families) and culture encompassing museums and monuments, particularly the Great Zimbabwe Monument (Muresherwa et al., 2022). According to the Zimbabwe Tourism Authority (ZTA) these groups of people represented in visits equate to the diverse nature of tourism as community, religious, and event tourism. Such visits have contributed to the growth of the tourism industry from 80,000 to over 2.5 million in 1980 and 2018, respectively.

This chapter seeks to parse out the opportunities that the tourism and hospitality industry in Africa, in particular Zimbabwe, presents in recovering the economy. Tourism and hospitality is the third largest sector after mining and agriculture (International Trade Administration, 2022). The chapter, therefore, aims to identify and discuss impediments to the involvement of communities in tourism in developing countries, with Zimbabwe's Community Based Tourism (CBT) projects, as a developing country, a case in point. This is key as the government of Zimbabwe (GoZ) embarked on establishing CBT projects in rural communities to advocate for their inclusion in tourism development and alleviate poverty through tourism in marginalised communities (Gohori and van der Merwe, 2021). Thus, this study aspires to assist developing countries such as Zimbabwe to promote community participation in tourism by promoting and implementing ideal policies.

The main aim of the review is to critically evaluate the contribution and relevance of the Zimbabwean perspective to the literature on the African tourism and hospitality experience, and to identify the areas for further research and development. Within the same vein the study aims to provide comprehensive insights into the wealth and uniqueness of the African tourism and hospitality experience in promoting inclusive and sustainable economic growth, decent work for all, and reducing inequalities in realising SDG 8. Likewise, the study seeks to propose policy recommendations and action plans for enhancing the wealth and uniqueness of the African tourism and hospitality experience. This will assist in maximising its contribution to the sustainable development of the continent and the world. In addition, the study suggests ways to improve the economic inclusion and employment prospects of low-skilled workers and other vulnerable groups (youths and women). This emanates from the unique history and natural wonders presented and gained in cultural, heritage and development tourism in enhancing economic inclusion, employment of low-skilled workers locally and other vulnerable groups (Signe, 2019). Hence, it is argued that the wealth and uniqueness of African tourism and hospitality signifies community consolidation through economic growth and local economic development that trickles down to poor communities, thus generating economic activities. Thus, the focal point of this enquiry stream is to show the uniqueness of African tourism and hospitality within the Zimbabwean context which links to how wealth is created in various sectors such as cultural heritage sites, various national game parks, hotels, food, and beverages among others. The chapter is guided by the following research questions:

Research Question 1: What are the main themes and debates in the literature on the wealth and uniqueness of the African tourism and hospitality experience, with a focus on the Zimbabwean perspective?

Research Question 2: How does the wealth and uniqueness of the African tourism and hospitality experience relate to the sustainable development goals, especially SDG 8?

The chapter proceeds as follows. The next section presents the literature discussion that underpins the study before presenting the methodological approach adopted in collecting data. Thereafter follows the empirical discussion on the wealth and uniqueness of Zimbabwean tourism and hospitality and the opportunities it presents to enhance economic recovery. The discussion is guided by perspectives drawn from local and foreign tourists, industry players, regulators, and policy makers. The chapter ends with a concluding discussion underscoring the need for sound economic inclusion, community and local economic development.

Literature Review

According to the tourist led growth theory, travel and tourism could play a key role in driving economic growth (Tang and Tan, 2018). There are two main categories of benefits that a nation might receive from tourism: non-economic and economic. The growth of business opportunities for the underprivileged, the expansion of employment and wages by guaranteeing commitments to local jobs and training of local residents, and the development of collective community income are some of the economic benefits of tourism (Bastidas-Manzano et al., 2021). Also, strengthening connections with other local economic sectors is key to maximising the total economic benefits of tourism. Strong economic ties to the tourism industry as well as other industries including agriculture and small and medium-sized businesses would strengthen the multiplier effect, increasing revenue retention and opening up job opportunities for the local population (Akama and Kieti, 2007). In addition, an increase in the demand for regional goods and services, the creation of new jobs, and the expansion of business prospects for regional companies are all outcomes of the strengthening of economic relations through tourist growth. To this end, non-financial advantages include reducing environmental effects, increasing access to services and infrastructure (such as transportation, security, health care, and water supplies), and training and capacity building (Lansing and Vries, 2007).

The mechanisms via which the benefits of tourism flow into the economy have been determined to be numerous. According to Ashley and Mitchell (2009), there are three primary channels: direct effects, indirect effects, and dynamic effects. Direct effects quantify the impact of a visitor's initial financial cycle (Adrian, 2018). In essence, this covers the earnings of people who work in the industry directly or are business owners. The tourism value chain generates indirect benefits (Vellas, 2011). A wide range of industries, including building, furniture, food and drink, transportation, and many more, provide inputs to the tourism industry. Indirect effects are, put simply, the subsequent business transactions that arise from direct tourism expenditures. Tourism development has dynamic effects that include how local households make a living, the business environment that supports the growth of small businesses, the growth patterns of the local or national economy, and the infrastructure and natural resource base of the destination (Shen et al., 2008). The total benefits that can be obtained by an economy through the growth of the country's tourism sector are equal to the sum of the three routes to tourism. The recognition of tourism as an in-situ export contributes to the sector's relevance in an economy (Sanderson and Leroux, 2017). This essentially indicates that, unlike other exports that require transportation to the final customer, visitors are the ones who visit the site of interest.

According to Awaritefe's study, the most common experiences, values, and reasons for travelling to a destination are leisure/recreation activities, cultural appreciation, and an appreciation of nature, history, and cultural artefacts. The quality of environmental aspects varies among tourists in different sites, and there are differences between domestic and foreign tourists in terms of their spatial behaviour, values, and motivations. The study has special implications for marketing and planning in the tourism industry, as well as a framework for comparing findings from related Western studies across cultural boundaries.

Theoretical Frameworks for Tourism and Hospitality Growth and Development

Tourism and hospitality development theories play a crucial role in understanding and guiding the growth of the sector. These theories can be regarded as frameworks for understanding the dynamics of the industry, and provide lenses through which the wealth and uniqueness of the African tourism and hospitality industry can be viewed. In addition, the application of these theories can significantly contribute to achieving Sustainable Development Goal (SDG) 8, which is to "Promote inclusive and sustainable economic growth, employment and decent work for all" (Strömstad, 2019). This chapter pivots on the lens of five related and complementary theories that advocate mainly for economic growth, employment, community participation and sustainable tourism practices. Butler's model outlines stages of tourism development, including exploration, involvement, development, consolidation, stagnation, and rejuvenation or decline (Butler, 2022; Karamustafa, 2023; McKercher, 2021). According to the model, proper management of tourism destination life cycles ensures sustained economic growth, employment, and community well-being, aligning with the goals of SDG 8 (Benner, 2020; Javed, 2020). The second theory that is relevant to this discussion is Community-Based Tourism (CBT), which, emphasises community involvement and empowerment, allowing local communities to actively participate in and benefit from tourism (Dolezal, 2022; Priatmoko, 2021; Zielinski, 2020). The chapter argues that by creating employment opportunities and promoting community engagement, CBT contributes directly to the goal (SDG 8) of decent work and economic growth in local communities. In addition, the Sustainable Tourism Development Theory, whose emphasis is the balancing of Economic, Social, and Environmental aspects of tourism informs the study (Mkono, 2020; Sharpley, 2020). Sustainable tourism development theory emphasises the importance of balancing economic benefits with social and environmental considerations (Sharpley, 2022). Sustainable tourism thus ensures that economic growth is achieved without compromising the well-being of current and future generations, which is also in alignment with the principles of SDG 8 (Jacobs, 2013; Sustainable Travel International, 2020; Torkington, 2020). In addition, the study deploys the Triple Bottom Line (TBL) theory that emphasises three key dimensions of sustainability: economic, social, and environmental. The theory also aligns with SDG 8 by promoting economic growth in the tourism sector while ensuring social inclusivity and minimising negative environmental impacts (Bui, 2021; Csikósová, 2020; Ozturkoglu, 2021). Lastly, the Pro-Poor Tourism (PPT) theory focuses on poverty alleviation and aims to create economic opportunities for marginalised communities and reduce poverty through tourism development (Ramkissoon, 2023,). This theory directly contributes to the goal of decent work and economic growth by promoting inclusive economic development and creating employment opportunities for the poor, as propounded by SDG 8 (Dada, 2022; Musavengane, 2019). The study submits that the wealth and uniqueness of African tourism can integrate various theoretical constructs in answering posed research questions.

The Future Tourism Picture in Africa

According to World Travel and Tourism, COVID-19 resulted in a 29.3% decrease in jobs and a 49.2% decrease in Africa's travel and tourism GDP contribution (WTTC, 2021). Before COVID-19, Africa was the world's second fastest growing tourist destination in 2018, with an 8.5% gain compared to 5.6% in Asia and a global growth average of 3.9% (Weigert, 2019). This rise is due to a 7% increase in tourism revenues in Sub-Saharan Africa and a 10% increase in tourism revenues in North Africa in 2018. In 2019, tourism contributed 7% of Africa's GDP (Monnier, 2021). However, as early as April 2020, most of Africa's key tourist attractions were closed due to suspended air transport, lockdowns, and social distancing regulations (Rogerson and Baum, 2020).

According to the Tourism 2020 Vision research, foreign tourist arrivals on the African continent are expected to expand at a rate above the global average, with an average yearly increase of 5.5% between 1995 and 2020 (Ishtiaque, 2013). However, the COVID-19 pandemic had a negative impact on the travel and tourism sector globally. All travel was restricted in order to slow the virus's spread, which led to a sharp drop in foreign arrivals and overall financial loss (World Travel and Tourism Council [WTTC], 2021; King et al., 2021). With the end of the COVID-19 epidemic, the global hospitality and tourism sector is starting to recover. Thus far, 2023 has yielded encouraging outcomes and opportunities for the industry. With a compound annual growth rate (CAGR) of 7%, the worldwide hospitality market has currently increased from USD4,390.59 billion in 2022 to USD4,699.57 billion, and this rise is predicted to continue (Hammanjoda et al., 2023). According to estimates from the International Aviation Transport Association (IATA), global aviation traffic reached 96.1% in the second quarter of 2023, a full recovery and a rise over the epidemic era (Abdelhady et al., 2022). In April 2023, the trip health index finally caught up with pre-pandemic levels, and in June 2023, it even exceeded them. By the end of 2023, the percentage of foreign tourism is predicted to reach 95% of 2019 figures. Even after COVID-19 requirements were lifted, the majority of travellers still place a high priority on health and safety precautions. Thus, excellent upkeep and sanitation are essential elements that hotels and resorts shouldn't ignore.

Domestic travel is considered to be the primary catalyst for short- to medium-term recovery in Zimbabwe since it can give the country's tourism firms the much-needed boost they need to survive (National Department of Tourism [NDT], 2020). As per the WTTC (2021), the tourism industry's share of the national GDP decreased to 3.7% in 2020 from roughly 6.9% in 2019. Domestic visitor expenditure fell quite significantly as well, from R160 billion in 2019 to R91.5 billion in 2020. As a result, a redesign of the industry's products is required. Even though COVID-19 is destructive, there are ways to change and reconsider tourism (Everingham and Chassagne, 2020). It was recommended that destinations evaluate their tourism policies critically and search for ways to strengthen the recovery after COVID-19 while also enhancing resilience (Organisation for Economic Co-operation and Development [OECD], 2020). Virtual tourism models were discussed as a "means to an end" and a means of survival, particularly in the short and near term (El-Said and Aziz, 2022).

Adapting to a world that is changing quickly is the driving force behind reimagining tourism sustainability. Utilising virtual tourism offers a creative approach to withstand change and adversity while ensuring the industry's survival after the epidemic. The most popular form of virtual tourism involves using virtual reality to create virtual tours. To give visitors a more in-depth and engaging experience, virtual tourism and tours can also be of advantage in augmented reality through the exposure of technology (Fotiadis et al., 2021). Through the promotion of virtual tours during the present pandemic, tourist locations like Cape Town, Harare, and New York have been able to earn cash and minimise employment losses, demonstrating the efficacy of virtual tourism to some extent. The tourist sector views augmented reality and virtual reality as practically indispensable, given their enhanced virtual experiences, numerous marketing advantages, and ability to give businesses a competitive edge (Lu and Xu, 2021).

An Overview of Zimbabwe's Tourism Industry

From the period of Zimbabwe's colonisation in 1890 to the present, the tourism industry has experienced significant changes and advancements. There are essentially two main periods that may be distinguished based on the acquisition of political and economic policies: the colonial phase and the post-colonial age. These have influenced the nation's tourism industry's growth and branding. Nonetheless, the main focus of this conversation is mostly on post-independence changes, particularly as they pertain to the establishment of two brands for Zimbabwean tourism. The reasons behind the change from "Zimbabwe: Africa's paradise" to "Zimbabwe: A World of Wonders" will be specifically highlighted. Soon after Zimbabwe gained its independence in 1980, the country's tourist sector faced numerous difficulties, one of which was central government indifference. This was mostly done in the immediate aftermath of the brutal independence war. Because of the urgency of the moment, the Zimbabwean government set out to rebuild the economy devastated by the war, giving top priority to the resuscitation of the mining, manufacturing, and agriculture sectors – which were regarded as the foundation industries for the resurgence and expansion of the economy. As a result, other industries, like tourism, were disregarded. The departure of important white businesspeople and workers, primarily to apartheid South Africa and Australia, as part of a broader wave of white emigration spurred by the uncertainties arising from the country's decision to grant independence to the black majority, exacerbated the situation for the tourism industry (Chibaya, 2013). As a result, only 100,000 white people were living in the country in 1980.

After realising the effects of emigration on the nation, the government turned its attention to the tourist sector to improve its reputation while protecting its workforce. The government established the Zimbabwe Tourism Development Corporation (ZTDC) in 1981. Its job was to invest on the government's behalf in the tourism sector. The Zimbabwe Tourism Authority (ZTA) was tasked with properly marketing Zimbabwe's tourism destinations locally, regionally, and worldwide after the ZTDC was abolished in 1996. As a result, the government began to own many tourism-related businesses, such as the hotel chain Rainbow Tourism Group. Some of the most popular tourist destinations in the nation, National Parks and Museums, were also controlled by the government. Additionally, the government purchased stock in a few hospitality businesses, such as the Zimbabwe Sun Hotels (Africa Sun Hotels).

The structure and makeup of Zimbabwe's tourist industry demonstrate the country's strength in this area. The four primary components of Zimbabwe's tourist product are its people, culture, and history; its built environment, which includes hotels and lodges; its natural resources, which include natural resources; and its distinctive scenery, which includes the Great Zimbabwe Monument and Victoria Falls. There was a significant drop in arrivals between 2000 and 2005 due to the political and economic crisis triggered by the land reform, which led to the large-scale eviction of commercial farms, ranches, and wildlife conservancies (Manwa, 2007). Wildlife, a critical backbone resource of Zimbabwe's tourism was no longer viable during this period as national parks and conservancies were invaded. Several hotels and restaurants closed or downsized, resulting in layoffs and the exodus of qualified and experienced professionals to neighbouring countries (Mpofu and Chitura, 2015). The global recession in 2008 exacerbated an already declining industry as hyperinflation eroded travellers' income. However, a new political dispensation of the Government of National Unity stabilised the economy and changed the narrative of the country resulting in the recovery of the tourism industry (Chibaya, 2013).

Wealth and Uniqueness of Zimbabwe's Tourism Industry

Zimbabwe, a country in Southern Africa, is home to five UNESCO World Heritage sites (Mana Pools National Park, Great Zimbabwe National Monument, Khami Ruins National Monument, Mosi-oa-Tunya/Victoria Falls, and Matobo Hills), making it one of the most visited tourist attractions on the continent. These sites have significant cultural and historical importance and UNESCO has recognised them as World Heritage Sites owing to their uniqueness and importance to the collective interests of humanity. Zimbabwe's increasing popularity might probably be attributed to the variety of experiences it provides to these kinds of tourists. Three characteristics that set this nation apart from others:

• Geography: Some of Southern Africa's most breathtaking vistas can be found in Zimbabwe. This is where the 355-foot waterfall, a portion of Victoria Falls, is situated. Hiking into the jungle will allow one to get up close and personal with the breathtaking vistas of this amazing monument, also known as "Smoke That Thunders." A rainbow that appears over the falls at specific times of the day completes the amazing scene. For those who love nature and photography, Victoria Falls is unquestionably a must-see destination (Dube, 2021).

- Wildlife: Zimbabwe boasts two significant national wildlife refuges, so visitors who love animals will have a blissful time there. The Zambezi River and more than 350 distinct plant and animal species can be found in Mana Pools, one of the most well-known parks in the nation. One might even be able to spot the Big Five animals: elephants, lions, buffalo, leopards, and rhinoceroses if they visit Mana Pools. Hwange National Park is another destination for travellers located in Zimbabwe. Along with 100 other mammal species, the Big Five mammals can be found in this 5,656-square-mile reservation (Kuguyo, 2022; Mapingure et al., 2019).
- Culture: Art and music play a big part in the Zimbabwean culture. The most commonly recognised art from the country is Shona sculpture. These stone sculptures are carved by members of the Shona tribe and are usually meant to depict characters from African folklore. Visitors can look out for the figurines of these characters, as they can always be found in local markets (Matura, 2022; Runyowa, 2017). Some of these cultural artefacts and traditional norms have become major sources of community-based cultural tourism attractions, for example, the Chesvingo Cultural Village in Masvingo (Matura, 2022), the KoMpisi Cultural Village in Victoria Falls (Runyowa, 2017).

Adding to the discussion the tourist sector in Zimbabwe has four primary chain groups and several individual businesses that function under the extensive regulatory oversight of the Zimbabwe Tourist Authority (ZTA). Great Zimbabwe and Victoria Falls are major factors in the sector's growth. The tourism industry in Zimbabwe is considered to be significantly influenced by the trinity of Victoria Falls, the Big Five, and Great Zimbabwe (Kuguyo, 2022; Mapingure, 2019). Beautiful terrain, untamed, highly regarded National Parks, distinctive flora and fauna, and, of course, some famous, globally recognised sights like the Great Zimbabwe Ruins, the Matobo Hills, Lake Kariba, and the magnificent Victoria Falls can all be found in Zimbabwe. It also has a fantastic climate and year-round outdoor activities (Mushawemhuka et al., 2021).

The landscape of Zimbabwe is characterised by sharp differences. The cooler, hillier Eastern Highlands with their apple orchards and tea estates; the drier, yet diverse, and unspoiled South East with the Rundi and Save Rivers and Gonarezhou National Park; the rugged Matobo Hills with its distinctive rock formations and San Bushmen paintings; the vast, varied terrain of the Hwange National Park; the powerful Zambezi River with its diverse animal and bird life; the immensely impressive Lake Kariba with its one-of-a-kind activities and breathtaking sunsets (Thondhlana et al., 2021).

Though it has much more to offer, Zimbabwe is most recognised for the magnificent Victoria Falls and Hwange Game Reserve. The world's most remarkable and well-preserved ancient towns are the Great Zimbabwe Ruins (Masvingo) and the Khami Ruins (Bulawayo) (Muresherwa et al., 2022). The Kariba Dam, at two hundred and forty kilometres in length, is a singular engineering achievement of its era and a menacing sight. It is a well-liked vacation spot with a variety of tourist amenities, such as houseboats, safari camps, fishing, and game viewing. Anglers continue to find it appealing, and the yearly Tiger Fishing Tournament, which is held in October, draws a large number of local and foreign participants. The competition is currently in its 55th year.

Zimbabwe is known for some of the best wildlife watching in Africa and is home to several ecologically diverse game parks. Wide swaths of the nation were designated as National Parks and Game Reserves because of the vision of the people in charge at the beginning of the 20th century.

There are 12 major parks located around the nation; the most well-known ones are Gonarezhou in the south-east, Matopos in the south-west, Hwange in the north-west, and Mana Pools in the north. Each is distinct and has something new to offer guests. There are roughly 199 known mammal species in Zimbabwe, and 694 known bird species, many of which are limited to particular regions of the country.

Over the years, the Zimbabwean government has made an effort to promote healthy growth in the tourism industry. In the 1990s, significant efforts were made to promote the industry on a broad scale, with the ZTA being designated as the primary marketing agency by an act of parliament. After that, legislation known as the two percent tourism charge was presented. To protect their interests in the profitable industry, the private sector also intervened and formed the Zimbabwe Council of Tourism (ZCT) and the Hospitality Association of Zimbabwe (HAZ). The industry received favourable treatment and was awarded an export zone status. Significant strides were achieved in enhancing the standard of the Zimbabwean tourism experience for foreign visitors as a consequence of these initiatives. However, much more remains to be done to develop Zimbabwe as a competitive international tourism destination.

A significant sector of Zimbabwe's economy is tourism. It has been designated by the government as one of the industries to help revive the flagging economy of the nation. According to the Zimbabwe Tourism Authority (ZTA) tourism directly provided 7.4% to employment, 8.4% to exports, and 5% to GDP in 2013, while in the same year, 1,832 million tourists spent USD856 million in the nation.

The Shape of Tourism in the Future

An assessment of relevant literature suggests that tourism in the future will have a greater focus on flexibility, segmentation, tailoring, and diagonal integration than it does today, with less emphasis on mass, rigidity, standardisation, and packaging. The 21st century will be shaped by several major megatrends (derived from Cooper et al., 1999; Hall, 2000): increased localisation and globalisation will force tourism stakeholders to "think globally, plan regionally, and act locally." Technology and telecommunications will dominate almost every aspect of tourism in the future (Gupta et al., 2023). All areas of travel will be prioritised based on speed, efficiency, safety, and convenience, with safety and convenience being especially important (Şimşekoğlu et al., 2015). Artificial intelligence (AI) will gain traction in the tourism industry in the future. The benefits of AI in tourism, such as lower prices, improved services and customer support, and the ability for customers to plan trips more quickly, will outweigh the challenges of AI, such as privacy and connectivity, which will be addressed as AI initially collaborates with tourism personnel (Bowen and Morosan, 2018).

There will be a growing divide between the type of tourism that people prefer (comfort versus adventure, for example), examining hotels and other facilities online, searching for last-minute deals via email, and finding hotels with inexpensive rooms on their websites. The effects of globalisation will be felt in a shrinking tourism world, with a greater emphasis on "off-beaten-track places" with targeted products (especially theme-based) that will primarily focus on entertainment, excitement, and education (Rasoolimanesh, 2021). As more destinations develop their "image" as a prerequisite to attracting tourists and diversifying their appeal, the importance of destination image development will increase (Heath, 2001). Consumption demands will clash with consumer desires as consumers become more aware of sustainable tourism development and ethical business practices (OECD, 2018). Increasingly, tourists in high-income countries are seeking meaningful holidays, such as agritourism, to foster human connections and contribute to sustainable tourism. According to Chatterjee and Prasad (2019) and Yamagishi et al. (2021), the concept of agritourism entails visiting a farm operation for relaxation, leisure, education, or participation in its activities. Zimbabwe's agro-based economy makes agritourism a natural next step for tourism.

Methodology

In this study, a qualitative inquiry was employed to explore the Zimbabwean tourism and hospitality landscape using the interpretivist method of inquiry. The qualitative research method was deemed more appropriate for phenomenological inquiry hence its application to this study, which sought to explore the wealth and uniqueness of the African hospitality and tourism experience based on an analysis of findings from previous studies, current developments, and forecasts for the future of the industry (Nunkoo, 2018; Olowoyo, 2021). The study drew its arguments from secondary data which encompassed documentary and conceptual analysis of authoritative sources to conceptualise and contextualise the wealth and uniqueness of African tourism. The significance behind the usage of this approach is that it gave leeway to focus on meaning and employ multiple methods to reflect different aspects of the issue of wealth and the uniqueness of African tourism. Hence, information from secondary sources was gathered from previously published academic research articles and unpublished documents, including websites, media reports, documentaries, local (the Ministry of Tourism and Hospitality Industry

(MoTHI) website, the Zimbabwe Tourism Authority (ZTA) website) and global tourism and hospitality reports by leading global institutions like the World Tourism Authority (WTO). All selected sources were genuine, authentic, credible, and meaningful to support the body of knowledge. The most appropriate sources were selected from various databases or search engines such as Google, DOAJ, Ebsco Host, Elsevier, Taylor and Francis, Science Direct, Emerald, Academic Search Premier using the terms "wealth and uniqueness of African tourism," "African tourism," "community-based tourism in Zimbabwe," "SGD8" and "tourism and hospitality." In total 30 documents were analysed using a conceptual analysis method from a selection of 75 documents after screening which was based on relevant documents which were answering the posed research questions. Thus, building on qualitative analysis of previous studies that have utilised past research in establishing the phenomena in tourism and hospitality and successfully established thematic trends and discovery of new knowledge in the process was adopted (Lee, 2022; Mura, 2018).

Empirical Evidence from the Broader African and Specific Zimbabwean Contexts

Africa is host to a myriad of tourist features that attract tourists locally, regionally, and internationally. Some of the major tourist attractions on the African continent include the Table Mountains and National Park, Robben Island in Cape Town, South Africa, the Victoria Falls, in Zimbabwe (Victoria Falls) and Zambia (Livingstone), the Omo River Region in Ethiopia, Mount Kilimanjaro in Tanzania, Maasai Mara National Park in Kenya, the Pyramids of Giza and the Great Sphinx in Egypt, the Virunga National Park in the Democratic Republic of Congo, the Okavango Delta in Botswana, the Zanzibar in Tanzania, and the Kruger National Park in South Africa. Evidence from previous research reflects the wealth and uniqueness of the African tourism and hospitality experience. In a study investigating the factors influencing the growth of the African tourism and hospitality market, Skobkin et al. (2020) analysed statistical data on the development of the tourism and hospitality industry, with a particular focus on the influence of cultural, historical, and modern political factors. The study findings reflect that the tourism and hospitality industry had vast growth potential, premised on Africa's strong historical and cultural foundations. The results of the study further confirmed, statistically, a growth trajectory of international tourist arrivals on the African continent, giving a strong basis for projecting further growth and development of the tourism and hospitality industry within the continent. An analysis of global trends reflects that in 2018, Africa was the second fastest growing tourism region, internationally, contributing 8.5% of the gross domestic product (GDP) equating to USD194 billion (Bama et al., 2022). In addition, the continent received 67 million international tourists during the same period (Mohrmann, 2019). African domestic tourism also had the highest expenditure by local visitors at 56% being 12% higher than international visitors who were at 44%. In 2019, tourism was one of the major contributory sectors towards the economic growth of African countries, contributing approximately 7% of the GDP in Africa (Monnier, 2021). In ranking African countries by tourist arrivals, Morocco (12.93 million) and South Africa (10.23 million) had the highest international tourist arrivals, while Mauritius, Seychelles, Namibia, Kenya, Tunisia, Cape Verde, Botswana and Tanzania were the other 8 comprising the top 10 in attracting international tourists (Statista, 2020) Africa also boasts a very wide range of renowned international brands in the hotel industry among them the Marriott International, Mangalis Hotel Group, Radisson Hotel Group, InterContinental Hotels Group, Accor, Best Western Hotels & Resorts, Hilton Hyatt International, Rotana Hotels, and Melia Hotels & Resorts.

Zimbabwe is home to five of the United Nations Educational, Scientific and Cultural Organization (UNESCO) World Heritage Sites and these are Mana Pools, Khami Ruins National Monument, Great Zimbabwe Ruins, Mosi-oa-Tunya/Victoria Falls, and Matobo Hills. The Great Zimbabwe National Monument, Khami Ruins National Monument, and Matobo Hills are cultural sites, while Mana Pools National Park and Mosi-oa-Tunya (Victoria Falls) are natural sites. The Victoria Falls is also one of the Seven Wonders of the World. These world heritage sites, together with other tourism and hospitality facilities, which include national parks, wildlife conservancies with vast species of flora and fauna, hotels, and lodges of international repute across the whole national divide, have proved to be a significant part of the wealth and uniqueness of the African tourism and hospitality experience. Zimbabwe has been experiencing a growing number of international tourist arrivals from 1,416,000 arrivals in 1995 to 2,580,000 arrivals in 2018, an 82% growth (Mundi, 2023). Income over the same period also increased by 97% from USD145 million in 1995 to USD285 million in 2019 (Macrotrends, 2023). While there was a significant decline in tourist arrivals due to travel restrictions imposed globally during the COVID-19 period, (639,000 arrivals in 2020, and a further decline to 380, 820 arrivals in 2021), there has been a resuscitation of the growth trajectory post COVID-19. Following the relaxation of COVID-19 related travel restrictions, Zimbabwe experienced an increase in tourist arrivals in 2022, a 174% increase from the 2021 arrivals (Zimbabwe Tourism Authority, 2023). This rebound and growth of the Zimbabwean tourism and hospitality industry have been supported by the existence of vast tourism attractions, hotel and other accommodation facilities, and other tourism and hospitality players that provide ancillary and complimentary services. In addition, the Zimbabwean tourism and hospitality industry has over the years gained great renown for high-quality food and beverages, accommodation facilities and excellent service at every encounter. A study by Mapingure et al. (2019) involving 869 tourists (both international and domestic), identified six competitiveness dimensions of Zimbabwe as a tourist destination. The tourists rated

Table 7.1 Major tourist attractions in Zimbabwe by province

Province	Major tourist attractions
Harare	National Heroes Acre, Mukuvisi Woodlands, Lake Chivero Recreational Park, Lion Park, Umfurudzi Game Park, Wild is Life Animal SanctuaryNational Gallery of Zimbabwe, Mbizi Game Park, Zimbabwe Museum of Human Sciences, Eco Nyathi Game Park, Tsindi Monuments, Chiremba Balancing Rocks
Bulawayo	Khami Monuments, Natural History Museum, Bulawayo Railway Museum, Lumene Falls, Tshabalala Game Sanctuary, Matobo National Park, Chipangali Wildlife Orphanage
Matabeleland North	Hwange National Park, Victoria Falls, Victoria Falls Bridge, Chizarira National Park, Zambezi National Park, KoMpisi Cultural Village
Manicaland	Bridal Veil Falls, Pungwe Falls, Vumba Botanical Gardens, Nyamgombe Falls, Mtarazi Falls
Mashonaland West	Lake Kariba Recreational Park, Darwendale Recreational Park, Mana Pools, Chinhoyi Caves, Matusadona National Park
Midlands	Antelope Park, Zimbabwe Military Museum, Naletele Monument
Masvingo	Save Valley Conservancy, Lake Mutirikwi National Park, Great Zimbabwe Monument, Chesvingo Cultural Village, Gonarezhou National Park
Matabeleland Sount	The Great Mapungubwe Transfrontier Conservation Area, Threeways Safaris, Nottingham Estates
Mashonaland Central	Domboshava Caves
Mashonaland East	Imire Game Park

Source: Zimbabwe Tourism Authority, 2023.

Zimbabwe as a competitive destination in relation to safety/relaxation opportunities, attractions and mix of activities, hospitality, and suitability of activities, while pricing/accommodation and facilities/accessibility were perceived as relatively less competitive (Mapingure et al., 2019). Each province within the country is endowed with vast tourist facilities. The table below summarises some of the major tourist attractions in each of the provinces in Zimbabwe (Table 7.1).

Zimbabwe also boasts a number of international brands in the hotel industry and some of the major hotels in the major cities and tourist destinations include (Table 7.2):

In addition to these hotels, there are other smaller hotels, lodges, chalets, campsites, restaurants and tourist facilities that offer excellent service to both local and international tourists. Zimbabwe also has a lot of registered travel agencies and tour guide operators that offer ancillary and complementary services to afford tourists an unparalleled experience. The table below summarises the UNESCO World Heritage Sites in Zimbabwe (Table 7.3):

Table 7.2 Major Zimbabwean hotels by city/tourism destination

City/Tourist destination	Major hotels
Victoria Falls	The Victoria Falls Hotel, Elephant Hills, Victoria Falls Safari Suites, The Victoria Falls Deluxe Suites, Palm River Lodge, Victoria Falls Safari Lodge, A'Zambezi River Lodge, The Kingdom Hotel, The Rainbow Hotel, Ilala Lodge, Manor Hotel
Harare	Rainbow Towers, Meikles Hotel, Holiday Inn, Bronte Hotel, Monomotapa Hotel, Cresta Lodge
Bulawayo	Holiday Inn, Rainbow Hotel, Cresta Churchill Hotel, Nesbitt Castle
Masvingo	Great Zimbabwe Hotel, Chevron Hotel, Flamboyant Hotel
Hwange	Hwange Safari Lodge, Baobab Hotel
Eastern Highlands	Troutbeck Inn, Montclair Casino, Manica Skyview, Hotel East Gate, Holiday Inn, Christmas Pass Hotel, La Rochelle

Source: Zimbabwe Tourism Authority, 2023.

Table 7.3 UNESCO world heritage site

Site (inscription date)	Description
Mana Pools National Park, Sapi and Chewore Safari Areas (1984)	Mana Pools, one of Zimbabwe's World Heritage Sites, is a wildlife conservation area comprised of three distinct areas: Mana Pools National Park, Sapi and Chewore Safari Areas. This is a secluded and spectacular spot in the Zambezi Valley, with wonderful views of the Zambezi River and the mountains of Zambia's Rift Valley.
Great Zimbabwe National Monument (1986)	Great Zimbabwe Ruins are the greatest stone buildings with gigantic curving walls made of millions of granite stones fitted together without cement. Great Zimbabwe which means "stone houses", is a historic city located 30 kilometres south of Masvingo.
Khami Ruins National Monument (1986)	After the Great Zimbabwe Ruins, the Khami Ruins are Zimbabwe's second most important archaeological site. Khami expanded and developed between 1450 and 1650, after Great Zimbabwe's capital was abandoned.
Mosi-oa-Tunya/ Victoria Falls (1989)	Victoria Falls is one of Zimbabwe's natural World Heritage Sites and one of the world's most stunning waterfalls. When the Zambezi River is at its peak, the Victoria Falls are 1708 m wide and drop 99 m, making it the world's greatest sheet of falling water. Victoria Falls (also known as Mosi-oa-Tunya, "the smoke that thunders") was named after Queen Victoria by David Livingstone, the first European to witness it.
Matobo Hills (2003)	The Matobo Hills, another World Heritage Site in Zimbabwe, are home to the biggest collection of rock art in southern Africa. The enormous boulders provide several natural shelters and have been linked to human occupancy from the early Stone Age.

Source: UNESCO World Heritage Convention, 2023.

Employment Creation and Community Benefits of Tourism and Hospitality in Zimbabwe

Bastidas-Manzano et al. (2021) assert that growth of business opportunities for the underprivileged, expansion of employment and wages by guaranteeing commitments to local jobs and training of local residents, and development of collective community income are some of the economic benefits of tourism. Zimbabwean citizens have enjoyed some of these benefits as the tourism and hospitality industry has facilitated employment creation and improvement of livelihoods for communities in some tourist areas. Some of the typical success stories include the CAMPFIRE programme with projects that run across the country, the Chesvingo Cultural Village in Masvingo (Matura, 2022), the KoMpisi Cultural Village in Victoria Falls (Runyowa, 2017) and the Save Valley Conservancy, which are all community based-tourism projects that have seen locals getting employment while also improving household incomes through shareholding in the projects.

Moyo and Tichawaa (2017) assert that although the poor benefit from employment opportunities and improvement of quality of life from tourism development, evidence from developing countries still shows that governments and external companies benefit disproportionately while local residents are paid low incomes. Moyo and Tichawaa (2017) also found that tourism in Zimbabwe is not consistent with the dictates of some organisations such as the World Bank who advocate for integration of all stakeholders in the development of tourism to obtain positive and sustainable outcomes. In terms of community involvement and participation, the findings of Moyo and Tichawaa (2017) and Gohori and Merwe concur in documenting how community members indicate the need for greater involvement in planning, decision-making and operational aspects. This will assist in community members in taking full responsibility for development that takes place within their residential areas. All these dictates will not materialise if local people are not aware of what the tourism industry constitutes, if they do not prioritise holiday opportunities, and if poor infrastructure and signage are the norm. To add, such benefits can be experienced if there is proper coordination between the ZTA and the private sector and through the inclusion of local people through provision of affordable services ranging from tourism activities (game viewing and sightseeing) and hospitality services (hotels, food, and beverages).

Evidence from interviews in a study by Shereni (2022) suggest five themes in which the hospitality industry can contribute to the achievement of the SDGs which include encouraging community participation, promoting gender equality, energy management initiatives, introducing policy framework(s) and the provision of decent jobs. For the purposes of analysis and benchmarking of this chapter three themes appear more relevant: encouraging community participation, introducing policy framework(s), and the provision of decent jobs. Linking them to SDGs, community participation contributes towards the achievement of social sustainability through following international mandates of reduced inequalities, poverty alleviation, and forging strong institutions. The findings of this study show that there is a need to promote the arts and crafts industry, buying products from locals to enhance self-sustainability and supporting local communities by providing incentives to enhance the supply of quality products. Evidence shows that community participation involves inclusion of locals in the entire value chain that ranges from buying from locals, Corporate Social Responsibility (CSR), promoting local arts, employing locals and supporting community-based enterprises. The participation of locals in tourism development and practices empowers them to have a voice in improving their own livelihoods.

Within this context of aligning tourism and hospitality industries to sustainable development goals 2030, empirical evidence from Shereni (2022) shows that the hospitality industry can contribute meaningfully towards SDG 8 on decent work, and SDG 1 on poverty eradication. Shereni (2022) shows that the hospitality industry provides low paying jobs (menial jobs) for the locals as a ruse for employment creation. For instance, in big resorts such as Victoria Falls foreign labour (expatriates) are more prevalent, especially for high positions with high incomes. This study contends that, to be recognised as contributing to SDG 8 which advocates for the creation of decent work and economic growth, local people must be capacitated to occupy higher and better paying positions. However, this works hand in hand with skills development, education and empowerment.

Shereni (2022) found that there is a lack of sustainability policies that guide operators in the hospitality industry which poses an impediment to achieving SDGs. The gap in the sustainability policy framework on SDG with regard to tourism and hospitality shows that the government has to play a major role in its crafting. Shereni suggests the Ministry of Environment, Tourism and Hospitality Industry (METHI) as a major stakeholder, needs to craft such a policy framework if SDGs are to be achieved in terms of economic growth. This study argues that the crafting of the sustainability policy framework assists in the implementation and enforcement of sustainability practices in the tourism and hospitality industry at all levels of government (local, national, and regional). Apart from crafting, implementing, and enforcing the sustainability policy framework in the tourism and hospitality industry by the government, funding and coordination of various stakeholders can play a crucial role in achieving SDGs.

This study contributes to the body of knowledge in three ways: Firstly, theoretically, it adds to the growing number of studies on the wealth and uniqueness of African tourism and other developing nations. Building on achieving SDG 8 to enhance human capital development the findings of this study build on existing theories in which each incorporates issues of economic growth, employment, community participation and sustainable tourism practices. Findings of this study focused on either one of the aspects on economic growth, employment, community participation or sustainable tourism practices.

Hence, the incorporation of five theories: The Community-Based Tourism CBT) theory; Butler's Model of Tourism Area Life Cycle; Sustainable Tourism Development Theory; Triple Bottom Line (TBL) theory and Pro-Poor Tourism (PPT) theory. These theories integrate all aspects, that is, economic growth, employment creation, community participation and sustainable tourism practices. What is lacking in all theories is the incorporation of legal practices in both sectors which will aid in advancing the theoretical framework in tourism and hospitality debates. Thus, a holistic and integrated theory could assist in answering multidimensional challenges that are faced in Zimbabwe. Secondly, methodologically it has added to the increasing research that uses qualitative conceptual analysis method in tourism studies. Finally, practically, the results may assist tourism stakeholders involved in various forms of tourism in Zimbabwe in crafting policies and strategies that promote the inclusion of local communities to enhance SDG 8. The results build on existing evidence from various scholars in the body of knowledge that present the data on how African tourism and hospitality industries contribute to the economic growth of a nation showcasing the wealth and uniqueness of such sectors. In this regard, empirical evidence shows that though tourism and hospitality can benefit local communities, in Zimbabwe they still face marginalisation, exploitation, poverty, insecurity, political instability, economic downturn, and inequalities since unemployment is still a major hindrance to human capital development hence affecting the achievement of SDG 8. Pro-poor approaches need to be integrated into tourism development, planning and management at local levels, integrating sustainability practices through crafting a sustainability policy framework. This points to the promotion, conservation and management of natural and cultural resources which will be of benefit as it leads to the inclusion and empowerment of local communities, specifically vulnerable groups like women and youths. However, for the sustainability policy framework to be effective in Zimbabwe there is a need for proper implementation and enforcement that incorporates the social, economic and environmental aspects so that future generations can benefit from cultural, natural and physical tourism sites. Closing the policy gap can capacitate the growth of such sectors, thereby leveraging on opportunities presented by the Agenda 2063, Africa Continental Free Trade Area (AfCFTA) and SDG 2030 that could promote regional integration, cooperation and development through learning from other international, regional and local tourism initiatives. While previous research has focused on the contribution of tourism and hospitality industries in economic growth, this inquiry shows that there is limited evidence of their contribution towards the focus on economic inclusion and employment opportunities for low-skilled workers, youths and women in the Zimbabwean context. To add on to that, innovation and diversification of the economy can foster tourism through developing new services, products and markets that will broaden and close the gap in competition harnessing the issue of affordability for local tourism.

On a sad note, tourism can have negative implications for SDG 8 in that workers might be exploited, there can be displacement of communities and degradation of resources in an environment that has various insecurities, inequalities and political instability. Aligning tourism policies and strategies with national and global development agendas together with adopting an integrated and holistic approach to tourism development, planning and regulation can assist in lessening the impact.

There remain some limitations to this study. The study involves a huge amount of secondary data from various documents which invites difficulties in selecting and summarising the findings. Secondly, the results of this study may not be generalised to all African and developing countries as they are based on the Zimbabwe case study. They are also based on documents found using imperfect search strategies for electronic documents. Therefore, the stated limitations of this study provide avenues for future research. There is a need for future research to focus on the perspectives of local people concerning the wealth and uniqueness of tourism in other parts of Zimbabwe which are not included in this study and how it contributes to enhancing their livelihoods. Future research may also seek the views of other stakeholders involved in various spheres of tourism and hospitality industries in Zimbabwe, for instance, development agencies, the private sector, government employees, other researchers, local and foreign tourists, industry players, regulators and policy makers (legal activists). These may provide fascinating findings that may contribute to contemporary knowledge in the field of tourism and hospitality.

Conclusion

Based on the empirical evidence above, the wealth and uniqueness of the African Tourism and Hospitality experience is undisputed. There is an abundance of flora and fauna coupled with high-quality accommodation, food and beverages, and excellent service delivery, which makes the African tourism and hospitality experience a marvel and a wonder. The continent is endowed with various renowned wonders of the world, world heritage sites (both cultural and natural) and very good climate conditions which make the continent a destination of choice globally. Tourism has great potential to support effective and sustainable economic growth for most African countries, Zimbabwe included, and can be instrumental in supporting recovery from the negative economic and socio-cultural effects of COVID-19. Based on the findings presented from the literature, tourism is partially an endogenous growth process, which, if supported through systematic allocation of resources by the government, has the potential to bring about sustainable growth of local economies. Through other lenses, the challenges and growth trajectory of the hospitality industry, as indicated in South Africa from 1994 to 2020, it can be acknowledged that basing on the South African government's policy interventions and the results that they yielded there is a need to increase efforts requisite for a more robust transformation of the industry to exploit its full potential. The view is, therefore, that the hospitality industry has a significant role in employment creation and as such there is a need to expedite human capital development as a capacity building intervention to accelerate socio-economic growth. Similarly, in Zimbabwe, tourism has been instrumental in driving economic growth and development, contributing an average of above 10% to GDP over the past two decades, facilitating employment creation, and also contributing to the improvement of livelihoods through community-based tourism initiatives. The abundance of world heritage sites, other tourist attractions, hospitable personnel, quality accommodation facilities, recreation, and adventure activities continue to position Zimbabwe as a prime tourist destination of choice for both international and domestic tourists.

Recommendations

While the African tourism and hospitality experience has its endowed wealth and uniqueness, it is not without its own challenges. There are a number of areas of improvement that could be implemented in making the African tourism and hospitality experience more attractive and enjoying sustainable growth going forward. More concerted efforts are needed in marketing the African tourist destinations for a wealth of unique hospitality experience. Tourism and hospitality operators can increase their online presence. The establishment and continuous update of high-quality websites and other social medial platforms such as Facebook, Instagram, X, and LinkedIn to market their products and services globally is essential. Internet connectivity generally remains poor in some African destinations. There is a need for greater investment in improving internet infrastructure for more stable connectivity. Liberalisation of the aviation market across the African continent could also significantly contribute to the improvement of Africa as a tourism destination. Some countries like Zimbabwe and the Democratic Republic of Congo also need to improve the quality of their roads as some of the tourists may need to travel by road from one destination to the other. Policy measures could also be implemented to increase the number of and improve the growth and profitability of community-based tourism projects so that tourism and hospitality benefits the local communities in more significant ways. Going into the future, the adoption of disruptive technologies such as artificial intelligence (AI), virtual reality, and 3D imaging may be explored to improve the quality of uniqueness of the African tourism experience. Future research could also focus on how these technologies could be explored to improve the growth and development of the tourism and hospitality industry.

References

Abdelhady, M., Ameen, F., and Abou-Hamad, M. (2022). The economic ramifications of COVID-19 pandemic on the global aviation industry. *Journal of Association of Arab Universities for Tourism and Hospitality*, *22*(2), 340–364.

Adeleye, B. N. (2023). Re-examining the tourism-led growth nexus and the role of information and communication technology in East Asia and the Pacific. *Heliyon, 9*(2).

Adrian, S. C. (2018). Measuring the effect of tourism propagation in the economy. *Ovidius University Annals, Economic Sciences Series, 18*(2), 359–362.

Akama, J. S., and Kieti, D. (2007). Tourism and socio-economic development in developing countries: A case study of Mombasa Resort in Kenya. *Journal of Sustainable Tourism, 15*(6), 735–748.

Aksoy, L., Choi, S., Dogru, T., Keiningham, T., Lorenz, M., Dan R., and Tracey, J. B. (2022). Global trends in hospitality. *Journal of Business Research, 142*, 957–973.

Ashley, C., and Mitchell, J. (2009). *Tourism and poverty reduction: Pathways to prosperity.* Taylor & Francis.

Authority, Z. T. (2023, 11 19). Statistics dashboard – Tourism arrivals trend. Retrieved from https://zimbabwetourism.net/.

Bama, H. N.-S. (2022). What innovations would enable the tourism and hospitality industry in Africa to re-build? *Worldwide Hospitality and Tourism Themes, 14*(6), 557–564.

Bastidas-Manzano, A. B., Sánchez-Fernández, J., and Casado-Aranda, L. A. (2021). The past, present, and future of smart tourism destinations: a bibliometric analysis. *Journal of Hospitality & Tourism Research, 45*(3), 529–552.

Benner, M. (2020). The decline of tourist destinations: An evolutionary perspective on overtourism. *Sustainability, 12*(9), 3653.

Bowen, J., and Morosan, C. (2018). Beware hospitality industry: The robots are coming. *Worldwide Hospitality and Tourism Themes, 10*(6), 726–733.

Bui, H. &. (2021). A recipe for sustainable development: assessing transition of commercial foodservices towards the goal of the triple bottom line sustainability. *International Journal of Contemporary Hospitality Management, 33*(10), 3535–3563.

Butler, R. (2022). COVID-19 and its potential impact on stages of tourist destination development. *Current Issues in Tourism, 25*(10), 1682–1695.

Chatterjee, S., and Prasad, D. M. V. (2019). The Evolution of Agri-Tourism practices in In-dia: Some Success Stories. *Madridge J Agric Environ Sci, 1*(1), 19–25.

Chibaya, T. (2013). From 'Zimbabwe Africa's Paradise to Zimbabwe A world of wonders': Benefits and challenges of rebranding Zimbabwe as a tourist destination. *Developing Country Studies, 13*(5), 84–91.

Cooper-Patrick, L., Gallo, J. J., Gonzales, J. J., Vu, H. T., Powe, N. R., Nelson, C., and Ford, D. E. (1999). Race, gender, and partnership in the patient-physician relationship. *Jama, 282*(6), 583–589.

Csikósová, A. J. (2020). Providing of tourism organizations sustainability through Tripple Bottom Line approach. *Entrepreneurship and Sustainability Issues, 8*(2), 764–776. doi:10.9770/jesi.2020.8.2(46).

Dada, Z. N. (2022). Pro-poor tourism as an antecedent of poverty alleviation: An assessment of the local community perception. *International Journal of Hospitality & Tourism Systems, 15*(1).

Dolezal, C., and Novelli, M. (2022). Power in community-based tourism: empowerment and partnership in Bali. *Journal of Sustainable Tourism, 30*(10), 2352–2370.

Dube, K. (2021). Sustainable development goals localisation in the hospitality sector in Botswana and Zimbabwe. *Sustainability, 13*(15), 8457.

El-Said, O., and Aziz, H. (2022). Virtual tours a means to an end: An analysis of virtual tours' role in tourism recovery post COVID-19. *Journal of Travel Research, 61*(3), 528–548.

Everingham, P., and Chassagne, N. (2020). Post COVID-19 ecological and social reset: Moving away from capitalist growth models towards tourism as Buen Vivir, *Tourism Geographies.* DOI: 10.1080/14616688.2020.1762119

Fotiadis, A., Polyzos, S. and Huan, T. C. T. (2021). The good, the bad and the ugly on COVID19 tourism recovery. *Annals of Tourism Research*, *87*, 103–117.

Gohori, O., and van der Merwe, P. (2021). Barriers to community participation in Zimbabwe's community-based tourism projects. *Tourism Recreation Research*, DOI: 10.1080/02508281.2021.1989654

Gupta, S., Modgil, S., Lee, C.-K., and Sivarajah, U. (2023). The future is yesterday: Use of AI-driven facial recognition to enhance value in the travel and tourism industry. *Information Systems Frontiers*, *25*(3), 1179–1195.

Hall, P. (2000). Creative cities and economic development. *Urban Studies*, *37*(4), 639–649.

Hammanjoda, K., Safiyanu, A., and Usman, J. (2023). Service quality in hotels: Understanding customers' perceptions for improved guest satisfaction. *Journal of Economics and Allied Research (Jear)*, *49*.

Heath, E. (2001). Globalisation of the tourism industry: Future trends and challenges for South Africa. *South African Journal Of Economic And Management Sciences*, *4*(3), 542–569.

International Trade Administration. (2022). Travel and tourism. Available at: https://www.trade.gov/country-commercial-guides/zimbabwe-travel-and-tourism (accessed on 6 November 2023).

International, S. T. (2020). Carbon footprint of tourism. Retrieved 01 04, 2024, from https://sustainabletravel.org/issues/carbon-footprint-tourism/

Ishtiaque, A. N. A. (2013). Tourism vision 2020: A case of Bangladesh tourism with special emphasis on international tourist arrivals and tourism receipts. *Journal of Business*, *34*(2), 13–36.

Jacobs, M. (2013). Green growth. In R.F. (ed.), *The Handbook of Global Climate and Environment Policy* (pp. 197–214). John Wiley & Sons. 10.1002/9781118326213.ch12

Javed, M. &. (2020). The role of government in tourism competitiveness and tourism area life cycle model. *Asia Pacific Journal of Tourism Research*, *25*(9), 997–1011.

Karamustafa, K., and Yilmaz, M. (2023). Destination life cycle: A conceptual approach based on the Turkish experience. *Journal of Tourism & Gastronomy Studies*, *8*(Special Issue 4), 439–451.

King, C., Iba, W., and Clifton, J. (2021). Reimagining resilience: COVID-19 and marine tourism in Indonesia. *Current Issues in Tourism*, *24*(19), 2784–2800.

Kuguyo, T. T., and Gandiwa, E. (2022). The influence of linking wildlife and non-wildlife tourist attractions on tourism marketing and performance in Zimbabwe. *Cogent Social Sciences*, *8*(1), 2044125.

Lansing, P., and Vries, P. D. (2007). Sustainable tourism: Ethical alternative or marketing ploy? *Journal of Business Ethics*, *72*, 77–85.

Lee, M. (2022). Evolution of hospitality and tourism technology research from Journal of Hospitality and Tourism Technology: A computer-assisted qualitative data analysis. *Journal of Hospitality and Tourism Technology*, *13*(1), 62–84.

Lu, J., and Xu, Z. (2021). Can virtual tourism aid in the recovery of tourism industry in the COVID-19 pandemic? *Travel and Tourism Research Association: Advancing Tourism Research Globally*.

Macrotrends. (2023, 11 19). Zimbabwe Tourism Statistics 1995-2023. Retrieved from https://www.macrotrends.net/countries/ZWE/zimbabwe/tourism-statistics.

Manwa, H. (2007). Is Zimbabwe ready to venture into the cultural tourism market? *Development Southern Africa*, *24*(3), 465–474.

Mapingure, C., du Plessis, E., and Saayman, M. (2019). Travel motivations of domestic tourists: The case of Zimbabwe. *African Journal of Hospitality, Tourism and Leisure*, *8*(2), 1–11.

Matura, P. (2022). Community-based tourism in Zimbabwe: The case of Chesvingo cultural village in Masvingo.

McKercher, B. &. (2021). Do destinations have multiple lifecycles? *Tourism Management, 83*, 1–5.

Mkono, M., and Hughes, K. (2020). Eco-guilt and eco-shame in tourism consumption contexts: Understanding the triggers and responses. *Journal of Sustainable Tourism, 28*(8), 1223–1244.

Mohrmann, B. (2019, 04 23). An analysis of Africa's tourism market for April 2019. Retrieved from https://www.atta.travel/.

Monnier, O. (2021), "A ticket to recovery: reinventing Africa's tourism industry", available at https://www.ifc.org/wps/wcm/connect/news_ext_content/ifc_external_corporate_site/newsþandþevents/news/reinventing-africa-tourism

Moyo, S., and Tichawaa, T.M. (2017). Community involvement and participation in tourism development: A Zimbabwe study. *African Journal of Hospitality, Tourism and Leisure, 6*(1), 1–15.

Mpofu, T. P. Z., and Chitura, M. (2015). An assessment of impact of the economic crisis on service delivery: A case study of the within the hospitality sector in at the resort town of Victoria Falls, Zimbabwe. *Greener Journal of Economics and Accountancy, 4*(1), 1–8.

Mundi, I. (2023, November 19). Zimbabwe – International tourism, number of arrivals. Retrieved from https://www.indexmundi.com/facts/zimbabwe/indicator/ST.INT.ARVL.

Mura, P. &. -L. (2018). Locating Asian research and selves in qualitative tourism research. *Perspectives on Asian Tourism*, 1–20.

Muresherwa, G., Tichaawa, T. M., and Swart, K. (2022). African journal of hospitality, tourism and leisure. ISSN: 2223-814X developing event tourism in Zimbabwe: Opportunities and challenges amid the covid-19 pandemic. *Pandemic. African Journal of Hospitality, Tourism and Leisure*, (113), 1259–12721259.

Musavengane, R., Siakwah, P., and Leonard, L. (2019). Does the poor matter in pro-poor driven sub-Saharan African cities? Towards progressive and inclusive pro-poor tourism. *International Journal of Tourism Cities, 5*(3), 392–411.

Mushawemhuka, W. J., Fitchett, J. M., and Hoogendoorn, G. (2021). Towards quantifying climate suitability for Zimbabwean nature-based tourism. *South African Geographical Journal, 103*(4), 443–463.

National Department of Tourism (NDT). (2020). Tourism Industry Survey of South Africa: COVID-19: Impact, mitigation and the future. Available at https://www.tourism.gov.za/CurrentProjects/Tourism_Relief_Fund_for_SMMEs/Documents/Tourism%20Industry%20Survey%20of%20South%20Africa%20-%20%20COVID-19.pdf.

Nunkoo, R. (2018). The state of research methods in tourism and hospitality. In Nunkoo, R. (ed.), *Handbook of Research Methods for Tourism and Hospitality Management* (pp. 3–23). Cheltenham: Edward Elgar Publishing.

OECD. (2018). Policy statement – Tourism policies for sustainable and inclusive growth. In *OECD Tourism Trends and Policies 2018* (pp. 11–14). OECD Publishing.

Olowoyo, M. R. (2021). Challenges and growth trajectory of the hospitality industry in South Africa (1994-2020). *African Journal of Hospitality, Tourism and Leisure, 10*(3), 1077–1091.

Organisation for Economic Co-operation and Development [OECD]. (2020). *Rebuilding Tourism for the Future: COVID-19 Policy Responses and Recovery.* Paris: OECD.

Ozturkoglu, Y. S. (2021). A new holistic conceptual framework for sustainability oriented hospitality innovation with triple bottom line perspective. *Journal of Hospitality and Tourism Technology, 12*(1), 39–57.

Priatmoko, S. K. (2021). Rethinking sustainable community-based tourism: A villager's point of view and case study in Pampang village, Indonesia 3245. *Sustainability, 13*(6), 1–15.

Ramkissoon, W. (2023). Perceived social impacts of tourism and quality-of-life: A new conceptual model. *Journal of Sustainable Tourism, 31*(2), 442–459.

Rasoolimanesh, S. M., Seyfi, S., Hall, C. M., and Hatamifar, P. (2021). Understanding memorable tourism experiences and behavioural intentions of heritage tourists. *Journal of Destination Marketing & Management, 21*, 100621.

Rogerson, C.M., and Baum, T., 2020. COVID-19 and African tourism research agendas. *Development. Southern Africa, 37*(5), 727–741.

Runyowa, D. (2017). Community-based tourism development in Victoria Falls, Kompisi Cultural Village: An entrepreneur's model. *African Journal of Hospitality, Tourism and Leisure, 6*(2), 1–7.

Sanderson, A. B. E. L., and Leroux, P. (2017). Tourism as an engine of wealth creation in Zimbabwe. *International Journal of Economics and Financial Issues, 7*(2), 129–137.

Sharpley, R. (2020). Tourism, sustainable development and the theoretical divide: 20 years on. *Journal of Sustainable Tourism, 28*(11), 1932–1946.

Sharpley, R. (2022). Tourism and development theory: Which way now? *Tourism Planning & Development, 19*(1), 1–12.

Shen, F., Hughey, K. F., and Simmons, D. G. (2008). Connecting the sustainable livelihoods approach and tourism: A review of the literature. *Journal of Hospitality and Tourism Management, 15*(1), 19–31.

Shereni, N.C. (2022). *Tourism and sustainable development goals in Zimbabwe: Contribution by the hospitality sector.* (Doctoral Thesis). Johannesburg: University of Johannesburg. Available from: http://hdl.handle.net/102000/0002 (Accessed: 10 January 2023)

Signe, L. (2019). Africa's tourism: A global destination for investment and entrepreneurship. Available at: https://www.brookings.edu/articles/africas-tourism-a-global-destination-for-investment-and-entrepreneurship/ (accessed 6 November 2023).

Şimşekoğlu, Ö., Nordfjærn, T., and Rundmo, T. (2015). The role of attitudes, transport priorities, and car use habit for travel mode use and intentions to use public transportation in an urban Norwegian public. *Transport Policy, 42*, 113–120.

Skobkin, S. S. (2020). The development of hospitality and tourism industry in Africa. *Journal of Environmental Management and Tourism, 11*(2(42)), 263–270.

Statista Research Development. (2023). Market size of the hospitality industry worldwide in 2023, with a forecast for 2027 (in billion U.S. dollars). Available at: https://www.statista.com/statistics/1247012/global-market-size-of-the-hospitality-industry/ (accessed 6 Novemeber 2023).

Statista. (2020). Selected African countries with the largest number of international tourist arrivals. Retrieved from https://www.statista.com/statistics/261740/countries-in-africa-ranked-by-international-tourist-arrivals/.

Strömstad, J. (2019). A global goal: Sustainable economic growth and decent work for all.

Tang, C. F., and Tan, E. C. (2018). Tourism-led growth hypothesis: A new global evidence. *Cornell Hospitality Quarterly, 59*(3), 304–311.

Thondhlana, T. P., Chitima, S. S., and Chirikure, S. (2021). Nation branding in Zimbabwe: Archaeological heritage, national cohesion, and corporate identities. *Journal of Social Archaeology, 21*(3), 283–305.

Torkington, K. S. (2020). Discourse(s) of growth and sustainability in national tourism policy documents. *Journal of Sustainable Tourism. Journal of Sustainable Tourism, 28*(7), 1041–1062.

UK Hospitality. (2018). The economic contribution of the UK hospitality industry 2018. Available at: https://www.ukhospitality.org.uk/page/EconomicContributionoftheUKHospitalityIndustry2018. (accessed 6 November 2023).

UNWTO. (2018). United Nations World Tourism Organisation, World tourism barometer.

Vellas, F. (2011, October). The indirect impact of tourism: An economic analysis. In *Third Meeting of T20 Tourism Ministers*. Paris, France.

Weigert, M. (2019). *Weigert* travel in Africa: Expanding the boundaries of the online travel agency business model. *Tourism Review, 74*(6), 1167–1178.

World Travel and Tourism Council (WTTC). (2021). South Africa 2021 Annual Research:

WTTC (2021). Global Economic Impact and Trends 2021 https://wttc.org/Portals/0/Documents/Reports/2021/Global%20Economic%20Impact%20and%20Trends%202021.pdf

WTTC. (2020). Recovery Scenarios 2020 & Economic Impact from COVID-19. Available at: https://wttc.org/Research/Economic-Impact/Recovery-Scenarios-2020-Economic-Impact-from-COVID-19. (accessed 6 November 2023).

WTTC. (2023). Economic impact research. Available at: https://wttc.org/research/economic-impact/recovery-scenarios-2020-econom (accessed 5 November 2023).

Yamagishi, K., Gantalao, C., and Ocampo, L. (2021). The future of farm tourism in the Philippines: Challenges, strategies and insights. *Journal of Tourism futures*. 10.1108/JTF-06-2020-0101

Zhou, Z. (2022). Critical shifts in the global tourism industry: Perspectives from Africa. *GeoJournal, 87*, 1245–1264.

Zielinski, S. K. (2020). Factors that facilitate and inhibit community-based tourism initiatives in developing countries. *Current Issues in Tourism, 23*(6), 723–739.

Zimbabwe Tourism. (2020). *Tourism trends & statistics*. Available at https://zimbabwetourism.net/download-category/tourism-trends-and-statistics/# (accessed 6 November 2023).

8 The State of Agritourism Development in Southern Africa

A Regional Analysis

Zibanai Zhou

Introduction

Tourism is one of the fastest growing economic sectors globally (Adeola et al., 2018; Matiza and Perks, 2021; Zhou, 2018), and is considered as a driver of economic development, diversification and catalyst for employment creation at community level. Emerging global trends indicate that the tourism market is increasingly becoming more inclined towards special interest tourism such as agritourism (Chase et al., 2018; Lak and Khairabadi, 2022). This is partly attributed to increased urbanisation coupled with growing fatigue of the global tourism market with traditional tourist attractions in favour of alternative less crowded and non-conventional forms of tourism in rural areas (Chase et al., 2018; Zhao et al., 2022). It is apparent that agritourism will continue to gain popularity as a form of special interest tourism worldwide (Back et al., 2019; Santeramo and Barbieri, 2017). In view of the increasing frequency of global health emergencies, Harari (2020) predicted that the global tourism market would recalibrate towards remote countryside tourist destinations, giving new impetus to special interest forms of tourism like agritourism. This is poised to become even more pronounced in a post COVID-19 period driven by the global travel market's heightened awareness of the risks associated with global health emergencies, and the need for authentic, unique and unspoilt vacation experiences in less crowded spaces (Zawadka et al., 2022). Chase et al. (2018) define agritourism as any farm activity that attracts visitors to the farm for experiences or product sales. Broadly, agritourism encompasses adventure tourism, fishing, food or wine tourism. The surge in agritourism has had profound socioeconomic impacts at local community, national and regional levels (Chiodo et al., 2019; Harari, 2020; Liao et al., 2022; Togaymurodov et al., 2023). Consequently, many southern African countries, after independence, promoted tourism as a driver for economic development through different forms of tourism (Rogerson and Rogerson, 2014). Despite the resurgence of agritourism, its growth and development is not uniform at national and regional levels as it is constrained by context specific challenges. In the context of southern Africa, agritourism development policy issues largely revolve around challenges relating to the need to build business skills and entrepreneurship

DOI: 10.4324/9781032696188-8

capabilities, farmer-to-farmer networking and connections, agritourism product development skill deficiencies, and to equip farmers with requisite tourism and hospitality skills (Rogerson and Rogerson, 2014). To this end, it is important to establish the level of agritourism development in order to plan for the sector's further development, and to devise ways of optimising socioeconomic benefits, and enhancing the competitiveness of southern Africa as an agritourism destination. Southern Africa is suited to become a prime agritourism destination given the region's predominantly agro-based economy. Most of southern Africa's population resides in the countryside, where agrarian activities are the main source of livelihood. Most tourist destinations in the Global South, including southern Africa are characterised by insufficient information on the level of agritourism development in their respective destinations. This makes tourism planning difficult, and often leads to the enactment of tourism development policies which are out of sync with practice. Consequently, most tourist destinations remain on the periphery of the booming agritourism market, thus losing out on the potential socioeconomic benefits.

Broadly, tourism has been significant to regional economies on many fronts: firstly, through creation of employment opportunities, infrastructural development and provision of social amenities in remote areas, generating foreign currency, preservation of resources, and multiplier effect within the economy (Dieke, 2013; Mutana, 2013; Shereni and Saarinen, 2020; Scheyvens and Monsen, 2020; Zhou, 2019). The potential benefits inspire countries and regions to develop the tourism industry; and southern Africa is not an exception. Agritourism entails recreational experiences which involve visits to rural settings or environments for the purposes of participating in activities and events not available in urban areas (Choo, 2012; Kunasekaran et al., 2011). Agritourism is viewed as a catalyst in revitalising farms, which face challenges like income stagnation, and it also supports the preservation of heritage resources (La Pan and Barbieri, 2013).

The present study aims to examine the contemporary state and level of agritourism development in southern Africa. It also seeks to establish the distribution of agritourism tourism resources, and identify the platforms for marketing and promoting southern Africa's agritourism products. Choo (2012) outlines that research on agritourism is still in the early stage of development and there is scope for further theoretical and conceptual advances mostly in developing countries. The study was inspired by the following research gaps observed in extant literature. The first theoretical gap is that Africa in general and southern Africa in particular for a very long time has been represented largely as an exclusive wildlife tourist destination (Zhou, 2019). The present study offers an alternative perspective. The second theoretical gap is that while research around agritourism is emerging in southern Africa, studies which offer a regional perspective remain limited as most previous studies focused on individual member countries, or a specific agritourism enterprise (Baipai et al., 2021; Chikuta et al., 2022; Myer and de Crom, 2013; Rogerson and Rogerson, 2014; van Zyl and van der Merwe, 2021). Hence, regional scale studies are limited, and the present study addresses this literature gap. Jeczmyk et al. (2015) acknowledged the limitation and called for more research in this regard in order to develop the sector. As a direct consequence of limited agritourism research at regional level, there is little information available on the state and level of agritourism needed to inform future tourism development and planning processes. The third theoretical gap is that in most emerging economies, the development of agritourism is often fraught with lack of proper coordination, insufficient knowledge and lack of government support (Askarpour et al., 2020; Kepe et al., 2001; Rogerson and Rogerson, 2014). Arguably, these variables further constrain southern Africa from achieving its full potential in agri-tourism. The study resonates with the United Nations World Tourism Organisation's (UNWTO) sustainable tourism thrust, which advocates for the development of alternative forms of tourism such as agritourism. However, very little information is available on the extent of special interest tourism develop-ment in emerging markets necessary to inform policy formulation. To address the research objectives, the following research questions are posed: What is the contemporary state of agritourism development in southern Africa? What is the spatial distribution of agritourism resources in southern Africa? How is southern Africa's agritourism product marketed and promoted? The study addresses these gaps identified in literature informed by the resource based theory. The use of the resource based theory would deepen the understanding of the various elements which influence agritourism tourism development in southern Africa. It is important to address the gaps identified in literature in order to enhance southern Africa's prospects as an agritourism destination. It is also envisaged that southern Africa's profile as an agritourism destination would be elevated to the wider agritourism market, increase the region's market share, and double agritourism's socioeconomic benefits. This would give new impetus to the development of a competitive regional agritourism sector. The study is significant given that the southern Africa region currently focuses on exploring other forms of tourism to diversify the regional economy. This could be an opportunity to develop the agritourism sector.

The study enhances the awareness of agritourism development among farmers and tourism policy makers in southern Africa as it highlights the need for capacity building for agritourism development. The study is important from a tourism planning, development and agritourism budget support perspective in southern Africa, especially in geographical areas with huge agritourism development potential. Despite an abundant agritourism resource base in southern Africa, the region largely remains unknown to the wider global agritourism market. Furthermore, in spite of the emerging global trends pointing to the growth of the agritourism segment, the current state of agritourism development in southern Africa and its ability to tap into this market remains relatively unknown. Consequently, it militates against formulation of requisite policies to spur the development of the promising sector. To this end, the region risks remaining on the fringe of agritourism, hence losing out on the associated socioeconomic benefits. The chapter is structured as follows. After the introduction, the next section examines discourses around agritourism. This is followed by an outline of the study's methodology. Subsequently, study findings are presented and discussed. Thereafter, the conclusion of the study is given.

Literature Review

The Concept of Agritourism

Agritourism is defined as visiting an active agricultural holding for leisure, entertainment, and recreational purposes or even educational programmes (Canovi and Lyon, 2020; Gil Arroyo et al., 2013; Tew and Barbieri, 2012). It revolves around experiencing various on-farm and off-farm attractions which exist mostly in remote rural areas (Askarpour et al., 2020; Krishna and Alok, 2020). Chase et al. (2018) posit that the production associated with farming is retained while tourism activities are an add on or embedded in the farming enterprise. Van Zyl and van der Merwe (2021) posit that agritourism is gaining fame worldwide, and has become a way of life in the US, Europe, and Asia. Most agritourism activities include farm stay, hiking, wildlife viewing, photographing, horse riding, and fruit picking (Van Zyl and van der Merwe, 2021). According to Back et al. (2019) other farm based activities include participating in festivals, dining, and wine and beer tasting. Similarly, Quella et al. (2021) and Fleischer et al. (2018) outline that agritourism includes core and peripheral activities which take place on a farm in the form of education, entertainment, outdoor recreation and hospitality. Agritourism is streamlined into agriculturally based recreational activities, agricultural education and rural based outdoor recreation and hospitality services. World over, agritourism is a significant source of income for the local communities, fosters farmers to practise sustainable agriculture, and helps preserve resources.

Agritourism is a growing tourism segment in both the Global North and Global South; however, little information is available on the current state of agritourism development in southern Africa. Information is paramount when planning tourism development. The unavailability of such information makes it difficult for policy makers to plan properly for the development of the sector. Similarly, it also affects the crafting of relevant regional agritourism development strategies, and marketing of agritourism products. Gao and Wu (2017) and Lane and Kastenholz (2015) frame agritourism as a subset of rural tourism, which is widely used as a means of rural regeneration and resource conservation. The unique, rich natural resources in the countryside are being turned into living spaces with multiple functions (Sardaro et al., 2020). The UNWTO (2022) launched 'Best Tourism Village' aimed at promoting rural culture and natural resource protection as well as sustainable social and economic development through tourism (Liao et al., 2022). Yang et al. (2022) posit that differences in tourism resources are the prerequisite and material bases for rural tourism development. This view corroborates Mitchell and Shinnon's (2018) observation that the spatial distribution characteristics of rural tourism resources influence the layout of the sector. To this end, Jonas-Berki et al. (2015) examined the characteristic features of health tourism destinations from a regional and spatial perspective; however, the study did not assess the state of agritourism development in southern Africa. It mainly focused on factors which influence agritourism development at country level, and agritourism development models at community level. There are aspects related to lack of robust agritourism development frameworks, and marketing and promotional strategies, which are of significant concern among tourism stakeholders. They influence agritourism development.

Drivers of Agritourism, and Socioeconomic Benefits

Several factors motivate farmers to engage in agritourism. These include rising farm input costs, unreliable rainfall, poor harvests, and food insecurity. In such circumstances, agritourism becomes an alternative source of livelihood and additional income stream (Bwana et al., 2015). The main drivers of agritourism development outlined in extant literature encompass economic decline, which causes income stagnation for farmers, the need to revitalise rural agrarian economies, heritage preservation, and improving the quality of life for communities (La Pan and Barbieri, 2013; Quella et al., 2021). Rogerson and Rogerson (2014) outline that agricultural sustainability, employment creation, additional revenue, recreational opportunities, and landscape management are other reasons often cited in support of agritourism development. Furthermore, agritourism is hailed for reducing rural-urban migration, providing incentives for the preservation of agricultural land and other heritage resources, and opening up vast opportunities for value addition at the farm. (Bwana et al., 2015). Bargi and Reeder (2012) assert that e-connectivity is another significant factor, which motivates farmers to participate in agritourism. Access to the internet increases farmers' information resources, as well as expansion of their market. In support of this perspective, Ferreira et al. (2020) state that the use of online marketing strategies has been harnessed to boost agritourism product sales. Tourism industry is one of the major economic activities, which contributes about 10% of the world's gross domestic product (GDP) (World Bank, 2019). In Africa, just like the trend in the Global South, rural communities have not benefited meaningfully from tourism because of low levels of tourism investment and inadequate promotion. The value of agritourism activities to the global economy stood at 79 billion dollars in 2019, before COVID-19 (FBI, 2020). According to Togaymurodov et al. (2023) agritourism is more developed in countries such as the US, UK, Germany, Spain, France, Malaysia, China, Poland, Italy, Indonesia, and India. In contrast, there is modest development of agritourism in countries in the Global South notably South Africa, Kenya, Tanzania, Namibia, and Morocco, which face a number of common challenges. However, in the case of southern Africa, the agricultural sector is the largest sector which supports the regional economy. It provides vast opportunities for farming communities to engage in agritourism activities as the main source of livelihood.

Resource Theory

Grant's (1991) resource-based theory is employed as the analytical lens in the study. The resource based theory is simply a strength and weakness analysis (Acede et al., 2006). The resource-based theory (RBT) entails that in order to maximise socioeconomic value, enterprises and tourist regions should not only possess great resources, they must also be able to leverage the uniqueness of the resources to transform the lives of local communities. The tenets of the resource-based theory are that an enterprise must assess its unique internal resources and determine its competitive advantage (Grant, 1991). Jeou-Shan and Chen-Tsang (2012) applied the resource-based theory to analyse the variables which determine the success of the development of culinary tourism in the Asia-Pacific region. The study found that the most important factor is to identify and capitalise on the core resources for tourism. In summary, it is important for an organisation to shift its focus from traditional and external positioning to the unique internal resources and capabilities in order to strengthen the organisation's core business and enhance its competitive advantage. In order for enterprises and tourist regions to hold sustainable competitive advantage, they must possess unique selling points which stand apart from competitors. To this end, tourist destinations must have unique immobile resources, which are scarce, not completely reproducible and irreplaceable. Enterprises should make use of resources and capabilities to develop competitive strategies. Consequently, in order for southern Africa to compete in the agritourism market, it must improve tourism planning, learn from best practices from leading agritourism destinations such as Israel, China, Germany, and the US and thus make regional agritourism planning a benchmark for global best practices (Hall and Sharples, 2003). There are synergies between the availability of agritourism resources, the distribution, and overall development of agritourism. A supportive agritourism development policy framework becomes an overarching variable. In addition, effective marketing and promotion of agritourism to the global tourism market leads to the optimisation of the socioeconomic benefits of agritourism. The resource-based theory embodies all these elements.

Southern Africa and Agritourism

In the context of southern Africa, most economies are agro-based, which provides opportunities for agritourism enterprising. However, southern Africa's tragedy is its overdependence on natural features and attractions such as wildlife, Table Mountain, Okavango Delta, Victoria Falls, beach tourism, Serengeti National Park, and Lake Malawi. The region has rich local African food systems, which largely remain untapped for tourism purposes.

The availability of such resources plays a critical role in agritourism development (Wang et al., 2022). Given that southern Africa is predominantly rural and its population rely on agriculture, it would benefit significantly from an agritourism-led rural development (Krishna and Alok, 2020; Togaymurodov et al., 2023). In view of this, agritourism offers huge benefits to southern Africa through provision of employment opportunities, which subsequently spur economic development and poverty alleviation. Additional benefits include the upgrading of social amenities and standard of living, boosting GDP, enhancing the region's competitiveness in the international tourism arena, diversification of the regional economy, and expanding the region's tourism product base. The significance of agritourism to southern Africa can be summarised as revolving around economic needs, employment creation opportunities, and social and environmental aspects. However, it must be stated that agritourism as an emerging eco-friendly form of tourism is yet to be fully promoted in southern Africa. Krishna and Alok (2020) pointed out a need to enact relevant tourism policies to provide a supportive regulatory environment conducive for tourism development in the Global South given that such policies are apparently lacking.

Constraints of Agritourism Development in Africa

Southern Africa is characterised by the neglect and marginalisation of small scale farm holders in economic and development policy, which contributes to the increasing vulnerability of rural communities, women, and youth. There is generally under-investment in agriculture, which in itself is increasingly being affected by climate change in southern Africa. The complexity of developing agritourism in southern Africa lies in the reality that most rural spaces are characterised by a general lack of proper facilities, underdevelopment, increasing climate change issues, and lack of investment in world-class tourism amenities, yet there are vast opportunities for the marginalised sections of society (women and the elderly) to benefit immensely from fishing, crafts, subsistence farming somehow related to agritourism (Cheteni and Umejesi. 2022). Rural spaces including farming communities, which characterise most of southern Africa, are associated with poverty; therefore, agritourism can play a significant role in achieving the sustainable development goals (SDGs), such as poverty reduction, inclusive development, economic growth and women empowerment. Central to tourism is its ability to alleviate poverty and creation of employment opportunities especially in remote areas. To this end, agritourism can be harnessed as a pathway to achieve SDGs 1, 8, and 13. It becomes important to explore possible ways through which agritourism can be developed sustainably for the benefit of mankind, environment and rural economies. Agritourism becomes critical for revitalising rural spaces through sustainable development (Ammirato et al., 2020). There are opportunities for agritourism development in the countryside given that the socioeconomic and cultural well-being of rural communities is intertwined with the tourism sector.

Agritourism can support entrepreneurship by residents of local communities (Swanson and DeVareaux, 2017). Despite the growth of agritourism globally, little is known about its extent, relationship between private and public tourism organisations and agritourism support measures in the context of southern Africa. Each southern Africa country has different agritourism framework conditions. There is still a lack of understanding of the status of the agritourism sector in less developed countries (Baipai et al., 2021), and African countries are yet to implement development policies for the agritourism sector.

The development of agritourism in southern Africa, which brings together the agrarian and tourism spheres, can contribute to the socioeconomic development of rural spaces (Litheko, 2022), and well-being of the rural population, thus contributing to achieving the SDGs such as poverty alleviation, empowerment, and inclusive development, among others. The unavailability of a well-constructed agritourism strategy for the region is a cause of concern holding back the development of the sector. There are limited economic opportunities for rural spaces, women, and entrepreneurs (Romanenko et al., 2020), of which agritourism development would go a long way to alleviate this. Agritourism development will create sustainable development for all stakeholders to benefit (Kubickova and Campbell, 2020; Backmatova, 2021; Maharjan and Dangol, 2018). Farms in most southern African countries are characterised by variations of highly capitalised and established farmers and small-scale farmers who may not have the resources to support their farm operations (Zantsi et al., 2019).

Agritourism has a positive effect on obtaining economic and social benefits. Overall, women are the main part of the labour force in the agricultural sector. Logically, it then follows that agritourism could be a pathway to empower women in farming communities and other rural spaces, thus promoting gender equality and empowerment of women in the work force. Women will get extra income from additional services for tourists. Agritourism can increase the employment of women, thereby reducing poverty. It also supports women's entrepreneurship, and this can vastly improve and elevate women's social, economic and cultural status (Vukovic et al., 2023). Agritourism opens opportunities for women to be empowered in poverty reduction through tourism. Empowerment is a critical aspect of gender equality (Mrema, 2015). However, despite these benefits, agritourism development is constrained by lack of funding at enterprise and government levels, as evidenced in Kenya, Tanzania, Namibia, South Africa and Morocco, whose potential for growth has been hampered by under-funding. Agritourism can positively contribute to rural development (Wijijayanti et al., 2023), and enhance the preservation of cultural heritage and natural resources (Roman and Grudzien, 2021; Ammirato et al., 2020). Agritourism can address SDGs, which are steeped in socioeconomic and environmental issues. SDG1 is poverty alleviation by providing communities with job opportunities to earn income and improving the quality of life. Agritourism also addresses SDG8, which is decent work and economic growth. It creates employment in rural spaces where job opportunities are limited. Agritourism addresses SDG13, which is climate action. As farming enterprises are increasingly being affected by climate change, agritourism can be a way to address climate change issues through agroforests, water conservation and use of renewable energy.

Tourism has been portrayed as a sector able to activate socioeconomic progress and gender equality and prone to generate income for some of the most remote destinations (Novelli et al., 2021). The demand for agritourism products gives way to the creation of new jobs, hence potentially contributing to poverty alleviation and reducing inequalities through entrepreneurship. Tourism can be a remedy for achieving all the 17 SDGs (Scheyvens and Hughes, 2019; Mutana and Mukwada, 2023). However, balanced collaboration in planning and implementing agritourism development activities is not always evident in southern African countries. Agritourism can address SDG1 of reducing poverty in all its forms by 2030 through inclusive considerations of groups of people that are easily left out, for example, women (Mutana and Mukwada, 2023). Hall (2019) and Dube (2020) acknowledged the interconnectedness of SDGs and tourism in general.

Methods and Materials

Study Site

The study aimed to examine the current state and level of agritourism development in southern Africa. Southern Africa is made up of 15 countries namely Angola, Botswana, Democratic Republic of Congo (DRC), Eswatini, Lesotho, Madagascar, Malawi, Mauritius, Mozambique, Namibia, South Africa, Seychelles, Tanzania, Zambia, and Zimbabwe. It is an economic grouping recognised by the acronym Southern Africa Development Community (SADC). The region depends on agriculture, export of raw minerals and tourism. Tourism as an economic sector has been targeted as a way of diversifying the regional economy (Zhou, 2019). Southern Africa has abundant swathes of land, farms, and wildlife reserves which can support agritourism. Currently, southern Africa is well known for wildlife tourism. The global tourism market has been concerned about southern Africa's rather restricted tourism resource base which revolves around African safaris or wildlife. Consequently, the development of agritourism in the region would partly resolve the restrictive tourism resource base concern as it would expand the region's tourism product.

Data Collection and Sampling Procedures

A qualitative research approach was adopted. Qualitative data collected before the survey were used to guide instrument development (Quella et al., 2021). The qualitative method involved interviews. Participants consisted of key informants who represented farmers, tour operators, hotels and national tourism organisations. Data were collected through interviews in line with similar studies conducted elsewhere (Brandth and Haugen, 2011; Mnguni, 2010; Myer and de Crom, 2013), which used interviews with farmer-owners and tourists. Askarpour et al. (2020) used face-to-face interviews with farmers to collect data. A semi-structured interview guide was designed to collect qualitative data. The interview guide was developed based on the works of Mnguni (2010), Myer and de Crom (2013), and Askarpour et al. (2020).

The findings of the study were based on data collected during the 2023 edition of the travel expo/Sanganai held in Bulawayo at the Zimbabwe International Trade Fair (ZITF) exhibition centre. Southern Africa countries were represented at the premier exhibition event. A database of exhibitors was obtained from the local national tourism organisation, Zimbabwe Tourism Authority (ZTA), which was then used to approach the respective targeted respondents. Initially, the researcher approached the respective organisations to make a formal booking for an interview. The aim of the study was explained, and after the interview booking was confirmed, data collection schedule was done. Interviews were conducted outside the exhibition hall, and data were recorded with the consent of the interviewees. National tourism organisations, Tour operators, hotels, and farmers were deliberately targeted. These were deemed to have adequate knowledge about agritourism activities in their respective countries, the potential of agritourism, its current state and level of development, spatial distribution of agritourism resources, as well as the marketing and promotion of agritourism products. In-depth interview data were augmented by an extensive review of extant literature around agritourism. Purposive and snow ball sampling strategies were employed to select the respondents. Eventually, 30 key informants were interviewed, which were distributed as follows: national tourism organisations-10; tour operators-13; farmers-5; and hotels-2. Interview questions focused on the general level of agritourism development in SADC countries, availability of a supportive agritourism development framework, geographical distribution of agritourism resources within SADC, and marketing and promotion of agritourism products. These themes were informed by the resource theory and were also developed from previous studies (Grant, 1991; Askarpour et al., 2020). A deductive analysis of data was concurrently done with data collection. A thematic content analysis was then employed for data analysis.

Findings and Discussion

Four major themes emerged from the dataset, which are presented and discussed below.

State of Agritourism Development

According to the interview narratives, the current state of agritourism development in southern Africa was overall described as on the back foot and not inspiring. Most respondents were of the view that although there is potential for a fully-fledged agritourism enterprise to emerge in southern

Africa, the overall development of the sector still lags behind as compared to other leading agritourism destinations mostly in North America, western Europe and northeast Asia (Krishna and Alok, 2020; Lv, 2020; Zoto et al., 2013). However, in comparison to other regions in continental Africa such as east, west, central and north Africa, southern Africa fared much better in terms of the development of its agritourism sector. This view is illustrated by the following interview excerpts:

Generally, agritourism is a very low tourism activity in southern Africa as compared to other forms of conventional tourism. Its level of development is very low, however promising (Participant 4).

Agritourism remains invisible on most southern Africa tourism packages and itineraries (Participant 27).

We have not yet reached the stage of countries such as Israel, China, the US, or Australia, which are leading agritourism destination globally (Participant 19).

The uninspiring state of agritourism is attributed to the general low level of awareness by the wider tourism market, inadequate supportive tourism infrastructure or amenities in farms, and lack of tourism and hospitality skills and knowledge on the part of farmers. This is exacerbated by under-investment, low priority and lack of support from government as highlighted in extant literature (Kepe et al., 2001; Rogerson and Rogerson, 2014). This also conforms to the general trend observed in the Global South as highlighted by Myer and de Crom (2013) in countries such as Kenya, Tanzania, and Morocco among others that there is still unutilised capacity in the agritourism sector, more so in developing countries. The low level of agritourism development is also a result of other systemic variables such as low awareness on the part of farmers that they can augment farming income with revenue stream from tourism activities, underinvestment in agritourism at farm level, and inadequate marketing of agritourism products at regional level. This is consistent with one pillar of tourism development described in the destination amalgam framework in which accommodation and avail-ability of supportive amenities are critical to stimulate the level of tourism development. This is unsurprising given that southern Africa has been focusing on conventional forms of tourism like wildlife tourism and other mega natural tourist attractions at the expense of other forms of tourism like agritourism. Another dimension which accounts for the current low level of development is the issue of farm owners who are unwilling and not committed to embracing agritourism as an add on to primary agrarian activities. This stems from a lack of knowledge on how to run tourism and hospitality services as highlighted in literature (Rogerson and Rogerson, 2014). Overall, southern Africa has huge potential in agritourism supported by an agro-based regional economy, and there is scope for the region to leverage the abundant agritourism resource base to compete at the level of

Israel, which is a leader in agricultural technologies in the Middle East region (Krishna and Alok, 2020) and China, which is also a leading agritourism destination in North East Asia region. As an example, in 2019, 3.2 billion tourists participated in agritourism in China alone (Lv, 2020).

Diversity and Distribution of Agritourism Resources

Findings showed an uneven distribution of a diverse range of agritourism resources in southern Africa. Participants observed overconcentration of resources in a relatively few member countries. A diverse range of agritourism resources was outlined encompassing vast rangelands, farm lands, forests, sugarcane, wine or grape, tea and coffee estates, citrus plantations, lucerne grass, and horse farming. This was regarded as of huge significance to the region, making it comparable with leading agritourism destinations worldwide. However, despite the availability of such an abundant base of agritourism resources, respondents were concerned about the mismatch with the provision of a supportive tourism infrastructure in terms of the requisite amenities. The concerns are consistent with previous studies such as Chikuta et al. (2022), who found a similar trend whereby agritourism development was curtailed by the unavailability of requisite amenities. The disproportionate distribution of agritourism resources partly explains the huge regional variation in terms of the overall agritourism development in southern Africa. This is in line with the destination amalgam theory that suggests that the availability of tourist attractions influences tourism development. As an example, successful tourist destinations with a well-developed tourism sector are characterised by a critical mass of requisite tourist attractions or resources. This is also in line with the tenets of the resource theory which suggests that destinations and communities with abundant resources tend to leverage such resources to stimulate socio-economic development. The finding also supports Wang et al. (2022) and Mitchell and Shinnon (2018), who argued that availability of resources influences the structural development of the sector. While tourism resource endowments provide important support for agritourism development, there is a need for a coordinated regional development approach. There is a need for more investment in tourism facilities on farms such as farm lodges, guest houses, wildlife sanctuaries, national monuments, snake parks, crocodile pens, gastronomy tourism, traditional meal cookouts, and traditional dishes, which have immense nutritional value and medicinal properties. The following quotes summarise participants' perspectives:

> *The most established agritourism centres in the southern Africa region include South Africa's Western Cape wine routes, Zimbabwe's eastern Highlands, Namibia's farms, and Botswana's cattle ranching* (Participant 3).
> *There is a diverse range of agritourism resources and activities in the form of crocodile farms, aquaculture, irrigation schemes, fisheries, cattle ranching,*

dairy farms, goat farming, bee keeping, poultry, hatchery farms, and ostrich farms, however, these are not evenly distributed (Participant 11).

Agritourism activities range from visiting farms, estates and these are concentrated in South Africa, Namibia, Zimbabwe, Botswana, Angola, and Tanzania which possesses a competitive agritourism resource base (Participant 7).

Southern Africa's economy is agro-based, therefore farm activities such as hayrides, farmscape observation, orchard tours, and food sales are popular (Participant 25).

There are different forms of agritourism activities which are prominent in southern Africa, which include visiting citrus plantations, orchards for bananas, avocadoes, olives, grapes, goats and sheep farms, maize fields, potatoes, horticulture, and bird watching. The outlined agritourism activities corroborate what was found by van Zyl (2019) that agritourism activities are diverse and expansive including visiting wine estates for wine tasting. This also conforms to the ideas of Xingping (2019) who framed agritourism activities as including exploring agricultural production plants, processing and sales factories. Other popular agritourism activities in southern Africa include outdoor recreation, educational experiences, entertainment, hospitality services and on-farm direct sales, bee production and rice fields in Malawi. These provide southern Africa with a head start in agritourism. This is also in line with what was established by Van Zyl and van der Merwe (2021) who outlined an expansive list of agritourism activities in South Africa ranging from on-site viewing of processing of agricultural produce, hiking, wildlife viewing, fruit picking, to horse riding. The findings are also similar to Back et al. (2019) who highlighted festivals, dining, wine and beer tasting. However, these are unevenly distributed therefore resulting in the development of an agritourism sector skewed in favour of only a few countries.

Agritourism Marketing and Promotion Platforms

Further analysis of data revealed that agritourism in southern Africa is marketed through various platforms and forums at local, regional and international levels. Respondents outlined a range of the most significant platforms employed to market agritourism. These are summarised in the following excerpts:

We market agritourism mostly at exhibitions, food cook out competitions, field days and through brochures, magazines, and pamphlets (Participant 15).

Promotion of agritourism products is mainly through digital marketing platforms, editorials, newspaper supplements, print and electronic media, billboards as well as overseas tourist offices (Participant 22).

Agritourism products are mostly marketed through themed events, wine tasting, and food expos, celebrities, agricultural shows, localised events, regional events and international platforms (Participant 6).

Tour operators, travel agencies, airline carriers, paraphernalia distributed through NTOs, and special features in newspapers are used to market agritourism (Participant 24).

The majority of participants believed that marketing of agritourism at international events such as Sanganai travel expo in Zimbabwe, Botswana Travel and Tourism expo, Africa's Travel Indaba in Durban, South Africa, as well as China's Shanghai World Travel Fair, and ITB in Berlin, Germany was deemed effective as it has a global reach as compared to using other platforms. These have been leveraged to market and promote southern Africa's agritourism. However, participants stated that marketing and promotion are restricted at an institutional level, given that most individual farmers lack the necessary resources to exhibit and market on such international platforms. This is attributed to lack of funding, a structural limitation which characterises agritourism development in the developing world (Rogerson and Rogerson, 2014). However, there is scope to cascade the marketing of agritourism through the tourism value chain by enlisting the services of leading airline carriers and tour operators, and harnessing online digital platforms to reach an even wider tourism market. This would go a long way to boost awareness of agritourism products in southern Africa to the global agritourism market. It was suggested that agritourism marketing platforms could be broadened to incorporate use of films, TikTok, documentaries and third party endorsement. As highlighted in literature, southern Africa's agritourism sector has not received sufficient recognition due to poor marketing. There is also a need to come up with joint regional agritourism itineraries, which cover the whole of southern Africa. This would resonate with the agritourism market, given the increasing multi-destination nature of visits by tourists. The use of effective marketing and promotion would optimise the socioeconomic benefits associated with agritourism development. The use of different marketing platforms is in line with international best practices as it mirrors global trends. In this day and age of information and communication technologies (ICTs), it is important to harness digital marketing platforms to broaden the marketing of southern Africa's agritourism. This is in consonance with the ideas of Bargi and Reeder (2012) who found that farmers who have access to the internet can increase the success of agritourism promotion. In addition, Ferreira et al. (2020) observed the increased use of digital marketing platforms for a global reach. Agricultural tourism promotion should not be limited to events like trade fairs and agricultural shows in order to put the region firmly on the global agritourism map. Some participants were sceptical about the effectiveness of the marketing and promotional strategies, and to back up their claims they cited southern African cuisines, local food systems, and traditional food baskets, which largely remain unknown. It was claimed that such agritourism products are underutilised. To this end, southern Africa should promote rural tourism as a way of broadening the regional economic base through development and unlocking of opportunities in rural areas which are often neglected from an economic development and provision of social services perspective. This would ultimately de-congest mega tourist centres, enhance inclusive development, and foster sustainable development.

Agritourism Development Frameworks

A recurring theme in the data set was the unavailability of a clear agritourism development policy framework, pointing to a structural challenge on the region's agritourism sector. According to respondents, agritourism development in southern Africa is uncoordinated at regional level. Overall, tourism development frameworks remain fragmented, in the sense that they do not incorporate other forms of tourism except the conventional forms of tourism. This curtails the development of agritourism as it is not prioritised in the current tourism development frameworks. This view is illustrated below:

Southern Africa does not have a specific agritourism development policy at regional level as this is devolved to member countries (Participant 13).

There is a need to make all forms of tourism prominent in tourism development master plans of SADC member countries (Participant 30).

There is a need to craft a tourism development policy framework that incorporates the interconnection between the various forms of tourism (Participant 2).

There is a need to unify agritourism with other tourism subsectors (Participant 26).

Tourism development policies are very broad, and not stream lined to speak to specific subsectors like agritourism (Participant 17).

This calls for a unified regional approach towards agritourism development within the southern Africa countries' membership. A regional institution is critical to spearhead agritourism development frameworks. There is also a need for member countries to review their tourism development policies, with a view to broaden and make them more inclusive, giving prominence to other forms of tourism such as agritourism. This is in line with shifting trends in the broader tourism industry in which tourists prefer to undertake vacation holidays in farm related settings. In the dataset, this aspect was recurring, which highlights the complexity of the challenges spawned by the fragmented nature of tourism development policy frameworks. This exacerbates the disconnect between agritourism development policies and practice. To this end, agritourism development models lack synergies with other tourism subsectors, an attribute which further weakens full-scale development. This particular finding is in line with Krishna and Alok (2020) who reiterated a need for policy makers to improve agritourism development in the Global South given that such policies are lacking.

Conclusion, Implications and Limitations

The study's purpose was to examine the contemporary state and level of agritourism development in southern Africa. The study found that despite southern Africa's huge potential, the current state of agritourism development in the region remains uninspiring. This has been attributed to multiple challenges, notably under investment, and lack of government support, which continue to hamstrung the region. Furthermore, the study established that southern Africa has a diverse range of agritourism resources, which are however, concentrated in very few countries. The level of agritourism development is directly related to the spatial distribution of agritourism resources. Furthermore, the study outlined the various marketing platforms through which agritourism is promoted namely travel expos, field days, food competitions, digital platforms, brochures and pamphlets. Additionally, findings revealed that southern Africa is characterised by fragmented and disconnected tourism development policy frameworks which constrain agritourism development. The study has implications on broad issues which include women empowerment, gender inequality, farming community livelihood, poverty alleviation, job creation, marketing and promotional strategies, and unavailability of clear agritourism development frameworks.

The study has shown a strong connection of agritourism with SDGs 1, 8 and 13, therefore agritourism aspects which support women empowerment, job creation, inclusive growth and development must be adopted and fully supported. It is important for southern Africa's tourism stakeholders to adopt and implement agritourism aspects, which uplift the vulnerable, women, address inequality, and job creation, in order for agritourism to be significant at both community and national levels when developing the agritourism sector. Agritourism development models must be remodelled and informed by these aspects. Agritourism offers opportunities to create jobs and social inclusion, economic empowerment and address gender inequality through income earnings (Vukovic et al., 2023). For example, women can engage in self-help groups which may help empower them by achieving economic emancipation, and greater control over resources such as information, data, and ideas. Women's entrepreneurship presents opportunities for economic empowerment, and involvement of women in agritourism through entrepreneurship is strongly connected with improving the status of women and community well-being (Haugh and Talwar, 2016). This can be achieved through establishing women's cooperatives and associations in the agritourism value chain. An agritourism support system collaborated by government and the private sector must back up these initiatives.

The current regional agritourism development policy framework needs review to facilitate linkages between agritourism farm enterprises and regional tourism institutions or organisations, tour operators, and the agritourism market. The study findings are significant to policy makers at regional and national levels, respectively, as the insights provide a better understanding of the impact of agritourism development frameworks on tourism development (Hollas et al., 2021). This would go a long to address the loose coordination between farmers and other stakeholders in the main stream tourism sector. The study employed Grant's (1991) resource-based theory to examine the state of agritourism development in southern Africa. This is a theoretical contribution as the RBT has not been applied in previous agritourism studies in the context of southern Africa. By employing the RBT, the study broadened the current understanding of the state of agritourism development by giving a fresh perspective of what authorities need to do to improve the development of the sector, and maximise agritourism's favourable impact in poverty alleviation, job creation, women empowerment, and gender equality. Through the resource-based theory, the study amplified the significance of southern Africa's agritourism resource base in influencing overall tourism development within the region. While the study reiterated the significance of the availability of agritourism resources, it also highlighted that it is not sufficient enough if the wider tourism market remains unaware of the resources. The study has shown that the availability of agritourism resources must be supported by investment in tourism and hospitality amenities. The findings of the study imply that agritourism development in southern Africa remains subdued, despite a favourable distribution of a diverse range of agritourism resources. There is huge potential for the development of a sustainable agritourism enterprise which, however, requires stakeholders to adopt more and expand effective marketing and promotional strategies for the agritourism product.

There is also a need to strengthen the link between tourism and agriculture in Africa. There are vast opportunities for the marginalised sections of society such as women to benefit immensely from fishing, crafts, and subsistence farming, which are related to agritourism (Cheteni and Umejesi 2022). The implication is that there is a need to design a regional tourism development policy framework that is inclusive of all the tourism subsectors to leverage on synergies and complementarity of the various tourism subsectors. However, there remains loose coordination between agritourism development at farm enterprise level and the wider tourism sector, and tourism bodies/institutions (Baipai et al., 2021), which presents challenges for southern Africa. This is an issue which policy makers need to address. The unavailability of an agritourism regulation framework and resulting variability in regulations throughout southern Africa creates a peculiar situation that needs to be addressed. This calls for regional oversight as most countries do not have an agritourism regulatory system, nor readily available support schemes (Saayman et al., 2018). The conclusion of agritourism development in southern Africa is one of qualified optimism. The natural resources of the southern Africa region match the growing tastes of the global agritourism market; however, there is a need to address issues around access and tourism infrastructural facilities. Generally, access to southern Africa countries is one of the most frequently cited difficulties (Dieke, 2013). There is a need for collaboration, and a move away from industry and sectoral fragmentation.

Limitations

The study findings relate to southern Africa, an emerging tourist destination. Further comparative studies may be replicated in advanced or fully developed tourist destinations to enable the possibility of examining the variation of results by distribution and diversity of agritourism resources, marketing and promotional strategies, and agritourism development models. This would allow comparisons and benchmarking of agritourism development processes. In addition, future studies should evaluate the extent to which agritourism contributes towards the achievement of SDGs by farming communities, particularly the role of women in agritourism development.

References

Acede, F.J., Barroso, C., and Galan, J. (2006). The resource based theory: Dissemination and main trends. *Strategic Management Journal*, *27*(7), 621–636.

Adeola, O., Boso, N., and Evans, O. (2018). Drivers of international tourism demand in Africa. *Business Economics*, *53*(1), 25–36.

Ammirato, S., Felicetti, A.M., Raso, C., Pansera, B.A., and Violi, A. (2020). Agritourism and sustainability: What we can learn from a systematic literature review. *Sustainability*, *12*(22), 9575.

Askarpour, M.H., Mohammadinejad, A., and Moghaddasi, R. (2020). Economics of agritourism development: An Iranian experience. *Economic Journal of Emerging Markets*, *12*(1), 93–104.

Back, R.M., Tasci, A.D.A., and Milman, A. (2019). Experiential consumption of a South Africa wine farm destination as an agritourism attraction. *Journal of Vacation Marketing*, *26*(1), 57–72. DOI: 10.1177/135676671985842.

Backmatova, G. (2021). Development prospect of agritourism and positive effects of tourism activities in rural regions. *Web of Conferences*, *273*.

Baipai, R., Chikuta, O., Gandiwa, E., and Mutanga, C.N. (2021). A critical review of success factors for sustainable agritourism development. *African Journal of Hospitality, Tourism and Leisure*, *10*(6), 1778–1793.

Bargi, F., and Reeder, R.J. (2012). Factors affecting farmer participation in agritourism. *Agricultural and Resource Economics Review*, *41*(2), 189–199.

Brandth, B., and Haugen, M. (2011). Farm, diversification into tourism-implications for social identity? *Journal of Rural Studies*, *27*(1), 35–44.

Bwana, M.A., Olima, W.H.A., Andika, D., Agong, S.G., and Hayambe, P. (2015). Agritourism: Potential socioeconomic impacts in Kisumu county. *IOSR Journal of Humanities and Social Sciences*, *20*(3), 1–11.

Canovi, M., and Lyon, A. (2020). Family centred motivations for agritourism diversification: The case of the Langhe region, Italy. *Tourism Planning and Development*, *17*(6), 591–610.

Chase, L.C., Stewart, M., Schilling, B., Smith, B., and Walk, M. (2018). Agritourism: Toward a conceptual framework for industry analysis. *Journal of Agriculture, Food Systems and Community Development*, *8*(1), 13–19.

Cheteni, P., and Umejesi, I. (2022). Evaluating the sustainability of agritourism in the wild coast region of South Africa. *Cogent Economics and Finance*, *1*, 2163542.

Chikuta, O., Baipai, R., Gandiwa, E., and Mutanga, C.N. (2022). Critical success factors for sustainable agritourism development in Zimbabwe: a multi-stakeholder perspective. *African Journal of Hospitality, Tourism and Leisure*, *11*(SEI), 617–632.

Chiodo, E., Fantini, A., Dickens, L., Arogundale, T., Lamie, R.D., Assing, L., Stewart, C., and Salvatore, R. (2019). Agritourism in mountaineous regions-insights from an international perspective. *Sustainability*, *11*(13), 3715.

Choo, H. (2012). Agritourism: Development and research. *Journal of Tourism Research and Hospitality, 1*(2).

Dieke, P.U.C. (2013). Tourism in sub-Saharan Africa: Production-consumption-nexus. *Current Issues in Tourism, 16*(7–8), 623–809.

Dube, K. (2020). Tourism and sustainable development goals in the African context. *International Journal in Economic and Financial Studies 12*(1), 88–102. 10.34109/ijefs.202012106

FBI. (2020). Agritourism Market Size, Share and Covid-19 impact analysis, by type (Direct-market, education, and experience, and event and recreation), and Regional forecast, 2020–2027. Available online: https://www.alliedmarketresearch.com/agritourism-market-A09097. Accessed 20 July 2023.

Ferreira, B., Morais, D.B., Szabo, A., Bowen, B., and Jakes, S. (2020). A gap analysis of farm tourism entrepreneurial mentoring needs in North Carolina, USA. *Journal of Agriculture, Food Systems, and Community Development, 10*(1), 83–99.

Fleischer, A., Tchetchok, A., Bar-Nahum, Z., and Talev, E. (2018). Is agriculture important for agritourism? The agritourism attraction market in Israel. *Euro. Review. Agricultural Economics, 45*(2), 273–296.

Gao, J., and Wu, B. (2017). Revitalising traditional villages through rural tourism: A case study of Yuanjila village, Shaanxi province, China. *Tourism Management, 63,* 223–233.

Gil Arroyo, C., Barbieri, C., and Rozier Rich, S. (2013). Defining agritourism: A comparative study of stakeholders' perceptions in Missouri and North Carolina. *Tourism Management, 37,* 39–47.

Grant, R.M. (1991). The resource-based theory of competitive advantage: Implications for strategy formulation. *California Management Review, 33*(3), 114–115.

Grillini, G., Sacchi, G., Chase, L., Taylor, J., Van Zyl, C.C., van der Merwe, P., Streifeneder, T., and Fischer, C. (2022). Qualitative assessment of agritourism development support schemes in Italy, the USA and South Africa. *Sustainability, 14,* 7903.

Hall, C.M. (2019). Constructing sustainable tourism development: The 2030 agenda and the managerial ecology of sustainable tourism. *Journal of Sustainable Tourism, 27*(7). 10.1080/09669582.2018.1560456.

Hall, M., and Sharples, L. (2003). The consumption experiences or the experiences of consumption. An introduction to the tourism taste. In: M. Hall, L. Sharples, R. Mitchell, N. Macionis. and B. Cambourne (Eds.), *Food Tourism around the world. Development, management and markets,* (pp. 1–24), UK, Butterworth-Heinemann.

Harari, Y.N. (2020). The world after coronavirus. *Financial Times, 20.*

Haugh, H.M., and Talwar, A. (2016). Linking social entrepreneurship and social change: The mediating role of empowerment. *Journal of Business Ethics, 133,* 643–658.

Hollas, C.R., Chase, L., Conner, D., Dickens, L., Lamie, R.D., Schidt, C., Singh-Knights, D., and Quella, L. (2021). Factors related to profitability of agritourism in the United States: Results from a national survey of operators. *Sustainability, 13*(23). 13334.

Jeczmyk, A., Uglis, J., Graja-Zwalinska, S., Mackowiak, M., Spychala, A., and Sikora, J. (2015). Research note. Economic benefits of agritourism development in Poland: An empirical study. *Tourism Economics, 21*(5), 1120–1126.

Jeou-Shan, H., and Chen-Tsang, T. (2012). Constructing indicators of culinary tourism strategies: An application of resource-based theory. *Journal of Travel and Tourism Marketing, 29*(8), 796–816.

Jonas-Berki, M., Csapo, J., Palfi, A., and Aubert, A. (2015). A market and spatial perspective of health tourism destinations: The Hungarian experience. *International Journal of Tourism Research, 17,* 602–612.

Kepe, T., Ntsebeza, L., and Pithers, L. (2001) Agritourism spatial development initiatives in South Africa: Are they enhancing rural livelihoods. *Natural Resources Perspectives*, *65*, 1–4.

Krishna, D.K., and Alok, K.S. (2020). Overview of agritourism in India and world. *Food and Scientific Reports*.

Kubickova, M., and Campbell, J.M. (2020). The role of government in agritourism development: A top-down bottom up approach. *Current Issues in Tourism*, *23*(5), 587–604.

Kunasekaran, P., Ramachandran, M., Yacob, M., and Shuib, A. (2011). Development of farmers' perception: Scale of agritourism in Cameroon Highlands, Malaysia. *World Applied Sciences Journal*, *12*, 10–18.

Lak, A., and Khairabadi, O. (2022). Leveraging agritourism in rural areas in developing countries: The case of Iran. *Frontiers on Sustainable Cities*, *4*, 863385.

Lane, B., and Kastenholz, E. (2015). Rural tourism: The evolution of practice and research approaches-Towards a new generation concept? *Journal of Sustainable Tourism*, *23*, 1133–1156.

La Pan, C., and Barbieri, C. (2013). The role of agritourism in heritage preservation. *Current Issues in Tourism*. DOI: 10.1080/13683500.2013.849667.

Liao, C., Zuo, Y., Law, R., Wang, Y., and Zhang, M. (2022). Spatial differentiation, influencing factors and development paths for rural tourism resources in Guangdong province. *Land*, *11*, 2046.

Litheko, A. (2022). Development and management of small agro tourism enterprises: A rural development strategy. *African Journal of Hospitality, Tourism and Leisure*, *11*(3), 1053–1069. 10.4622/ajhtl.19770720.275.

Lv, J. (2020). Leisure agriculture and rural tourism policy guidance. *Leisure Agriculture and Rural Tourism Policy Interpretation: Agricultural Resources and Regionalisation in China*, *41*(3), 35–35.

Maharjan, S.K., and Dangol, D.R. (2018). Agro-tourism education and research in Nepal. *Agricultural Research and Technology 14*(5), 001–005.

Matiza, T., and Perks, S. (2021). The tourism-foreign direct investment nexus: Empirical evidence from Zimbabwe (2009-2015). *Tourism Economics*, 13548166211007603.

Mitchell, C.J.A., and Shinnon, M. (2018). Exploring cultural heritage tourism in rural Newfoundland through the lens of the evolutionary economic geographer. *Journal of Rural Studies*, *59*, 21–34.

Mnguni, K.I. (2010). *The socioeconomic analysis of agritourism in two rural communities in the Limpopo province*. University of South Africa.

Mrema, A.A. (2015). Tourism and women empowerment in Monduli district, Arusha-Tanzania. *African Journal of Hospitality, Tourism and Leisure*, *4*(2), 1–14.

Mutana, S. (2013). Rural tourism for pro-poor development in Zimbabwean rural communities: prospects in Binga rural district along Lake Kariba. *International Journal of Advanced Research in Management and Social Sciences*, *2*(4), 147–164.

Mutana, S., and Mukwada, G. (2023). SDGs as indicators of holistic small town tourism development. A case for Philthadithaba South Africa. In: A. Membretta et al. (Eds.), *Sustainable Futures in southern Africa's mountains*. Sustainable Development Goals series. 10.1007/978-3-031-15773-8_10.

Myer, S., and de Crom, E. (2013). Agritourism activities in the Mopani district municipality, Limpopo, South Africa. Perceptions and opportunities. *TD: The Journal for Transdisciplinary Research in Southern Africa*, *9*(2), 295–308.

Novelli, M., Adu-Ampong, E., and Ribeiro, M.A. (2021). Prospects and future for tourism in Africa. In: M. Novelli, E. Adu-Ampong, and M.A. Ribeiro (Eds.), *Routledge handbook of tourism in Africa*, (pp. 491–498). Routledge.

Quella, L., Chase, L., Conner, D., Reynolds, T., Wang, W., and Singh-Knights, D. (2021). Visitors and value: A qualitative analysis of agritourism operator motivation across the US. *Journal of Agriculture, Food Systems and Community Development* *10*(3), 287–301.

Romanenko, Y.O., Boiko, V.O., Shevchok, S.M., Barabanova, V.V., and Karpinska, N.V. (2020). Rural development by stimulating agritourism activities. *International Journal of Management*, *11*(4), 605–613.

Rogerson, C.M., and Rogerson, J.M. (2014). Agritourism and local economic development in southern Africa. *Bulletin of Geography. Socioeconomic Series, 26*, 93–106.

Roman, M., and Grudzien, P. (2021). The essence of agritourism and its profitability during the coronavirus (COVID-19) pandemic. *Agriculture*, *11*(5), 458.

Saayman, M., van der Merwe, P., and Saayman, A. (2018). The economic impacts of trophy hunting in the South African wildlife industry. *Global Ecology Conservation, 16*, e00510.

Santeramo, S.G., and Barbieri, C. (2017). On the demand for agritourism: A cursory review of methodologies and practice. *Tourism Planning and Development*, *14*(1), 139–148.

Sardaro, R., La Sala, P., and Rosell, L. (2020). How does the land market capitalise environmental, historical and cultural components in rural areas? Evidence from Italy. *Journal of Environmental Management, 269*, 110776.

Scheyvens, R., and Hughes, E. (2019). Can tourism help to end poverty in all its forms everywhere? The challenge of tourism addressing SDG1. *Journal of Sustainable Tourism*, *27*(7). 10.1080/09669582.2018.1551404.

Scheyvens, R., and Monsen, J.H. (2020). Tourism and poverty reduction: Issues for small islands states. *Tourism Geographies*, *10*(1), 22–41.

Shereni, N.C., and Saarinen, J. (2020). Community perceptions on the benefits and challenges of community-based natural resources management in Zimbabwe. *Development Southern Africa*, 1–17.

Swanson, K.K., and DeVareaux, C. (2017). A theoretical framework for sustaining culture: Culturally sustainable entrepreneurship. *Annals of Tourism Research, 62*, 78–88.

Tew, C., and Barbieri, C. (2012). The perceived benefits of agritourism: The provider's perspective. *Tourism Management*, *33*(1), 215–224.

Togaymurodov, E., Roman, M., and Prus, P. (2023). Opportunities and directions of development of agritourism: Evidence from Samarkand region. *Sustainability*, *15*, 981.

UNWTO (2022). Tourism Best Village. https://www.unwto.org>news>best> tourism>village.

Van Zyl, C.C. (2019). *The size and scope of agritourism in South Africa. Master's degree.* Potchefstroom, North West University.

Van Zyl, C.C., and van der Merwe, P. (2021). The motives of South African farmers for offering agritourism. *De Gruyter, Open Agriculture, 6*, 537–548.

Vukovic, D.B., Petrovic, M., Maiti, M., and Vujko, A. (2023). Tourism development, entrepreneurship and women's empowerment. Focus on Serbian countryside. *Journal of Tourism Futures*, *9*(3), 417–437.

Wang. H., Hollas, C.R., Chase, L., Conner, D., and Kolodinsky, J. (2022). Challenges for the agritourism sector in the US: Regional comparisons of access. *Journal of Agriculture, Food Systems and Community Development*, *11*(4), 61–74.

Wijijayanti, T., Salleh, N.H.M., Hashim, N.A., Mond Saukani, M.N., and Abu Bakar, N. (2023). The feasibility of rural tourism in fostering real sustainable development in host communities. *Journal of Tourism and Geosites*, *4*(1), 336–345.

World Bank (2019). Available online. https://data.worldbank.org/indicator/SP.RUR. TOTL.ZS?locations=UZ. Accessed 30 August 2023.

Xingping, L. (2019). Exploring in the approach to rural revitalisation in view of green development. In *IOP Conference Series. Earth and Environmental Science, 310*(5), 052074. IOP Publishing.

Yang, Q., Li, J., and Tang, Y. (2022). The dilemma of the development of rural tourism from the sustainable environment perspective. *Journal of Environmental and Public Health*, 7195813.

Zantsi, S., Greyling, J.C., and Vink, N. (2019). Towards a common understanding of emerging summer in a South African continent using data from a survey of three district municipalities in the Eastern Cape province, South Africa. *Journal of Agriculture Extension, 47*(2), 81–93.

Zawadka, J., Jeczmyk, A., Wojcieszk-Zbierska, M.M., Niedbaka, G., Uglis, J., and Pietrzak-Zawdka, J. (2022). Socioeconomic factors influencing agritourism from stays and their safety during the Covid-19 pandemic: evidence from Poland. *Sustainability, 14*(6), 3526.

Zhao., Z., Xue, Y., Geng, L., Xu, Y., and Meline, N.N. (2022). The influence of environmental consumer intentions to participate in agritourism. A model to extend TPB. *Journal of Agricultural and Environmental Ethics, 35*, 15.

Zhou, Z. (2019). Tourism progress in the SADC region: Postcolonial era milestones. In: M. Mkono (Ed.), *Positive tourism in Africa* (pp. 137–146). London: Routledge.

Zhou, Z. (2018). The Tourism Sector: A bright light in Zimbabwe's depressed economic environment. *African Journal of Hospitality, Tourism and Leisure, 7*(1), 1–19.

Zoto, S., Qinii, E., and Potena, E. (2013). Agritourism-A sustainable development for rural area of Korea. *European Academic Research, 1*(2), 209–223.

9 Bibliometric Analysis of Agritourism for Rural Transformation and Poverty Reduction Policy Implications

Francis Aron Mwaijande

Introduction

Agritourism is conceptualized to link agriculture and tourism to social, economic and environmental sustainability of the rural areas, whereby farmers earn extra income from touristic related activities. Agritourism is also conceptualized as farm tourism or rural tourism which happens when local or international visitors go to farms for leisure and learning. (Rogerson, 2012; Cheteni and Umejesi, 2023) argue that this type of tourism helps farmers in achieving agricultural viability and diversifying rural economies when providing farm-based tourist experiences for visitors interested in traditional rural hospitality, nature access, outdoor activities, and cultural experiences. Similarly, (Walke and Kumar, 2015) define agritourism as farm enterprises designed for the enjoyment of visitors that generate on-farm income. It also refers to the act of visiting farms or any agricultural, horticultural, or agribusiness operations for the purpose of enjoyment, education, or active involvement in the activities of the farm that also adds to the farm economic base.

While agriculture is the main source of living for about 78% of the population in Tanzania (URT, 2021) contributing to 26% of the Gross Domestic Product (GDP) (URT, 2017), tourism contributes 8.1% of Africa's GDP (Bolaky, 2016). However, the on-farm incomes are declining due to multiple factors such as high prices of inputs, low prices of farm products, and climate change (Myer and De Crom, 2013). In order to sustain farmers' livelihood, agritourism may complement the rural economy by increasing farm attractions, hunting, camping, and on-farm accommodation (Mahaliyanaarachchi et al., 2019).

Scholars such as (Blake et al., 2003; Torres, 2003; Rogerson, 2012) have studied the economic potential and backward linkages of tourism on employment and poverty reduction globally; however, there are few complementarities of operations between agriculture and tourism in many African countries including Tanzania. To effectively increase linkages of agriculture and tourism sectors, innovation in agriculture policy is required for opening up new agritourism enterprises. Agritourism represents an innovation for rural transformation approach for minimizing dependency on farm economy to alternative off-farm activities through rural tourism destinations. The

DOI: 10.4324/9781032696188-9

transformation toward agritourism is important because it provides alternative sources for farmers' declining income due to unreliable production caused by drought, pests, low technology, and market prices. Following the uncertainties of agriculture, farmers' livelihoods are affected and agritourism takes on an economic importance for rural transformations by supplementing farmers' income. This is an innovative model of "rural entrepreneurship" focused on farm diversification, employment creation and increased productivity as a direct response to rural policy (De Rosa et al., 2019) by linkages of agriculture to the tourism sector.

The motivation toward agritourism is due to the increasing demand for food supply caused by the expanding tourism and hospitality business. Following the expanding tourism sector there should be a relative supply increase for the local agricultural produce to the tourism industry. When farmers supply food and or provide touristic activities (on-farm bed and breakfast), they create opportunities for earning tourist dollars. The on-farm activities range from food and beverage, landscape scenery attractions, animal farms, hunting and biodiversity with a focus of diversification of on-farm economic activities.

Literature Review

The assumption is that for rural transformation to occur, on-farm income activities have to be diversified with tourism to earn tourism dollars. The diversified dollar income is important for reducing rural poverty. Sustainable rural development is attained when nature is conserved while it meets the current needs without compromising future use. The important thing to take note of is for agritourism to be connected to the environmental values of rural regions for income generation and economic growth. Having a diversified economy for the rural areas is critical at the moment because agricultural on-farm incomes are declining due to low prices of agricultural commodities, rising input costs, environmental and climate change pressures (Braun et al., 1999; Mahaliyanaarachchi et al., 2019). Agritourism is a viable option for diversified rural socio-economic development and transformation for enhancing economically and socially distressed rural areas (Morris, Henley and Dowell, 2017; Cheteni and Umejesi, 2023). Many approaches for rural poverty reduction (Ellis and Freeman, 2004; Saleem and Donaldson, 2016) have been attempted, yet it has not been effectively achieved. Rural poverty and unemployment challenges have not been addressed as expected. Scholars (Cheteni and Umejesi, 2023) argue that rural tourism can increase employment, income distribution, and the preservation of the countryside's natural beauty, thus contributing to Sustainable Development Goals. Literature on agritourism indicates that the tourism sector can provide a potential market for the local agricultural sector if food requirements in the tourism sector are provided by the local agricultural producers. Apart from the economic benefits of agritourism to farmers and the hospitality industry, there are also socio-ecological benefits that the society gains.

Theoretical Perspective

The underpinning theory of rural development refers to policies that aim to boost the economy, society, and environment, including to promote poverty eradication and pro-poor planning in rural areas (Pimid et al., 2023). The contention is on the nexus of local development and poverty reduction through agritourism. It goes without saying that agritourism is essentially a Western model for agricultural development as (Bhatta, 2020) found that farm diversification on tourism activities has become an important economic activity in European countries. Agritourism offers farmers additional and diversified income through on-farm touristic activities in order to help supplement their low agricultural income (Sachaleli, 2020; Back, Lowry and Higgins, 2021). Agritourism or rural tourism is a viable rural development pathway for improving livelihood for rural residents (Pimid et al., 2023) as agritourism encompasses the hospitality industry at tourism businesses and services; it renders visitors or tourists in a form of hotel accommodation, food and drinks, recreation, shopping, and local travel connects agriculture for extra earning. Tourism industry entities include primary products whose businesses include traveling, food eating and experiencing tastes, entertainment, and site scenery viewing. Agriculture on the other hand encompasses food production and maintenance of the environment and natural resources. The agritourism enterprises therefore are the farm value-added activities that provide pleasure, recreation, information, education, or other experiences for which the tourist pays admission to participate in purchase of agricultural products, thus creating linkages to hotels and restaurants and promotion of local food business for tourism purposes.

Scholars use agritourism interchangeably to refer to "farm tourism" (Nguyen and Binh, 2022; Cheteni and Umejesi, 2023; Chikuta, 2023), or "agrotourism" or "agriculture-tourism linkages", while others mean "country hospitality". Nguyen and Binh (2022) define farm tourism as any tourism activity on farms from as little as selling farm produce to countryside visitors that is increasingly becoming an important rural policy intervention in many European countries where it promotes farm income diversification (Hjalager, 1996). It is argued that agritourism provides multiplier effects such as minimizing rural-urban migration, supporting local services and helping to maintain farm landscapes for beautification. Agriculture-tourism linkages represent an important alternative farm enterprise for increasing economic diversification of the rural population by creating supplemental income (Yuan et al., 2017).

Shrinking farmers' income is caused by unreliable production due to drought, pests, low technology, and market prices. Following the uncertainties of agriculture and their effects on farmers' livelihoods, agritourism should take place to maintain the rural livelihoods. This is further confirmed in a study by (Vyslobodska et al., 2022) who found that the diversification of activities and the development of green tourism in rural areas have a positive

Table 9.1 Linkage effects of tourism in the Tanzanian economy

Sector	Output linkages		Employment linkages	
	Backward	*Forward*	*Backward*	*Forward*
Agriculture	0.801	1.036	2.096	12.373
Manufacturing	1.076	0.889	0.738	0.389
Tourism	1.155	1.139	0.847	1.476
Other services	0.968	0.936	0.319	0.956

effect on the socio-economic situation of the territories, and is a catalyst for the development of infrastructure and the growth of employment of the population in rural areas. Table 9.1 shows that tourism has the highest level of linkage effects with the rest of the economy compared to other sectors.

In the changing agrarian economy, agritourism integrates agricultural production into tourism chain to benefit the rural economy. This is central to addressing the bottlenecks of the agricultural sector, such as markets for locally produced food produce. Farmers are likely to produce and access the tourism market needs. The demand for food supply and new natural recreational destinations creates a forward linkage between agriculture and tourism. However, the agritourism linkage is constrained by the mismatch between the demand and supply of quality and quantity of locally produced agricultural produce to the tourism sector, of which scientific research ought to inform the rural transformation policies.

Methodology

The study employed a mixed research method by employing triangulation methods to answer the questions: how can farmers increase on-farm income through tourism, how does tourism intersectionality promote rural transformations, and what literature provides evidence for agritourism for rural transformation and poverty reduction? Data were collected using desk-top review of secondary data, whereby a bibliometric analysis was adopted. The bibliometric analysis is an approach that uses a set of quantitative methods to measure, track, and analyze scholarly literature on a specified study area by identifying scholarly publications, methodologies and conclusions obtained (Rojas-Sánchez et al., 2023). A bibliometrics is a set of statistical and mathematical methods used to measure and analyze quality and quantity of articles and books for a better understanding of the scientific research in a study area. This method was adopted to enable a better understanding of the status of agritourism in Africa in scientific literature to inform policies (Büyükkidik, 2022). We used the appropriate Boolean operators and key search words on "agritourism" AND "Africa" AND "Rural" from Scopus database. Only 14 scientific and scholarly studies were found, which were firstly, reviewed and later exported to R-Studio software for bibliometric analysis. The adopted method enabled a metadata review and generated evidence to inform agritourism for rural transformation for poverty reduction in Africa. This method contributes to a better understanding of the scientific research in agritourism research which synthesize past and current research findings (Büyükkidik, 2022).

An interview was conducted with the late Prosper Ngowi about his farming activities. He translated his agribusiness enterprise with a pig farm, reptiles, rabbit farming and other animals as the main base for attractions. The farm cottage is where the "wazungu" – typical tourists in Dar Es Salaam – can come for a picnic or overnight stay.

The picture shows an agritourism enterprise whereby local and foreign visitors can be included in a combined tourism package and visitation to national parks and farm visitation. In so doing, the visitation generates tourist dollar expenditure on the farm, thus diversifying on-farm income (Ilbery et al., 2015).

Results and Discussion

With the shrinking farm size due to low prices, markets and climate change, farmers' income from agricultural activities also declines. That is to say innovative rural policies for addressing the challenges of farmers and needs of the poor. Over 75% of the rural population in Africa and in Tanzania in particular are engaged in agricultural production. Agriculture has become the principal supplier of food for the domestic and hospitality industries. Tourism, on the other hand is an important source of income for many developing countries that provides a market for the agricultural produce (Fleischer and Tchetchik, 2005). It can therefore logically be argued that, agritourism is able to drive the rural economy as it is considered as an innovative pro-poor political economy that offers opportunities for rural transformations (Srisomyong and Meyer, 2015).

This chapter presents a view that agritourism ought to create sustainable rural incomes through diversified economic activities. Thus, farming activities should also include tourism attractions and destinations in the form of agritourism approach by producing environmental, socio-cultural, and economic benefits. However, agritourism is not well developed in Africa compared to the Western and Caribbean countries due to inadequate knowledge, entrepreneurship skills, technology, working capital and enabling environmental policy.

This study observes that there are few African countries that have adopted agritourism for rural development (Baipai et al., 2023; Chikuta, 2023); however, even for the few African countries that have adopted, there are multiple challenges that prevent adoption of agritourism for rural transformation including rural road infrastructure, electricity, attractions, government support and scientific evidence. However, agritourism in Africa is relatively new and requires more scientific evidence to inform rural transformation policies. The scientific knowledge production to inform on agritourism policy decisions in Africa is very scanty. The results of the conducted bibliometric analysis show

Table 9.2 Local impact index

Element	h_index	g_index	m_index	TC	NP	PY_start
RAMAANO AI	2	3	0.66666667	14	3	2022
ADEDIBU BA	1	1	0.25	21	1	2021
BACK RM	1	1	0.25	8	1	2021
CHETENI P	1	1	0.5	6	1	2023
HIGGINS LM	1	1	0.25	8	1	2021
LOWRY LL	1	1	0.25	8	1	2021
NNABUKO JO	1	1	0.25	21	1	2021
OKOLO-OBASI EN	1	1	0.25	21	1	2021
ONODUGO VA	1	1	0.25	21	1	2021
ROGERSON CM	1	1	0.09090909	59	1	2014
ROGERSON JM	1	1	0.09090909	59	1	2014
UDUJI JI	1	1	0.25	21	1	2021
UMEJESI I	1	1	0.5	6	1	2023

Source: Study result.

scientific production on agritourism in Africa is a recent phenomenon beginning in 2014. Nevertheless, it is encouraging that it has been increasing since 2019.

Scholarly evidence on agritourism for rural transformation and poverty reduction is generating 0.25 impact (Table 9.2) as only diversified economic activities can drive rural transformations (Orboi, 2012).

African scholars within and outside African Universities have begun publishing to provide scientific evidence.

It is encouraging that scientific citations on agritourism have been increasing since 2000. However, it is still a challenge for African scholars to generate evidence of agritourism as a strategy for rural transformation and poverty reduction in Africa.

With the exclusion of the period of COVID-19, the volume of tourists in Africa showed an increasing trend from 64 million in 2018 to 69 million in 2019 who contributed to foreign exchange in Africa. In specific countries like Tanzania, tourist arrivals increased from 295,312 in 1995 to 782,699 in 2010, with tourism earnings increasing from USD259.44 million to USD1.2 billion. The recent trend also shows a foreign exchange earnings increase from USD2,412 million in 2018 to USD2,528 million in 2022 (URT, 2022) whose effect should trickle down to farmers engaged in agritourism (Figure 9.1).

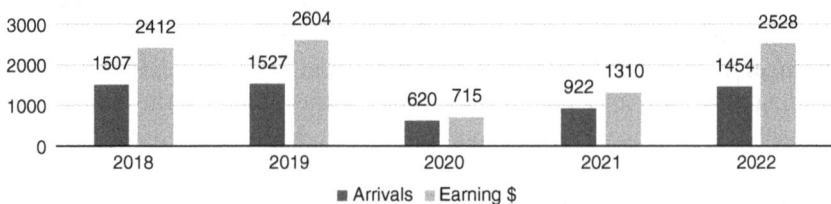

Figure 9.1 Tourism arrivals in 000 and earnings in million.

Necessary Conditions for Developing Agritourism

We examine and identify the necessary conditions for developing agritourism enterprises as means for creating alternative farm income that will contribute to rural poverty reduction. We also examine ways for establishing rural tourism destinations in order to promote agricultural growth. The flourishing of the tourism industry depends very much on the functioning of other sectors such as agriculture, tourism marketing, transportation, communication, energy. For agritourism enterprises to grow, rural road, electricity, and water infrastructure must be improved along with landscaping.

Examples of Agritourism for Rural Transformation in Africa

Agritourism is an emerging important economic activity in Africa where they have realized the potential linkages between agriculture and tourism to benefit the rural agrarian economies. There are a few examples of agritourism in Kenya, Tanzania, and South Africa. South Africa and Nigeria are the countries with the highest agritourism activity for diversified farm incomes that contribute to rural transformations.

South Africa Agritourism

The Western Cape province is the leading destination and national core region for agritourism in South Africa, accounting for a 42% share of all accommodation establishments. Other provinces with developed agritourism destinations are found in Free State, Eastern Cape, KwaZuluNatal, Mpumalanga and Northern Cape (Rogerson, 2012, 2016). The most outstanding agritourism attractions comprise wine and ostrich farms, visits to banana plantations, citrus, olive, avocado, sheep and cattle farms as well as maize and potato farms and macadamia farms. The most popular farm picking and food tasting experience include strawberry picking, cheese tastings, horse riding, fishing and bird watching (Rogerson, 2016) (Table 9.3).

Table 9.3 Examples of agritourism in South Africa

Town	Type of agritourism
Stellenbosch	Wine, olives, vinegar, proteas, and roses, horse riding and farm animal petting and feeding
Montagu	Wine, fruit, apricots, olives, stud farm, horse riding, fishing, farm animals
Worcester	Grape picking, fishing, bird watching
Tulbagh	Olive and wine farming, fishing, orchards, horse riding
Oudtshoorn	Ostrich farming, horse riding and bird watching
Knysna	Fishing, berry farms, horse riding, jersey herd, collecting farm eggs
Citrusdal	Citrus farming, bird watching, fishing, horse riding
Dullstroom	Trout fishing, horse riding, cattle farming
George	Fishing, horse riding, strawberry picking, animal feeding
Memel	Fly fishing, cattle, maize, and potato farms farming, bird watching
Parl	Wine, fruit and olives, bird watching, fishing, horse riding

Source: Rogerson, 2016.

Tunisia Agritourism

Tunisia represents one of the unique gender-based agritourism enterprises in the rural setting as a means to diversifying the economy. Their agritourism is defined as: nature-based, environmental conservation, environmental education, sustainability, and ethically responsible enterprises. Tourism done in rural settings is vividly seen through bed and breakfast lodges that offer employment for women in the fields of hospitality industry and management providing services to tourists in the form of tour guides, cooks, caterers, and selling food products and local handicrafts which result in earning income and the increase in female labor force participation. Other agricultural based activities include processing of agricultural products and animal husbandry (Khazami et al., 2023).

Kenya-Kericho Agritourism

Kenya in East Africa has also become an important rural tourism destination. Obonyo and Fwaya studied strategies for integrating agritourism strategies in Western Kenya and argued that Kenya is confronted with the challenge of improving the lives of its rural population including in the context of poverty, food insecurity, and unemployment. Kenyan agritourism enterprises in Kericho encompass environmental conservation attractions and tea farms, which economically, have led to the reduction of poverty levels through the creation of employment (Kipkorir et al., 2022).

Zimbabwe Agritourism

The Zimbabwe agritourism case presents a framework for developing a successful agritourism. It proposes critical stages for sustainable agritourism development which include planning, development and implementation. It also provides the main guiding principles as the environmental scanning for enablers, multistakeholder engagement, identification of potential farms that meet the requirements for agritourism and identification of critical success factors at farm level (Baipai et al., 2023).

Practical Implication

Tanzania Government has established a Build Better Tomorrow project for youth engagement into agriculture. One of the farming activities is grape growing in Dodoma. Wine and food tourism is one of the fastest-growing rural tourism niches, with effects on the orientation of food systems, the livelihoods of producers, and the viability of rural communities (Robinson, 2021). This is one of the potential agritourism enterprises in which youth could get employment as it creates the interrelationships Between Tourism and Agriculture. Innovative agritourism in Tanzania may include visitation by tourists where farm activities such as palm juice drinking, coffee or tea picking, wine tasting, fishing, meat roasting, or horse riding, and picnics could be integrated.

Implication of Agritourism on Rural Transformation

The implication of agritourism is the relative mutual benefit of agriculture from tourism and vice versa. In order to capture benefits from the expanding tourism market and business, the agricultural sector must identify the needs of the tourism sector and overcome the constraints. On the other hand, the tourism sector would establish new tourism destinations. By so doing the country would contribute to poverty reduction. Tanzania faces a challenge to reduce the number of people living in poverty and attain the Sustainable Development Goals by 2030. According to the Tanzania Human Development Report, over 51% of the population lived on less than USD1 a day, and about 42% of these lived in absolute poverty of less than USD0.75 a day. About 24% of the 11,021,000 urban dwellers live in poverty. The Labor Force Survey established rural unemployment at 7.4%, whereas urban unemployment was even worse at 11.0%. Youth unemployment in rural areas stood at 4% in rural areas and 5% in urban areas, indicating a young labor force unemployment at 9.2% national wide (NBS, 2021). One of the macro-causes of poverty in Tanzania is the weak incentives in the agricultural sector to attract the young labor who could create agriculture-tourism forward and backward linkages to create employment and address poverty.

By introducing agritourism, it is likely to attract the young labor force who migrate into urban areas. Urban sociologists explain rural young men and women migrate into cities mainly due to low economic-activities in the rural areas. The pull-push theory (Todaro, 1997) explains the causes for the rural-urban migration as that the migrating youths are pushed by the so deplorable and unfavorable socio-economic conditions in the rural areas that do not promise employment. The increasing rural-urban migration is attributed to the unfavorable socio-economic factors in the rural areas, because the current farm activities do not attract the interests of the young population who are mostly attracted by urban employment and modernity. According to Labor Force Survey (NBS, 2021) According to Tanzania's population census (2022), 61.9% of the population are employed in the agriculture sector; nevertheless rural unemployment is 7.2%. Agritourism offers room for decent job creation of agribusiness, tour operators, etc.

Lessons from the Chinese agricultural reforms and spatial development policy indicate that well designed agricultural development in the suburban areas is likely to absorb surplus labor in cities by creating small urban cities in the city's vicinity. Lo observed that along with the village small town urbanization, the Chinese Spatial development policy promoted industries in villages that provided employment and income. Lo argued that these industries absorbed surplus labor released from improved agriculture pro-duction (p. 299) and promoted urban agricultural growth. In the same way, agritourism in rural areas has the potential to absorb the migrated urban population or halt the rural-urban migration by providing employment or income generating activities in rural areas.

Table 9.4 Tourism and rural employment

	2005	2006	2007	2008	2009
Tourism share of all employment	8	8.5	8.8	8.1	6.7
Real growth	3.1	10.3	5.8	−6.2	−15.1
Direct employment	300.1	342.3	356.7	325.5	259
Total employment ('000)	694.7	766.5	811.1	760.3	645.1

Given the projection that half of Tanzania's population will have moved to urban areas by 2030, a need for reversing the trend by retaining the urban population is implied. People desire to escape rural deprivation and rural livelihoods. By developing agritourism, rural areas will experience growth and attract a working-age population. The potential of tourism linkages to agriculture and other sectors to provide rural employment is also acknowledged in Tanzania, whereby direct employment in tourism increased from 300,000 in 2005 to over 350,000 in 2007 (Table 9.4).

The linkages between tourism and agriculture provide multiple benefits across socio-economic sectors. Some of the benefits that Parrot, Boyne and Hall found include stronger agriculture-tourism linkages that reduce economic leakages, and create new employment and markets for local products. The benefits accrue from building policy linkages between tourism and agriculture as experienced in the European Union countries. Agriculturally based economies need to put the linkages between agriculture (food) and tourism at the top of the policy agenda whereby strategies for increasing agricultural production and capacity for agro-industries must be taken.

Agriculture-Tourism Interface

The section examines the interface between agriculture and tourism. It highlights policy issues, objectives and statements that display the agritourism interface. Much of the available literature supports the view that agriculture-tourism linkages in the form of agritourism activities provide opportunities for agricultural growth (Orboi, 2012; Ilbery et al., 2015; Walke and Kumar, 2015) and create alternatives to non-farm economic base. The review provides the foundation for understanding agritourism as an alternative for rural transformation and poverty reduction. In the attempt to stimulate agricultural growth, some scholars have argued that tourism could act like an economic engine that moves up agriculture. Such approaches advocate for strong linkages between agriculture and tourism (Rogerson, 2012).

Policy Issues

The 2013 National Agriculture Policy's cross cutting policy issue is employment. It recognizes the need for creating decent employment in agriculture. One way of creating decent employment is through agritourism activities (Figure 9.2).

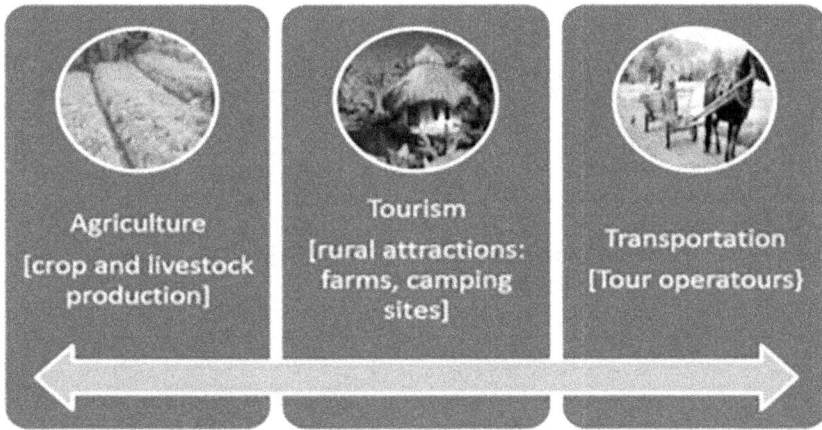

Figure 9.2 Agritourism policy issues.

Agriculture Policy Issue – Decent Job and Employment

The question is how can agritourism create decent jobs and employment? Creating decent job and employment opportunities is of critical importance in the contemporary times of massive youth unemployment. The agriculture sector has not been attracting many youth for many reasons as it is considered labor intensive. One way of attracting youth into agriculture is to create an interface with tourism sector activities; that is agritourism jobs such as tour operations for farms and on-farm activities of high value productions. Tanzania's agriculture policy states that "the Government recognizes the importance of decent employment in agriculture and its central role in the achievement of sustainable agricultural growth and rural development". Economic sustainability will have a fast and long-term impact on poverty alleviation if the created employment potential enables poor people to raise their income, either through reduced unemployment (Knowd, 2015) (Table 9.5).

Tourism Policy Issue

The National Tourism Policy seeks to assist in efforts to promote the economy and livelihoods of the people, essentially poverty alleviation through encouraging the development of sustainable and quality tourism that is culturally and socially acceptable and environmentally sustainable. Since 75% of the Tanzanian population's livelihoods depends on agriculture, the tourism policy goal targets farmers, whose activities ought to integrate into the tourism economy in a more sustainable manner. For many years tourism products have concentrated on wildlife and beaches. The tourism policy strategy for growth is to diversify its operations. The policy strategy is

Table 9.5 Agritourism policy issues

Sector	Policy issue	Intervention
Agriculture	Food production for the tourism market Farms oriented for tourism diversified activities; Farmers' capacity	Increase crop production; Orient farmers for tourism diversified economy; Orientation of farmers for agritourism enterprises
Tourism	On-farm attractions; On-farm accommodation; Human resource capacity; Diversified tourism	Coordination with agriculture stakeholders; Training for agritourism; Developing agritourism products and cuisine
Transport	Rural roads	Improve passage of roads all year round

Source: Knowd, 2015.

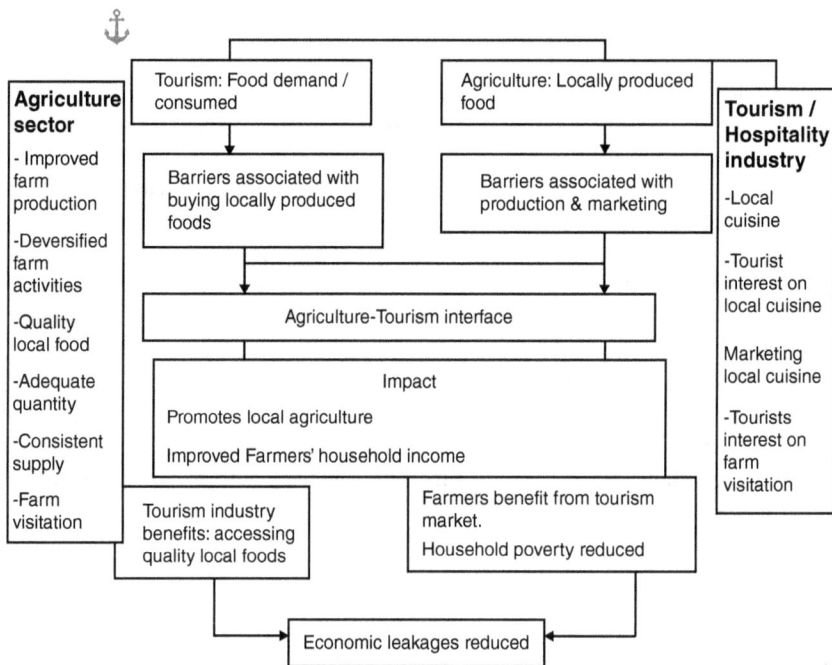

Figure 9.3 Agriculture-tourism interface.

to diversify tourism attractions such as on-farm tourism (agritourism) in farm visitations (Figure 9.3).

As the tourism sector is increasingly becoming an important source of foreign exchange in many countries, it promises reliable markets for locally grown food products. The interface between agriculture and tourism can become more reciprocal if the agriculture sector provides adequate and quality foods to the tourism and hospitality industry. The reciprocity would then create impact on farmers' livelihoods by increasing household incomes and reducing poverty, because "investments in a tourist attraction benefit not only its promoters, but also those in accommodation, food and drinks" and "tourism creates market opportunities for farmers' produce, tourism expansion should ideally generate agricultural earnings".

Many studies, mostly conducted in European countries, have found complex relationships ranging from competition for land and labor, to symbiotic relations where the tourist industry buys local agricultural products. Torres and Momsen identified diverting land and water resources away from agriculture to support tourism as a negative impact of pro-poor tourism on local agriculture. However, there is a great opportunity for complementarity between the two sectors as Klejdzinski argues "tourism needs a healthy agriculture, similarly agriculture needs tourism". Studies like "Our Countryside: The Future. A Fair deal for England" argues that tourism businesses are well placed to promote food and drink products benefiting local agricultural producers; while the study "Forward Strategy for Scottish Agriculture" acknowledges there is potential for the farmers to earn from tourism. Boyne, Hall and Williams carried out a quite detailed study and found that "high-quality food and beverage products can enhance the overall tourism"; on the other hand "tourism-related spending on locally produced goods provides economic stimuli" Boudy. Klejdzinski (1990) observes that tourism provides a direct source of income to farmers by selling farm products; it also contributes to the agricultural related agritourism and farm employment (Figure 9.4).

Boyne et al. argue that local foodstuffs enhance and strengthen tourism products while tourists provide a market for these products. While tourism spending on the locally food is envisaged to promote agriculture growth, the approach to create synergy faces implementation problems related to (a) lack of

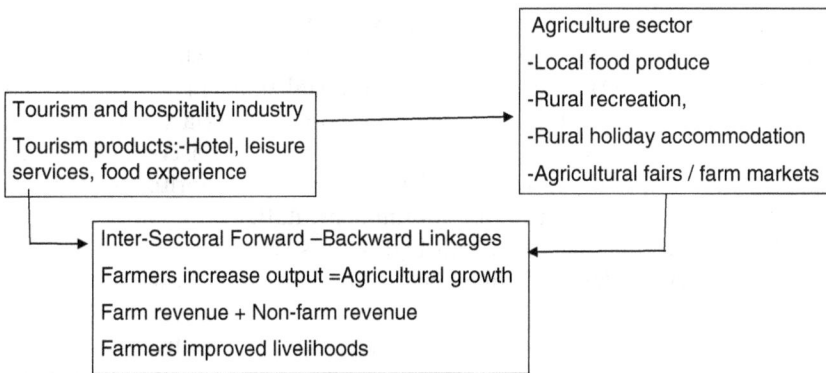

Figure 9.4 Agritourism nexus for rural transformation.

adequate understanding of consumer behavior as regards to local food and (b) consumer attitude regarding local food production. This is contrary to a survey made of tourists' attitude toward local food in the Isle of Arran Taste Trail in Scotland which found that tourists were "prepared to spend more money on meals consisting of locally produced food, more inclined in the future to purchase locally produced groceries, and that the quality of Arran's food would be a positive factor in their decision to make a return visit to the Islands".

Based on the academic research findings, policy documents such as "Tomorrow's Tourism" require tourism operators to source and promote locally produced foodstuff. Boyne et al.'s study provides basic inference to the study "Trade, Tenure and Tourism in the U.S Virgin Islands: Understanding the Policy Frameworks", as it singles out policy barriers related to tourists' quality assurance systems. This can be resolved by creating websites with tourist information that not only provide information on local food menus, but also the recipes that may be considered critical food information for tourists. Strong and organized tourist boards could provide a strong network for tourism business decision making that could also be emulated by the farmers who seem to be less organized and lack networks amongst themselves.

The significance of tourism to agriculture development is also studied in the context of the "economic benefits" gained from the tourism businesses. The economic benefits are derived from tourists' expenditure on the goods and services they procure during their stay. For example, Belisle studied "tourism and food imports" in Jamaica where he found that tourism plays an important role in economic development by generating investment opportunities, employment and foreign exchange. The direct benefits for the agriculture sector are derived when tourists purchase farmers' produce at the market or agriculture fair, whereas the indirect benefit is derived from the tourists' expenditures on food and drinks in hotels and restaurants.

Mapping Stakeholders for Agritourism for Rural Transformation

Stakeholder analysis is a mapping matrix used to identify and align the interests and commitments of stakeholders. The Research and Policy in Development (RAPID) developed a tool for stakeholders' analysis called "Alignment-Interest and Influence Matrix (AIIM)". It is a framework for identifying policy stakeholders working together toward a common policy goal. The tool is useful during the identification and prioritization of the target policy engagement, influence and implementation. It comprises of the vertical and horizontal axes, low and high points (Figure 9.5). Vertical axis measures the general level of alignment[1] while horizontal measures level of interest[2] toward a policy goal.

With the help of the Alignment-Interest and Influence Matrix (AIIM), the following stakeholders are likely to be in the agritourism development. The list of stakeholders is not exhaustive; these include the Ministry of Agriculture

High
	MoA, MoT, MoFL, Farmers, Hoteliers
MoT, Transporters	
MoI, TANAPA	Universities, Tourism college, Tour operators

Low

Low   High
Interest

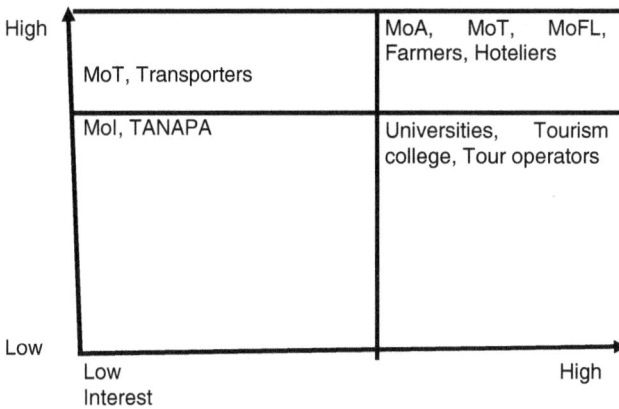

Figure 9.5 Mapping stakeholder by AIIM matrix.

Source: Study result.

Food and Cooperatives, Ministry of Tourism, Ministry of Livestock and Fisheries Development, Tanzania National Parks (TANAPA), Ministry of Infrastructure (Transport), Academic, Research, and Training institutions, Local Government Authorities, Private Sector. Mapping of stakeholders must be guided by interest and alignment in developing and sustaining agritourism. The roles played by each stakeholder could be different but complementing each other directly or indirectly.

Social Network Analysis (SNA) and the associated measurements are a methodological approach of policy analysis and research analytical tool for studying ties and social setting among individuals groups or organizations. Policy networks are defined as interactions among policy actors for accomplishment of a common goal. Policy networks facilitate exchange of information, expertize and or resources among nodes that are necessary but cannot be unilaterally obtained. On the other hand Social Network Analysis (SNA) is a specified set of nodes representing actors such as individuals, teams, agencies, or nations of which at least two are connected by a specified set of ties representing a class of relations such as resource transaction, affiliation, authority, economic exchange, or technological diffusion (Ties can be directed or non-directed, and dichotomous or valued represented as a matrix or as a sociogram that explains relationships between and among social entities. It begins with the premises of business or policy communities called Nodes interacting or making ties to make business or policies happen. SNA is defined by actors (called Nodes) and by relations (called ties). Social networks are defined as social structures made up of Nodes which are connected by one or more Ties of interdependency or relationships. Social Network Analysis (SNA) is therefore the analysis of nodes and ties (also called edges, links, or connections) measured in terms of centrality, betweenness, cohesion, etc.

(Tabassum et al., 2018). We begin with a description of basic elements in the Social Network Analysis. Nodes are the individuals, organizations, or entities of the network, whereas ties are the number of relations that particular nodes have. The relations of nodes could be one directional or non-directional, each giving a specific meaning and purpose.

In the directed relations the arrows or edges show K relates to G and [B]; G relates to [F]; B relates to [FAM and F]; but FAM relates to none. The more ties the node has, the more influence and capacity for policy implementation could be as the node have allies for support. These can be visualized in a netdraw and calculated to determine some measurements such as betweenness, centrality, and structural cohesion.

In the non-directional node relations, the actors' relations do not show power relations, that is, they show reciprocal relations. Social Network Analysis borrows mathematical and graphical terms such as Degree measure in SNA is the number of links to a vertex (indegree, and outdegree), whereas degree of centrality in SNA means the number of direct connections or ties a node has showing the most active or connected node in the network. These are the people or organizations that can influence effective implementation of policies.

Closeness centrality in SNA means the shortest paths to all others, these are nodes that are close to everyone else. They are useful nodes for policy monitoring and information sources in the network. Betweenness centrality indicates the number of shortest paths it passes the node. Because of its centrality positions, such nodes play the brokering role in the network. Density on the other hand is the sum of tie values divided by the number of possible ties (Tabassum et al., 2018).

Many public or private organizations are facing enormous challenges in improving efficiency and effectiveness in performance. Effectiveness is measured in terms of increasing capacity for the utilization of productive resources, whereas efficiency is seen in terms of reducing transactional unit costs. Tourism involves networked operations of business actors who could be geographically dispersed; these may include among others, national tourism boards, local and international tourism associations, travel agents, airlines, tour operators, hotels and restaurants. Looking at the operations of the actors indicates that no single node can make the tourism industry complete. There is so much interdependence of the businesses and actors in this industry. This has necessitated the adoption of Social Network Analysis for policy network as an analytical tool and a new approach of understanding policy implementation that enhances outcome results.

Formulating a development problem in the context of agritourism requires a multidimensional approach involving multiple actors. Social Network Analysis is useful for studying wicked policy problems that would require compliance to local government regulations, while others may require financial support, community engagement, working infrastructures, or promotional activities. Different scholars use social networks for different purposes. Social networks are used in many complex ways across disciplines as used in computer sciences, public policy, economics, anthropology, and sociology. Policy scholars use policy networks as an interactive relationship between policy structures and agencies, structural linkages between public and private actors for the purpose of increasing civic engagement and participation, means for reducing transactional costs, as well as a means for increasing policy implementation.

Social scientists' theories use analytical frameworks for understanding mechanisms for mutual inter-sectoral linkages. For example, Kritsh and Kauffeld-Monz argue that because partners in an innovation network tend to have closely related interests, the chances of gaining valuable information and knowledge in such a network are relatively high. The importance of social networks in social sciences "is the notion that individuals are embedded in thick webs of social relations and interactions" (Wasserman and Faust, 1994). In the same vein, individual sectors such as agriculture or tourism may increase their outputs through interactions.

SNA is a tool of policy analysis for examining and understanding relations among policy actors. It involves relational thinking and understanding the social systems among entities called nodes in the language of social network analysis. Nodes represent actors with specific attributes that characterize them in terms of physical or functional descriptions. The relationships among nodes are defined as "ties"; thus, a tie between the Ministry of Agriculture and the Ministry of Tourism can be characterized as inter-dependence of government entities, or business whereby one sector provides services to another sector. A prominent hypothesis put forward is based on the idea that "strong ties" characterize a dense network of actors who are mutually connected to each other (Tabassum et al., 2018).

Doing agritourism business requires a lot of information and information sharing among key actors. This entails a need for a strong tie among actors in the network; this is important as argued that the actors of sub-cluster tend to interact frequently and share information in the social system.

The social network is usually considered to operate among human beings because they have social interaction attributes. The SNA has borrowed the concept to the non-human relational analysis, such as organizational, companies, cities or nations' networks.

A relationship between actors is called a tie at the very basic level in SNA. The tie could be a pair or multiple actors. When a pair of actors form ties between them, they form a dyad. Dyads are defined as the most basic units in the statistical analysis of social networks (Wasserman and Faust, 1994). We would examine the common ties in agritourism among actors such as farmers, hotels/restaurants, tour operators/transports and airlines, fishermen, food sellers, and banks. Our examination of agritourism network is centered on the assumptions that: a) actors in networks i.e the agriculture and tourism can benefit more in the presence of strong ties than if they have weak ties; b) network cohesion or connectedness of the network actors would have a positive effect from the information sharing on the available goods and services to each other.

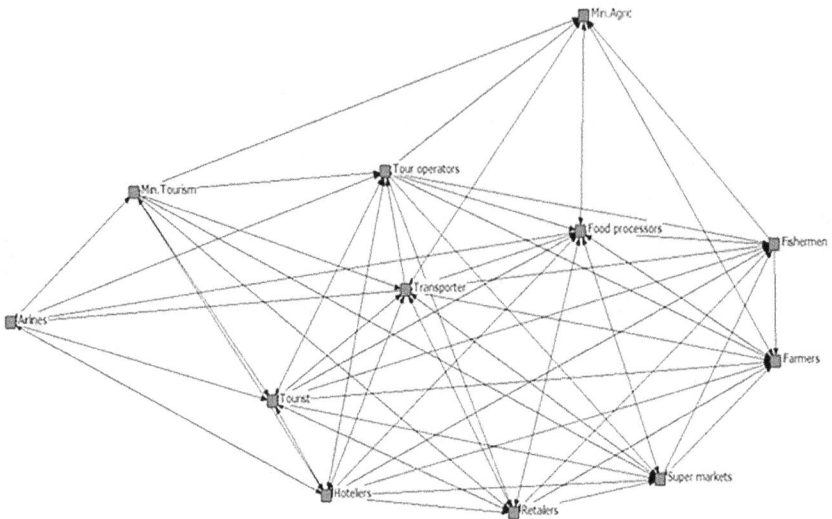

Figure 9.6 Visualization of key nodes for agritourism network in Tanzania.
Source: Study result.

In a very specialized language, a network is "a set of discrete elements called "vertices" and a set of connections called "edges" that link the actors (Borgatti et al.). These elements and their links can consist of communication lines, people and their organizations. These may be visualized using the graph theory.

Agricultural-tourism network analysis begins at the inputs supply to consumers, such that agriculture becomes the supply side to the tourism consumers. We have taken the view that farming activities provide key inputs to the tourism sector. Farmers or producers of cereals, vegetables, fruits, oil seeds, livestock and fish keepers are important components in the network. Other important nodes in the network are the marketers who constitute those who bring products to consumers. This category includes processors, wholesalers, retailers, and traders. The processes involved play roles in increasing opportunities for growth across the board (Figure 9.6).

At this juncture, it is important to define key terms of SNA. According to Borgatti et al., Centrality is the measurement of the social power of a node based on how well it is connected in the network. There are two measurements of centrality that is betweenness, and Closeness.

Degree Measure of Centrality in Agritourism

In the network analysis (SNA) perspectives, measures degree of centrality is defined as the nodes or actors that feature prominently, and have greater control or influence on major policy decision making. In other words, the degree of centrality is the number of ties of a given actor has to influence change or policy direction. The Ministry of Natural Resources and Tourism is implementing anti-poaching policy following the increased poaching of wildlife. According to the Ministry of Natural Resources and Tourism policy objective 1; they have an obligation of protecting the Wildlife by preventing illegal use of wildlife through appropriate surveillance, policing and law enforcement. It is very much understood that the ministry in isolation cannot afford protecting the wildlife; policy objective 2 therefore gears toward the involvement of conservation stakeholders within national, regional, and international boundaries. The policy recognizes that a successful anti-poaching policy implementation requires the concerted efforts of all stakeholders. We use the Stakeholders Analysis for identifying the Stakeholders in the Anti-Poaching policy implementationand and we use the SNA for estimating the degree of centrality of stakeholders in the anti-poaching policy implementation.

The Ministry of Natural Resources and Tourism is implementing anti-poaching policy following the increased poaching of wildlife. We wanted to understand how big is the problem? The key informants at the ministry acknowledge that the poaching problem is alarming and it requires joint interventive and preventive actions by all stakeholders. According to the Ministry of Natural Resources and Tourism policy objective 1, You have an obligation of protecting the Wildlife from illegal use of wildlife by taking appropriate surveillance, policing and law enforcement. We also inquired how you have implemented the policy. It was informed that policy implementation is done through the deployment of wildlife sub-sector patrols, surveillance operations and prosecution for those who are caught. The rule of law requires that the culprits be prosecuted and brought before the court of justice. Together with communities, the Ministry through the village game scouts and private sector participate in the anti-poaching strategies by enforcing the WCA No. 5 of 2009 and TANAPA Cap 2820 of 2002.

It is well recognized that in the promotion of Tourism, the ministry in isolation cannot overcome anti-poaching as it has become more complicated and involves many crimr networks. One of the specific implementation strategies is to create linkages with institutions related to tourism; such as protection of wildlife – elephants, rhino, etc. The main objective in the anti-poaching policy network implementation is to protect the natural resources, the wildlife so that tourism continues to contribute to the national economy. However, there are policy implementation problems such as: inadequate funding of the massive anti-poaching campaigns and operations; inadequate knowledge and awareness of roles to play among stakeholders and conflicting laws and policies. The following national, regional or international institutions work or have agreed to work together in overcoming anti-poaching in Tanzania. The process requires coordination; we use social network analysis for establishing nodes, ties and centrality (Figure 9.7 and Table 9.6).

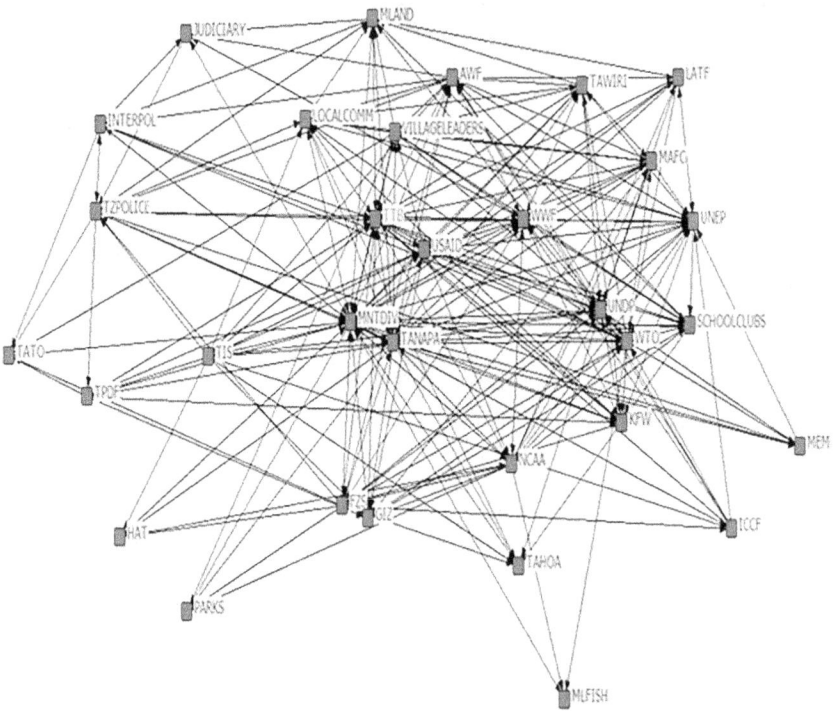

Figure 9.7 Visualization of key nodes for agritourism network in Tanzania.
Source: Study result.

It is important to define and examine key terms of SNA. According to Borgatti et al., Centrality is the measurement of the social power of a node based on how well it is connected in the network. There are two measurements of centrality that is betweenness, and Closeness. Centralization is the difference between the numbers of links for each node divided by maximum possible sum of differences. A centralized network will have many of its links dispersed around one or a few nodes, while a decentralized network is one in which there is little variation between the numbers of links each node possesses. Whereas, Closeness is the degree an individual is near all other individuals in a network (directly or indirectly). It reflects the ability to access information through the grapevine of network members. Thus, closeness is the inverse of the sum of the shortest distances between each individual and every other person in the network.

Cohesion measures the degree to which actors are connected directly to each other by cohesive bonds. Groups are identified as 'cliques' if every individual is directly tied to every other individual, 'social circles' if there is less stringency of direct contact, which is imprecise, or as structurally cohesive blocks if precision is wanted, and density is the degree a respondent's ties know one another/proportion of ties among an individual's nominees.

Table 9.6 List of abbreviations for organizations

Abbreviation	Organization
MNTDIV	Ministry of Natural Resources Tourism Division
TTB	Tanzania Tourist Board
TATO	Tanzania Association of Tourism Operators
TAHOA	Tanzania Hunting Operators Association
WTO	World Tourism Organization
HAT	Hotel Association of Tanzania
FZS	Frunkfurt Zoological Society
WWF	World Wildlife Fund
PARKS	Parks Congress
AWF	African Wildlife Foundation
UNEP	United Nations Environmental Program
UNDP	United Nations Development Program
TZPOLICE	Tanzania Police Force
JUDICIARY	
LATF	Lusaka Agreement Task Force
INTERPOL	
LOCALCOMM	Local community
ICCF	
TPDF	Tanzania Peoples Defense Force
SCHOOL CLUBS	Primary school MaliUhai Club of Tanzania
TANAPA	Tanzania National Parks
NCAA	Ngorongoro Conservation Areas Authority
TIS	Tanzania Latelight System

Network or global-level density is the proportion of ties in a network relative to the total number possible.

Policy Implication and Recommendations

This chapter implies that rural transformation calls for innovations to overcome rural poverty in countries whose agriculture is the main economic base. It is postulated that for rural socio-economic transformation to occur, agriculture must create strong linkages with tourism which has the potential of bringing in foreign money. This implies innovative ways for building farmers' capacity to engage in transforming their farms and farming activities into agritourism ventures. However, decisions for such transformation ought to be informed by scientific research. Unfortunately, there are few scientific publications on agritourism in Africa, and certainly fewer by African scholars and universities. This implies that more scholarly and scientific evidence are required to inform evidence-policy decisions for the African governments to adopt the agritourism pathway for rural transformation and poverty reduction. It is recommended for countries to support this research agenda on the role of agritourism for sustainable rural development and poverty reduction.

The chapter also implies that building a strong intersectoral linkages between agriculture and tourism requires a careful understanding of forward and backward linkages and the associated challenges. The identified poor rural infrastructure such as on-farm accommodation, transportation, electricity and communications requires a strong government inter-sectoral coordination. The current Public-Private Partnership framework provides room for design and build, design, build, and operate, design, build, operate and maintain, build, operate and transfer, build, lease, and transfer, design, build, finance and operate, design, build, finance and maintain, build, own and operate, and buy, build and operate agritourism. Whereas, the observed weak capacity of agriculture sector to supply the required quantity and quality of food to the tourism sector, implies an urgent need for capacity building for improving the quality and supply of farm products to the tourism industry, improving the value chain of farm products to meet the tourism industry demands. This shall require developing a specialized hospitality curriculum in the training institutions to meet the needs of the new tourism product. The chapter implies entrepreneurial skills for our farmers whose resource and incomes are shrinking. The proposed innovative model of agritourism entrepreneurship must be taken for on-farm income diversification as a rural policy response to overcome poverty.

Notes

1 *Do they agree with our approach? Do they agree with our assumptions? Do they want to do the same things that we think need to be done? Are they thinking what we are thinking?*
2 *Are they committing time and money to this issue? Do they want something to happen (whether it is for or against)? Are they going to events on the subject? Are they publicly speaking about this?*

References

Back, R.M., Lowry, L.L. and Higgins, L.M. (2021). 'Exploring a wine farm micro-cluster: A novel business model of diversified ownership', *Journal of Vacation Marketing*, 27(1), pp. 103–116.
Baipai, R. *et al.* (2023). 'A framework for sustainable agritourism development in Zimbabwe', *Cogent Social Sciences*, 9(1), pp. 1–18.
Bhatta, K. (2020). 'A review of quantitative studies in agritourism: The implications for developing countries', pp. 23–40.
Blake, A. *et al.* (2003) 'The economic potential of tourism in Tanzania', *Journal of International Development*, 15(April), pp. 335–351.
Bolaky, B. (2016). 'Tourism for economic development in Africa', *Journal of Research in Business, Economics and Management*, 7(4), pp. 1222–1248.
Braun, O.L. *et al.* (1999). 'Potential impact of climate change effects on preferences for tourism destinations. A psychological pilot study', *Climate Research*, 11(3), pp. 247–254.
Büyükkidik, S. (2022). 'A bibliometric analysis: A tutorial for the bibliometrix package in r using IRT literature', *Journal of Measurement and Evaluation in Education and Psychology*, 13(3), pp. 164–193.

Cheteni, P. and Umejesi, I. (2023). 'Evaluating the sustainability of agritourism in the wild coast region of South Africa', *Cogent Economics and Finance*, 11(1). Available at: 10.1080/23322039.2022.2163542.

Chikuta, O. (2023). 'Agritourism in Botswana: Challenges and Prospects', *Conference Paper*, 16(3), pp. 216–217.

Ellis, F. and Freeman, H.A. (2004). 'Rural livelihoods and poverty reduction strategies in four African countries', *Journal of Development Studies*, 40(4), pp. 1–30.

Fleischer, A. and Tchetchik, A. (2005). 'Does rural tourism benefit from agriculture?', *Tourism Management*, 26(4), pp. 493–501.

Hjalager, A.M. (1996). 'Agricultural diversification into tourism: Evidence of a European community development programme', *Tourism Management*, 17(2), pp. 103–111.

Ilbery, B. *et al.* (2015). 'Farm-based tourism as an alternative farm enterprise: A case study from the', (February).

Khazami, N., Nefzi, A. and Yahyaoui, A. (2023). 'Role of rural women on the agritourism entrepreneurial behavior in Tunisia', *Cogent Business and Management*, 11(1), pp. 1–19.

Kipkorir, N., Twili, N.S. and Gogo, A. (2022). 'Effects of agritourism development on the local community in Kericho County, Kenya', *Journal of Tourism, Culinary and Entrepreneurship (Jtce)*, 2(1), pp. 34–53.

Knowd, I. (2015). 'Tourism as a mechanism for farm survival', (January 2006). Available at: 10.1080/09669580608668589.

Mahaliyanaarachchi, R.P. *et al.* (2019). 'Agritourism as a sustainable adaptation option for climate change', *Open Agriculture*, 4(1), pp. 737–742.

Morris, W., Henley, A. and Dowell, D. (2017). 'Farm diversification, entrepreneurship and technology adoption: Analysis of upland farmers in Wales', *Journal of Rural Studies*, 53, pp. 132–143.

Myer, S.L. and De Crom, E.P. (2013). 'Agritourism activities in the Mopani District Municipality, Limpopo Province, South Africa: Perceptions and opportunities', *The Journal for Transdisciplinary Research in Southern Africa*, 9(2). Available at: 10. 4102/td.v9i2.208.

NBS (2021). *Integrated labour force survey 2020/21*. Dodoma: National Bureau of Statistics.

Nguyen, C. and Binh, T.C. (2022). 'Agritourism and the sustainable development in the mekong delta of Vietnam', … *Symposium on Sustainable Development In …* [Preprint], (December). Available at: https://papers.ssrn.com/sol3/papers.cfm?abstract_id=4300973%0Ahttps://www.zbw.eu/econis-archiv/bitstream/11159/608409/1/EBP091272866_0.pdf.

Orboi, M.D. (2012). 'Development of rural communities by diversification of rural economy in the context of sustainable development', *Animal Science and Biotechnologies*, 45(1), pp. 450–453.

Pimid, M. *et al.* (2023). 'Rural development: Motivational factors impacting community support for rural tourism', *Planning Malaysia*, 21(6), pp. 408–424.

Robinson, D. (2021). 'Rural food and wine tourism in Canada's south okanagan valley: Transformations for food sovereignty?', *Sustainability (Switzerland)*, 13(4), pp. 1–19.

Rogerson, C.M. (2012). 'Tourism-agriculture linkages in rural South Africa: Evidence from the accommodation sector', *Journal of Sustainable Tourism*, 20(3), pp. 477–495.

Rogerson, C.M. (2016). 'Climate change, tourism and local economic development in South Africa', *Local Economy*, 31(1–2), pp. 322–331. Available at: 10.1177/0269094215624354.

Rojas-Sánchez, M.A., Palos-Sánchez, P.R. and Folgado-Fernández, J.A. (2023). *Systematic literature review and bibliometric analysis on virtual reality and education,*

Education and Information Technologies. Springer US. Available at: 10.1007/s10639-022-11167-5.

De Rosa, M., McElwee, G. and Smith, R. (2019). 'Farm diversification strategies in response to rural policy: a case from rural Italy', *Land Use Policy*, 81(November 2018), pp. 291–301.

Sachaleli, N. (2020). 'Agritourism as a business in regional rural development', (1), pp. 89–100.

Saleem, Z. and Donaldson, J.A. (2016). 'Pathways to poverty reduction', *Development Policy Review*, 34(5), pp. 671–690.

Srisomyong, N. and Meyer, D. (2015). 'Political economy of agritourism initiatives in Thailand', *Journal of Rural Studies*, 41, pp. 95–108.

Tabassum, S. *et al.* (2018). 'Social network analysis: An overview', (January), pp. 1–21. Available at: 10.1002/widm.1256.

Todaro, M.P. (1997). 'Urbanization, unemployment, and migration in Africa: theory and policy', *Reviewing Social and Econimic Progres in Africa*, 104(104), p. 54.

Torres, R. (2003). 'Linkages between tourism and agriculture in Mexico', *Annals of Tourism Research*, 30(3), pp. 546–566. Available at: 10.1016/S0160-7383(02)00103-2.

URT (2017). *Agricultural sector development programme phase II*. Dar Es Salaam: Ministry of Agriculture.

URT (2021). 'National five year development plan: Realising competitiveness and industrialisation for human development', *Dodoma: Ministry of Finance and Planning*, pp. 106–107.

URT (2022). 'The 2014 international visitors' exit survey report', *National Bureau of Statistics*, p. 72. Available at: http://www.nbs.go.tz/nbs/takwimu/trade/International_Visitors_Exit_Survey_Report_ 2014.pdf.

Vyslobodska, H., Brychka, B. and Bulyk, O. (2022). 'Rural tourism as an alternative direction of activity diversification of agricultural products producers', *Scientific Messenger of LNU of Veterinary Medicine and Biotechnologies*, 24(99), pp. 10–14.

Walke, S.G. and Kumar, A. (2015). 'Agritourism: Supplementary business for farmers in Maharashtra state', 2(6), pp. 480–485.

Wasserman, S. and Faust, K. (1994). 'Social network analysis: Methods and applications'. Cambridge University Press. https://doi.org/10.1017/CBO9780511815478

Yuan, P. *et al.* (2017). 'A study on farmers' agriculture related tourism entrepreneurship behavior', *Procedia Computer Science*, 122, pp. 743–750.

10 Agritourism Digital and Social Media Marketing in Africa
The Case Study of Zimbabwe

Sinothando Tshuma, Mercy Dube, and Phillip Dangaiso

Introduction and Background

Agritourism is a subsector of the tourism industry and a subset of rural tourism that uses a diverse array of farms as tourist destinations. It supports farm activities targeting tourists, including farm demonstrations, farm visits, farm training, as well as on-site value-adding and sales of farm products. Agritourism occurs on a working farm or agricultural plant and is conducted for the visitors' enjoyment and to generate supplemental income for farmers. Agritourism and nature tourism farms might include a diverse array of activities such as outdoor recreation, farming experiences, cultural entertainment, hospitality services, and on-farm direct sales. As an emerging wing of the tourism industry in Zimbabwe, digital and social media marketing has a role to play in order to increase the visibility of the sector. In the African context, the sector is still lagging as far as the adoption of digital and social media marketing is concerned.

Objectives

i To identify the digital and social media channels that are available for marketing Agritourism products in Africa.
ii To analyse the challenges faced by marketers in Agritourism digital and social media marketing.
iii To recommend solutions to challenges faced by marketers in Agritourism digital and social media marketing

Literature Review

The yardstick for the effectiveness of any agritourism marketing campaign is its ability to inform and persuade consumers to consume the offer (Joyner et al., 2018) regardless of consumers' needs and wants (Miller et al., 2021). The strength of a marketing campaign from a tourist perspective is the ability of the tourism resort to deliver leisure and experience beyond what the firm has communicated to the customer (Liang et al., 2020; Suhartanto et al., 2020; Miller et al., 2021). The rise in technological advancement has led to the

DOI: 10.4324/9781032696188-10

development of new marketing platforms that the agritourism industry should embrace to effectively reach its target market. One of the most used platforms has been nominated as a social media platform like Alexa because of its low or zero expenditure. The rise in the need for agritourism marketing is customer boredom (Comen, 2017) due to experiential tourist resorts that offer the same customer menu every day (Chikuta and Makacha, 2016) with no tourist connection with the environment or its species (Comen, 2017; Chikuta and Makacha, 2016). The major drawback that has led to the low uptake of agritourism products in Zimbabwe is the lack of stakeholder input, which in turn leads to limited or nil marketing of agritourism sites in Zimbabwe. Effective agritourism marketing in Zimbabwe can turn around the economic well-being of a farmer and the entire country (Chikuta and Makacha, 2016), as is the norm in developed countries like the United States of America with an income above USD3.5 billion from farm agritourism sites (Chase et al., 2018). Recent research conducted by Baipai, Chikuta, Gandiwa, and Mutanga (2022) noted with concern that farmers have not been empowered to redesign their farming calendars and embrace agritourism for economic development, which in turn makes the sector docile to generate income. The major recommendation from the study conducted was that farmers should design marketing strategies for their farms and encompass various farming activities that increase tourist appeal to visit the farms including developing artificial tourist sites within farms. For farmers to fully embrace the digital and social media marketing strategies at a low cost in the Zimbabwe environment to boost agritourism, they should register with the Zimbabwe Tourism Authority and attend scheduled in-house training sessions. According to Zvavahera and Chigora (2023), the transformation of Zimbabwe's socio-economic state lies in the effective use of marketing strategies such as online marketing; this helps to broaden the agritourism prospective customers to visit the rural agro-based farms (Chigora et al., 2020), as consumers can subscribe to their alternative tourist site at the comfort of their homes. One of the enemies of the low uptake of agritourism in Africa has been identified as ineffective marketing (Zvavahera and Chigora, 2023). The marketing of agritourism products awaits success if well implemented as Zimbabwe has rich natural resources that require limited or no modification to be tourist sites than just marketing (Chikuta and Makacha, 2016).

Overview of Agritourism in Africa

The backbone of economic survival of many African economies is their agricultural resources (BFAP, 2018). With the passage of years, the sector has been infiltrated by changes in climatic conditions which have created many challenges for the survival of the sector. In South Africa, 1 in 23 individuals employed on the farms are directly employed in the agritourism sector (STATSSA, 2023); this has led to an increase in rural GDP even in seasons of drought where direct farm output is low and increasing community loyalty to their cultural beliefs and values (Barbieri, 2010). It has also been noted that besides rural development, agritourism will help in the urbanisation of rural farms, on the other hand curbing rural-to-urban migration and ensuring equity distribution of the country's generated revenue (Cheteni and Umejesi, 2023). The Western Cape province has been noted as the province with the highest uptake of agritourism activities in South Africa.

To effectively implement Agritourism in Tanzania, also termed Pro-Poor Tourism (PPT) and meant to alleviate poverty in the Tanzanian rural farms, the country has identified nine regions documented in their ACT magazine (ACT, 2023). The main PPT regions encompass Arusha, Kilimanjaro, Iringa, Mbeya, Coast, Morogoro and Dodoma. The main activities of Agritourism were noted from various rural farms (ACT, 2023). In Arusha, Tourists are given the platform to taste the home-brewed coffee on the farm, tour the coffee plantation and even participate in the roasting and grinding of coffee beans and finally are introduced to the local culture as they drink coffee. Furthermore, in Ndombo-Mfulony-Nkoarisambu they are given the taste of Banana beer or Mbege from the plantations. In the Kilimanjaro region, tourists are given the platform to learn more about coffee brewing and, cultural origins surrounding the coffee estate and during their spare time they tour the mountainous slopes in the region. In summary, the Iringa region gives the agri-tea tour, Dodoma gives the agri grapes and horticulture tour, Morogo gives the horticulture agri tour, and in Mbeya agri cocoa tour, tea, coffee, banana and avocado tours. Also to mention is the spice farms in Zanzibar. These regions could have been earmarked to be huge revenue centres if they are policy-guided, as to date United Republic of Tanzania has launched and published an official policy for PPT or Agritourism and its linkages to the tourism value chain (ACT, 2023).

Moving on to other African countries, Lesotho as a landlocked country of South Africa, still suffers from low GDP with many of its citizens migrating to South Africa instead of the country utilising its lucrative agritourism rural farms (Mpiti and Harpe, 2015). It was noted that the area has natural farms with tourism artefacts but does not have ICT resources to lure the tourists to respective farms. In the context of Morocco, the concept of agritourism has been encompassed as the country's farms are still using traditional farming methods that also serve as a tourist experience. Many tourists opt to visit the rural farms and they participate in land tilling, planting and harvesting using traditional methods. Tourists are also given an appreciation of testing the farm produce and they are inducted of the benefits of the menus they consume on the farms. According to the European Training Journal (2021), more jobs have been created mainly by farmers who have embraced agritourism concepts. This is because the government has provided financial and social support to farmers to embrace the concept.

In Kenya, many farmers are mainly involved in gardening and farming. Tourists are given tours around the farms and they have a chance to participate in experiential tourism. In this scenario, tourists partake in fruit picking and drinking coffee. Some tourists who visit animal-rearing farms, are shown brands of animals and some milking methods (Ajayi and Gündüz, 2020). In the Central and Rift Valley, tea and coffee drinking are the main farm tourism activities. The Kenyan government has invested in the sector by holding some farmer training to appreciate agritourism and this has increased the country's revenue from the sector (Agayi and Gündüz, 2020; Khanal et al., 2019).

In the Zimbabwean economy, the country lost its bread basket of agricultural produce after land reforms in the year 2000, and this has led to low income from the sector (Guvamombe, 2019). More than 7 million hectares of land are farm plots, with many farmers weeping of low produce (Guvamombe, 2019), yet there have been recommendations to embrace agritourism (Chikuta and Makacha, 2016), which farmers are still hesitant to embrace with main reasons being lack of frameworks to fully roll the agritourism concept for sustainable economic development (Baipai et al., 2023). On another note, a few farmers have come on board to embrace the agritourism concept in their farms with many now building tourist lodges to come and tour some farms (Arru et al., 2019) as the country has five regions with vibrant diversifications in farming. This in turn gives tourist options on which region to visit depending on their affiliations. The full commercialisation of agritourism in Zimbabwe has the capacity to develop the rural population and at the same time lead to economic development through ministry of agriculture field days (Wijijayanti et al., 2023). Agritourism in Zimbabwe is also most visible in Msinje farm with intense agriculture tourism learning activities (Matarise, 2023). The vibrancy of the sector rests in ZTA implementation policy with the Ministry of lands and also of Agriculture.

Shift towards Agritourism in Africa

In European countries, the rural sphere also faces similar struggles as in developing nations, which encompass slow development initiatives; the current trending initiative to develop these communities according to the European Union is investing in agritourism through finance and training because people love to spend on outgoing activities (Bacsi and Szálteleki, 2022). The rise in agritourism in Europe has helped to alleviate poverty in the rural prone families. In countries like France and the United States of America, tourists now participate in harvesting gatherings and seasonal festive celebrations during the time of touring (Akwii and Kruszewski, 2021). The main thrust of community alleviation in these communities is that farmers also sell their produce to tourists, thereby increasing their marketing linkages. In Japan, some revenue is also generated from viewing rice planting and harvesting (Akwii, and Kruszewski, 2021; Zvavahera and Chigora, 2023). The rise in tourism especially in the United States is based on the need to broaden the tourism industry scope, keep the small farms relevant and surviving and preserve the rural community cultures and ethics (Ollenburg and Buckley, 2007; Veeck et al., 2006). The main creativity in tourism is to make agriculture a tourist attraction that increases productivity on the farm and improves the lives of various stakeholders. According to Zvavahera and Chigora (2023), the agritourism initiative is rapidly being appreciated by various African countries such as South Africa, Kenya, Morocco, and Tanzania, Lesotho, among other countries; on another aspect, Zimbabwe is still shaping its sector through the implementation of some frameworks for sustainable development (Baipai et al., 2023).

Revolutionising Agritourism through Digital Marketing

According to Stelzner (2018), some sectors have diverted their marketing activities to using digital platforms; it has been noted that about 90% of successful firms use Facebook and 64% of them are also found on Instagram. This gives the agritourism farmers to change their perception towards digital marketing for economic returns. Many successful farmers in Oklahoma alluded that the main root of effective agritourism marketing is Facebook marketing and all platforms that are to be used in future should always direct users to the Facebook page regardless of farm location or tourism experiences (Miller et al., 2021). It also alluded that the effectiveness of the print or traditional marketing tools saved their purpose in alerting customers about agritourism resorts in specific delimitated areas as they cannot harness prospective customers outside the reach of the media (Miller et al., 2021). The upcoming agritourism sites alluded that the main reason that has made many small farmers embrace digital marketing tools is based on surveys they conducted which alluded that many tourists got to know them through social media platforms rather than conventional platforms. According to Orias and Borbon (2022), the upsurge of COVID-19 also turned around the agritourism landscape. Many farms were forced to aggressively adopt digital marketing to remain relevant in the tourism sector as many customers were locked indoors yet bottled with wishes to tour. It was noted that digital marketing platforms provided platforms for farmers to boost sales and make the community aware of the existence of agritourism resorts. The online platform can help to establish customer consciousness to visiting sites as displayed digitally. The current trending effective platforms include the use of the internet, social media sites as well as mobile-based technologies (Yasmin et al., 2015). The rise in the use of social media sites is their ability to reach a wider market at the lowest possible cost. The Department of Tourism and Food and Agriculture in the Philippines took it upon themselves to revive the agritourism sector after COVID-19 by massively embracing digital marketing platforms (Choenkwan et al., 2016). The digital marketing platform has emerged to reach even the poor tourist who wishes for luxury but cannot afford expensive tourist sites and they can digitally access and book tourism space on farms (Tew and Barbieri, 2012).

2.3 Enhancing Consumer Experience in Agritourism through Digital Marketing

According to Zvavahera and Chigora (2023), one of the key elements needed to transform the agritourism industry in Zimbabwe is the need for effective marketing as indicated in the figure that follows (Figure 10.1).

The main conception of the relevance of marketing in a business entity is its ability to bring customers and keep them for organisational revenue generation. There are many marketing strategies that can be used by marketers ranging from the traditional to the electronic or online marketing channels. The effectiveness of any marketing campaign according to Zvavahera and Chigora (2023) lies in its capacity to educate the consumers about the tour offering and build site brand equity. The power of agritourism marketing should create a customer experience that leads to easy retrieval of information about the resort site and linking it with its tour artefacts (Figure 10.2).

According to Sanderson, Nyamadzawo, Nyaruwata and Moyo (2013), the rise of the tourism industry in Zimbabwe is based on the above factors. The major factors that were raised through consultation with various stakeholders included investment in agriculture and ICT. These two major elements are the thrust to agritourism and the implementation of agritourist sites through the use of information technological tools. The policy articulated noted that these enablers should be implemented for the effectiveness of the tourism industry. One of the noted concerns is that 'Tourists are financially rich and time poor', as a result they need quick platforms to explore agritourism sites.

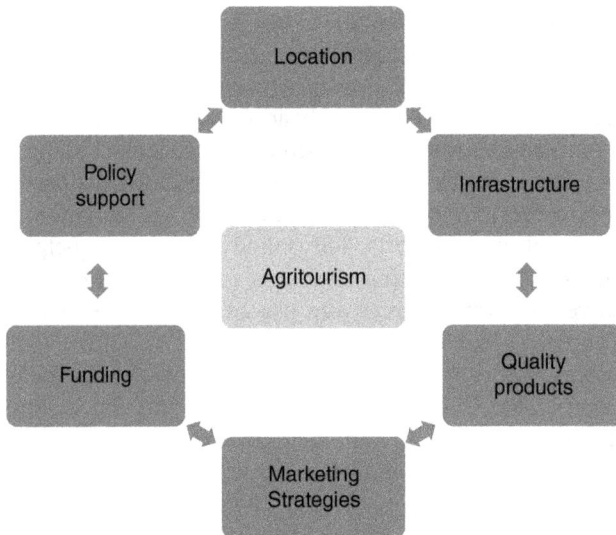

Figure 10.1 Model of agritourism transformation.

Source: Adapted from Zvavahera and Chigora, 2023.

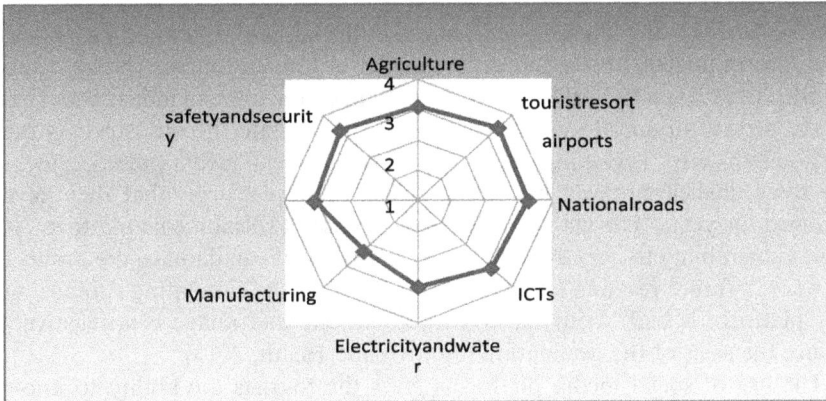

Figure 10.2 Key enablers for tourism growth.

Social Media Marketing for Sustainable Agritourism

The use of online tools has been widely used in the tourism sector, ranging from the use of digital tools and social media tools to access tourism products and ensure effective online bookings (Buhalis, 2013). The continuous evolution of technology has led to the increasing use of social media platforms in tourism marketing which include Facebook, Linkedin, MySpace, blogs, YouTube, Scribd, Flickr, microblogs, and Twitter, among other platforms. The social media platforms have the highest number of subscribers at about 4.4 billion users. In the international space, the most used platform is Facebook with the highest number of users at about 2.9 billion users monthly. In the tourism sector, social media platforms have the highest number of users with many evaluating resorts and making online bookings using Facebook post promotions. The major rise of Facebook as a prominent platform is its ability to generate user content, instant touch with consumers and its closeness to the social reality of consumer affiliations. Within Africa, limited studies have been done on Agritourism marketing, creating literature gaps for more primary research in future (Boateng and Hinson, 2007; Wiig, 2003).

According to Dube, Kupika and Nyaruwata, the growth in rural tourism is mainly being limited by limited exposure of these tourist resorts in the digital and social media platforms as is the norm with bigger tourist resort areas. It should not be undermined that the rise in digital technology is transforming every sector globally even through the use of geolocation (Shin and Perdue, 2022) and giving every organisation or tourist resort the opportunity to gain market share and compete globally (Zheng et al., 2020). The use of content marketing and community engagement, including blogging and video content, can significantly educate and engage potential agritourism visitors (Mathew and Soliman, 2021).

Challenges of Social Media Marketing in Agritourism

The challenges in the tourism sector began at the onset of the country events of land reform in 2000 and reduced the number of foreign tourists (Ndlovu and Heath, 2013). Then COVID-19 put a grip again on the tourism industry and led to the growth of social media marketing to resuscitate the industry. As the agritourism sector is gaining momentum through social media presence, there are some challenges faced by farmers using these platforms that they have shunned for years. The main drawback to the use of social media platforms is their vulnerability to negative reviews by tourists that can damage the product or resort's future revenue (Chigora et al., 2021). If the circulating rumour on any platform is bad about the tourist resort, the bad image can negatively reduce the sales of the destination (Ndlovu and Heath, 2013).

The use of social media platforms gives the tourists the ability to know more about the agritourism site than the farmers or owners of the site; the platforms that create customer to consumer marketing can de-market the tourist site if the past experiences of the customers were not pleasant (Chigora et al., 2021). The major challenge in the use of social media platforms in tourism is the ability of a marketer to present the resort offering using the media trusted by the intended customers and blending the mix correctly to avoid misconceptions (Chigora et al., 2021).

The major challenge in the uptake of agritourism social media marketing in Zimbabwe is the lack of a marketing framework (Baipai et al., 2023). The farmers who are to partake in marketing their agritourist resorts have not been capacitated in terms of hands-on skills, ICT network linkages and financial investments to invest in these platforms (Chikuta and Makacha, 2016; Guvamombe, 2019). The authors recommended farmer collaboration to share marketing costs and exposure on the social platforms.

The other concern by farmers is that the sector is prone to drought and floods, fluctuating produce and climate dependant farming activities which later create a gap in the marketing of the sector (Mutami, 2015), making the marketing platforms chosen docile when the sector is not lucrative for tourists. According to Baipai, Chikuta, Gandiwa, and Mutanga (2022), the success of any marketing strategy is the support by the guiding sectoral ministry; as a result, the low uptake of social media marketing in the agritourism sector in Zimbabwe is lack of consideration by the sector as a vital strategy in alleviating the struggling growth of the agritourism sector. (Chikuta and Makacha, 2016) noted in their tourism article concern that although the agritourism sector has growth potential, it is stifled by the concentration of all digital marketing efforts on the old tourism sites. Many farmers in Manicaland and Mashonaland West alluded that they could have started digital marketing of their agritourist farms, but they lack the skills to undertake the marketing efforts (Baipai et al., 2022), and the Ministry representative of Tourism in Zimbabwe noted that ZTA still has an obligation of training and educating farmers about the marketing options they can use. It was heart felt that some farmers noted from interviews carried out that the agritourism digital marketing concept is new, they have never heard of it and how it can be implemented to increase their profits (Baipai et al., 2022).

Opportunities for Social Media Marketing in Agritourism

According to Nyahunzwi (2019), the main advantage of encompassing online marketing strategies in the tourism sector is to blend with the traditional platforms and trap the technologically savvy customer in the digital platforms. The author further articulated that the Zimbabwe Tourism Authority is encouraging the use of dual marketing platforms as both traditional and online platforms offer diverse opportunities when used correctly in the tourism sector. The main reason that has led to the rise in the use of online-based platforms like the social media platform is its ability to cut marketing costs, target the right market and communicate to the market effectively.

Social media marketing also has some opportunities in the agritourism sector such as, offering live snapshots of the resort area to prospective tourists, timely communication between farmers and prospective customers, offering practical science-driven sites, helps on collaboration with various stakeholders, reviving life in rural communities virtually, online resort bookings, timely address of queries and updated clientele information (Vuori and Okkonen, 2012; Baipai et al., 2022; Chikuta and Makacha, 2016; Guvamombe, 2019).

The use of digital and social media platforms also offers organisations the opportunity to establish profitable customer based relationships (Hofacker et al., 2020), encourages customer based marketing and two-way conversations with the company in comparison with traditional channels; the content created can easily convince a sale due to blending of images, audio, videos to attract attention to sale, making it easy for tourists to choose places they will enjoy before making disappointing bookings and offering live customer set ups as they can film their adventures and make lifetime selfie pictures (Castillo and Jimber, 2018; Harrigan et al., 2017).

Methodology

This study adopted a systematic literature review as a research method. This methodology was used because it allowed large quantities of information to be assessed, analysed and reduced into smaller useful information. The approach allowed for the integration and analysis of information to produce themes that were used for selecting the articles. A systematic literature review is considered a reliable and valid method for synthesising research findings in a specific field (Wong et al., 2013). The search was conducted on multiple databases, including Scopus, ResearchGate, and Sage Publications to ensure a comprehensive review of the literature. The themes used in searching for the articles are digital marketing and agritourism, social media marketing and agritourism, digital marketing channels, social media channels and sustainable agritourism. The search yielded 36 journal articles and 16 were included in the study. The types of publications were quantitative, qualitative as well and conceptual articles. The inclusion criteria for the articles selected for review were that they had to be published in English and had to focus on digital marketing in agritourism, social media marketing in agritourism, digital marketing channels and sustainable agritourism marketing. The articles had to be published in peer-reviewed journals, conference proceedings, or ebooks. For the exclusion criteria, the authors discarded all articles irrelevant to the chapter title, articles published in non-peer-reviewed journals or magazines, and articles published in languages other than English. The data extraction process involved reading the abstract and full text of the articles selected for review. The articles were then analysed to identify key themes and concepts related to agritourism digital and social media marketing. The themes were then categorised into digital marketing channels for agritourism, social media channels for agritourism, challenges for digital and social media marketing in agritourism and sustainable agritourism through digital and social media marketing. The data synthesis process involved analysing the themes and concepts identified in the articles and synthesising them into a coherent narrative (Table 10.1).

Table 10.1 List of scientific papers on enhancing consumer experience in agritourism through digital marketing in Zimbabwe (2013–2023)

Number	Author	Journal	Title	Strategies and best practices for enhancing consumer experience in agritourism through digital marketing in Zimbabwe
1	Zvavahera and Chigora (2023)	Qeios	Agritourism: a source for socio-economic transformation in developing economies	Developed a model of agritourism transformation for the agritourism sector in Zimbabwe, the model encompassed the need to intensify marketing strategies as one of the drivers
2	Nyamadzawo, Nyaruwata and Moyo (2013)	Nathan Associates Inc.	USAID Strategic Economic Research And Analysis – Zimbabwe (Sera) Programme Positioning The Zimbabwe Tourism Sector For Growth: Issues And Challenges	Developed key enablers for tourism growth in Zimbabwe, major drivers being investment in the tourist resort outlook, electricity and water and ICT

Discussion of Findings

From the two models developed by (Zvavahera and Chigora, 2023; Nyamadzawo et al., 2013) for the Zimbabwean landscape, the authors proposed the following marketing model for agritourism sustainable digital and social media marketing in Zimbabwe: (Figure 10.3)

From Table 10.2 above, many agritourism businesses in Zimbabwe are not fully utilising popular social media platforms such as Facebook, Instagram, Twitter, and Pinterest to showcase their offerings due to lack of digital marketing knowledge of farmers, although the platforms allow them to share visually appealing content, including images and videos of their farms, lodges, and experiences, to attract potential visitors. Content marketing is also another strategy the African countries can use to promote agritourism. Content marketing, including blogging and video content can significantly educate and engage potential agritourism visitors (Mathew and Soliman, 2021). Blog posts, YouTube videos, and Instagram stories can be used to tell the story of the farm, its products, and the unique experiences visitors offer (Buhalis, 2013). These are some of the emerging social media marketing strategies that the Zimbabwean Agritourism sector should grab and implement to boost revenue from the sector.

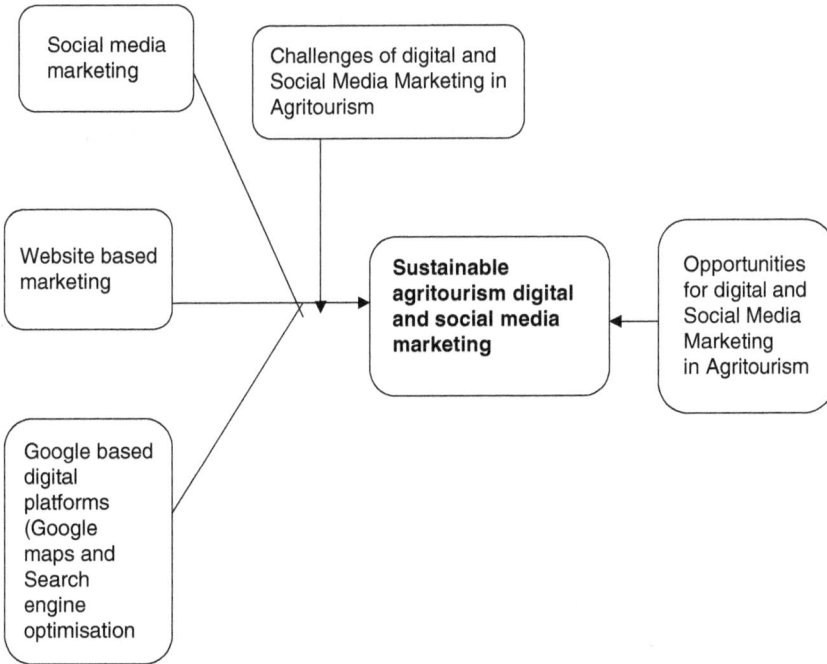

Figure 10.3 Model for agritourism sustainable digital and social media marketing in Zimbabwe.

Source: Researcher's construct, 2023.

Table 10.2 List of scientific papers on social media marketing for sustainable agritourism (2012–2023)

Number	Author	Journal	Title	Strategies of social media marketing for sustainable agritourism in Zimbabwe
1	Buhalis (2013)	UNWTO seminar on Tourism and New Technologies	e-Tourism trends and challenges in the social media era	Articulated that digital marketing strategies have been widely used in the tourism sector, ranging from the use of digital tools and social media tools to access tourism products and ensure effective online bookings
2	Fotis, Buhalis and Rossides (2012)	www.academia.edu/1324948	Social media use and impact during the holiday planning process	Articulated the need to embrace social media platforms noting the need to use Facebook as it has the highest number of users with about 2.9 billion users monthly
3	Dube, Kupika and Nyaruwata (2023)			In the Zimbabwean context the growth in rural tourism is mainly being limited by limited exposure of these tourist resorts in the digital and social media platforms as is the norm with bigger tourist resort areas.
4	Shin and Perdue (2022)	International Journal of Hospitality Management	Hospitality and tourism service innovation: A bibliometric review and future research agenda	The rise in digital technology is transforming every sector globally even through the use of geolocation
5	Mathew and Soliman (2021)	Journal of Consumer Behaviour	Does digital content marketing affect tourism consumer behaviour? An extension of technology acceptance model	Content and community engagement should be used in digital marketing
6	Zheng et al. (2020)	Journal of Cleaner Production	From digital to sustainable: A scientometric review of smart city literature between 1990 and 2019	Digital marketing helps organisations to gain market share and compete

The other digital strategy African countries can utilise to promote agritourism is Community Engagement, a key strategy in engaging with local and international communities through social media (Mathew and Soliman, 2021). Agritourism businesses often interact with followers, answer questions, and share user-generated content. This helps build a loyal customer base and encourages word-of-mouth marketing. The high penetration rate of mobile devices should be beneficial to the Agritourism business in Zimbabwe. Given the widespread use of mobile devices in Zimbabwe, agritourism businesses needed to ensure that their websites and social media content were mobile-friendly. Some agritourism businesses integrated booking and reservation systems into their websites and social media profiles. This makes it easier for potential visitors to plan and book their trips. Collaborating with social media influencers and bloggers, especially those with a focus on travel and lifestyle has become a popular trend (Shin and Perdue, 2022). These influencers could reach a wider audience and provide authentic reviews and recommendations. Some agritourism businesses use geolocation services to help visitors find their farms or lodges. Maps and navigation tools can be integrated into their websites and social media profiles.

Email newsletters and marketing campaigns are used to keep past visitors and interested parties informed about special offers, events, and updates. Online Reviews positive reviews and ratings on platforms like TripAdvisor and Google Maps were influential in attracting new visitors. Businesses often encourage satisfied customers to leave reviews. Language Localisation is another strategy the Agritourism sector uses to encourage the uptake of their services. Given the linguistic diversity in Africa, agritourism businesses often have multilingual websites and social media content to reach a broader audience. It's important to note that the digital and social media landscape is constantly evolving. To stay competitive, agritourism businesses in Africa should adapt their marketing strategies to the latest trends, emerging platforms, and changing consumer preferences. It is important to note that the popularity of specific platforms and marketing strategies may vary by region and target audience (Table 10.3).

The adoption of different digital and social media strategies also comes with challenges. The agritourism sector is not spared from the different challenges to effectively implement digital and social media marketing to gain a competitive urge in the industry. Some of the challenges are highlighted below:

In Zimbabwe, there is limited access to the internet, and when it is available, it can be unreliable (Baipai et al., 2023). This poses a significant challenge for digital marketing, as it hinders the reach and engagement of potential customers in rural areas where agritourism destinations are often located. Language and Cultural Diversity is another challenge faced by marketers in Agritourism digital and social media marketing. Zimbabwe is a country of diverse languages and cultures. Marketers in agritourism need to address these differences to effectively communicate with and appeal to a wide range of potential visitors, even foreign visitors from other nations.

Table 10.3 List of scientific papers on challenges of social media marketing in agritourism (2013–2023)

Number	Author	Journal	Title	Challenges of social media marketing in agritourism
1	Ndlovu and Heath (2013)	Academic Journals	Rebranding of Zimbabwe to enhance sustainable tourism development: Panacea or villain	Challenges in the tourism sector began from the time Zimbabwe made a fast track land reform and this reduced the number of foreign tourists, also noted that the social media platform can circulate false or bad rumours about the resort areas which might de-market the place's attractiveness
2	Chigora, Ndlovu, and Zvavahera (2021)	Journal of Sustainable Tourism and Entrepreneurship	Zimbabwe tourism destination brand positioning and identity through media: A tourist's perspective	The main drawback to the use of social media platforms is their vulnerability to negative reviews by tourists that can damage the product or resort's future revenue. Also in the Zimbabwean context, the major challenge in the use of social media platforms in tourism is the inability of a marketer to present the resort offering using the media trusted by the intended customers and blending the mix correctly to avoid misconceptions
3	Baipai et al. (2023)	Cogent Social Sciences	A framework for sustainable agritourism development in Zimbabwe	The major challenge in the uptake of agritourism social media marketing in Zimbabwe's lack of marketing framework. The farmers who are to partake in marketing their agritourist resorts have not been capacitated in terms of hands-on skills, ICT network linkages and financial investments to invest in these platforms
4	Chikuta and Makacha (2016)	Journal of Tourism and Hospitality Management	Agritourism: A Possible Alternative to Zimbabwe's Tourism Product?	Farmer collaboration to share marketing costs and exposure on the social platforms
5	Guvamombe (2019)	Herald	Farm tourism is trendy, and refreshing	Farmer grouping in the area to share huge marketing costs and exposure on the social platforms that they could not afford individually

6	Mutami (2015)	The Journal of Sustainable Development	Smallholder Agriculture Production in Zimbabwe: A Survey	The agritourism sector is prone to drought and floods, fluctuating produce and climate dependant farming activities which later create a gap in the marketing of the sector making the marketing platforms chosen docile when the sector is not lucrative for tourists
7	Baipai, Chikuta, Gandiwa, and Mutanga (2022)	African Journal of Hospitality, Tourism and Leisure	Critical Success Factors for Sustainable Agritourism Development in Zimbabwe	The low uptake of social media marketing in the agritourism sector in Zimbabwe is lack of consideration by the sector as a vital strategy in alleviating the struggling growth of the agritourism sector and the sector personnel also lack the digital marketing skill to market the agritourism sector
8	Chikuta and Makacha (2016)	Journal of Tourism and Hospitality Management	Agritourism: A Possible Alternative to Zimbabwe's Tourism Product?	The growth of the agritourism sector is stifled by the concentration of all digital marketing efforts on the old tourism sites

Poor infrastructure, including roads and transportation, can make it difficult for tourists to access remote agritourism destinations. Marketers must consider how to promote these locations while acknowledging the logistical challenges visitors may face. Not all potential customers in Africa may be digitally literate. This can make it challenging for agritourism marketers to engage with and convert their target audience through digital and social media platforms (Baipai et al., 2023; Baipai, Chikuta, Gandiwa and Mutanga, 2022). With the growth of agritourism in Africa, marketers face increasing competition. It can be challenging to differentiate one's agritourism offering from others in the market (Baipai et al., 2023). Many agritourism businesses in Africa, Zimbabwe not spared, are small and may lack the resources to invest in sophisticated digital marketing campaigns. This limitation can impact their ability to reach a wider audience. Agritourism often promotes environmentally responsible practices. Marketers must navigate the challenge of communicating their commitment to sustainability while also appealing to tourists. In many African countries, access to digital payment systems can be limited. Marketers must find ways to address payment challenges when promoting their agritourism offerings, as many marketers still use traditional marketing practices only to market the sector. Political and economic instability in some African regions can affect the marketing and perception of agritourism destinations, making it challenging to attract tourists (Mutami, 2015; Chikuta and Makacha, 2016). Marketers must address concerns related to limited budgets through corporative marketing with competitors (Guvamombe, 2019) (Table 10.4).

The use of digital and social media platforms creates more opportunities in the marketing field of agritourism products. These opportunities include, creation of a competitive advantage, improving resort attractiveness, solving customer queries instantly, online booking platforms, recorded tour scenarios and reviving the docile rural communities in Zimbabwe as tabulated above.

Discussion

Agritourism is a growing industry in Zimbabwe, and digital and social media marketing can be powerful tools to promote agritourism destinations and experiences. However, several challenges need to be addressed in the Zimbabwean context. Here are some recommended solutions to agritourism digital and social media marketing challenges in Zimbabwe:

Partner with local internet service providers and government initiatives to improve internet infrastructure in rural areas, making it more accessible for agritourism businesses. Additionally, focus on creating mobile-friendly content that can be accessed with slower internet connections. Create multilingual content to reach a wider audience. Collaborate with local translators and use translation tools to ensure your content is available in various African languages.

Invest in training and development programmes for local farmers and agritourism operators to build their digital and social media marketing skills.

Table 10.4 List of scientific papers on opportunities for social media marketing in agritourism (2013–2023)

Number	Author	Journal	Title	Opportunities for social media marketing in agritourism
1	Nyahunzwi (2019)	Journal of Tourism and Hospitality	The Online Marketing Strategies of the Zimbabwe Tourism Authority (ZTA) and South Africa Tourism (SAT): A Comparative Study	Zimbabwe Tourism Authority is encouraging the use of dual marketing platforms as both traditional and online platforms offer diverse opportunities when used correctly in the tourism sector such as to cut marketing costs, target the right market and communicate to the market effectively
2	Guvamombe (2019)	Herald	Farm tourism is trendy, and refreshing	Social media marketing also has some opportunities in the agritourism sector such as, offering live snapshots of the resort area to prospective tourists
3	Vuori and Okkonen (2012)	VINE	Refining information and knowledge by social media application: Adding value by insight	Offering practical science-driven sites
4	Baipai et al. (2022)	African Journal of Hospitality, Tourism and Leisure	Critical Success Factors for Sustainable Agritourism Development in Zimbabwe: A Multi-Stakeholder Perspective	Helps in collaboration with various stakeholders, reviving life in rural communities virtually, online resort bookings
5	Chikuta and Makacha (2016)	Journal of Tourism and Hospitality Management	Agritourism: A Possible Alternative to Zimbabwe's Tourism Product?	Timely address of queries and updated clientele information
6	Hofacker et al. (2020)	EJM	Digital marketing and business-to-business relationships: a close look at the interface and a roadmap for the future	Creates a platform to establish profitable customer based relationships
7	Castillo and Jimber (2018)	Journal of Place Management and Development	Quality, satisfaction and loyalty indices	Encourages creation of content that can easily convince a sale due to blending with mages, audio, videos to attract attention to sale
8	Harrigan et al. (2017)	Tourism Management	Customer engagement with tourism social media brands	Offer live customer set ups

Consider collaborating with local universities and institutions to offer training programmes. Implement cost-effective marketing strategies, such as utilising free social media platforms, building a website with open-source tools, and leveraging user-generated content. Seek out grants and funding opportunities for agritourism development.

Showcase the authenticity of agritourism experiences through user-generated content, testimonials, and real stories. Highlight sustainable and ethical practices to build trust among potential visitors. Plan and promote year-round agritourism experiences to attract visitors during different seasons. This may include season-specific activities, festivals, and workshops. Agritourism players must partner with local financial institutions to offer secure and convenient payment options for domestic and international visitors, including mobile money and digital wallets.

To increase awareness and brand awareness, Agritourism players must collaborate with regional and national tourism boards to promote agritourism as a part of the broader tourism industry. Participate in international travel and trade shows to showcase African agritourism offerings. They must also invest in high-quality photography, videography, and storytelling to capture the unique and captivating aspects of agritourism experiences. Share behind-the-scenes stories, farm-to-table journeys, and cultural experiences.

Involve local communities in marketing by featuring their cultural elements, traditions, and craftsmanship. Foster a sense of community ownership and pride in agritourism activities. Emphasise sustainability and responsible tourism practices in your marketing efforts. Highlight eco-friendly initiatives, support local conservation efforts, and engage with eco-conscious travellers. They must also form partnerships with other local businesses, tour operators, and travel agencies to cross-promote agritourism experiences and create package deals for visitors.

The Government also through ZTA should aid in resuscitating the life of the sector through sustainable funding towards sustainable digital and social media marketing.

By addressing these challenges and implementing these solutions, agritourism businesses in Zimbabwe can effectively use digital and social media marketing to attract visitors, boost local economies, and promote sustainable tourism practices.

Conclusion

In conclusion, digital and social media marketing can transform the agritourism sector by enhancing the visitor experience, expanding the reach of agritourism businesses, and increasing operational efficiency. These innovations can enable agritourism operators to adapt to changing consumer preferences and offer unique and engaging experiences to tourists interested in rural and agricultural experiences.

Area of Future Studies

Future research can be conducted on how to integrate digital and social media marketing strategies with conventional marketing strategies especially in remote farms where digital platforms are ineffective when targeting the local community. Research can also be conducted on tourist perception towards digital and social media marketing strategies on their purchase behaviour of agritourism products.

References

Agricultural Council of Tanzania (ACT) (2023). *Agritourism Regions for Enhancing Linkages between Tourism and Sustainable Agriculture in the United Republic of Tanzania*. Dar es Salaam: ACT.

Ajayi, C.O. and Gündüz, E. (2020). An evaluation of rural tourism potential for rural development in Kenya. *International Journal of African and Asian Studies*, 63, 35–46.

Akwii, E. and Kruszewski, S. (2021). *Defining and Regulating Agritourism Trends*. Royalton, VT: Center for Agriculture and Food Systems at Vermont Law School.

Arru, B., Furesi, R., Madau, F.A. and Pulina, P. (2019). Recreational services provision and farm diversification: A technical efficiency analysis on Italian agritourism. *Sustainability*, 9(2), 42.

Bacsi, Z. and Szálteleki, P. (2022). Farm profitability and agritourism in the EU – Does size matter? *Deturope*, 14(2), 152–171.

Baipai, R., Chikuta, O., Gandiwa, E. and Mutanga, C. (2022). Critical success factors for sustainable agritourism development in Zimbabwe: A multi-stakeholder perspective. *African Journal of Hospitality, Tourism and Leisure*, 11(SE1), 617–632.

Baipai, R., Chikuta, O., Gandiwa, E. and Mutanga, C.N. (2023). A framework for sustainable agritourism development in Zimbabwe. *Cogent Social Sciences*, 9(1), DOI: 10.1080/23311886.2023.2201025

Barbieri, C. (2010). An importance-performance analysis of the motivations behind agritourism and other farm enterprise developments in Canada. *Journal of Rural and Community Development*, 5, 1–20.

Barbieri, C., Stevenson, K.T. and Knollenberg, W. (2019). Broadening the utilitarian epistemology of agritourism research through children and families. *Current Issues in Tourism*, 22(19), 2333–2336.

BFAP (2018). *Baseline Agricultural Outlook 2018–2027. Outlook 2018–2027*. Retrieved from: http://www.bfap.co.za/wp-content/uploads/2018/08/BFAPBaseline-2018.pdf

Bhatta, K., Itagaki, K. and Ohe, Y. (2019). Determinant factors of farmers' willingness to start agritourism in rural Nepal. *Open Agriculture*, 4(1), 431–445.

Boateng, R. and Hinson, R. (2007). Perceived benefits and management commitment to e-business usages in selected Ghanian firms. Electronic Journal of Information Systems in Developing Countries. *Tourism Management*, 59, 597–609.

Buhalis, D. (2013). e-Tourism Trends And Challenges in The Social Media Era; UNWTO seminar on Tourism and New Technologies: 14-15 May 2013, San Jose, Costa Rica.

Castillo, C.A.M. and Jimber, J.A.D.R. (2018). Quality, satisfaction and loyalty indices. *Journal of Place Management and Development*, 11(4), 428–446.

Chase, L., Stewart, M., Schilling, B., Smith, B. and Walk, M. (2018). Agritourism: Toward a conceptual framework for industry analysis. *Journal of Agriculture, Food, Systems, and Community Development*, 8(1), 13–19.

Cheteni, P. and Umejesi, I. (2023). Evaluating the sustainability of agritourism in the wild coast region of South Africa. *Cogent Economics and Finance*, 11(1), 2163542.

Chigora, F., Mutambabra, E., Ndlovu, J., Muzurura, J. and Zvavahera, P. (2020). Towards establishing Zimbabwe tourism destination brand equity variables through sustainable community involvement. *African Journal of Hospitality, Tourism and Leisure*, 9(5), 1094–1110.

Chigora, F., Ndlovu, J. and Zvavahera, P. (2021). Zimbabwe tourism destination brand positioning and identity through media: A tourist's perspective. *Journal of Sustainable Tourism and Entrepreneurship*, 2(3), 133–146.

Chikuta, O. and Makacha, C. (2016). Agritourism: A possible alternative to Zimbabwe's tourism product? *Journal of Tourism and Hospitality Management*, 4(3), 103–113.

Choenkwan, S., Promkhambut, A., Hayao, F. and Rambo, A.T. (2016). Does agrotourism benefit mountain farmers? A case study in PhuRuea District, Northeast Thailand. *Mountain Research and Development*, 36(2), 162–172.

Comen, T. (2017). Critical success factors for agritourism entrepreneurs. *The 2nd International Congress on Marketing, Rural Development, and Sustainable Tourism*, 91(June 13– June 15), 399–404.

European Training Foundation (2021). *The future of skills A case study of the agri-food sector in Morocco.* Retrieved from: https://www.etf.europa.eu/sites/default/files/2021-04/future_of_skills_agri-food_sector_in_morocco.pdf

Guvamombe, I. (2019, August 3). Farm tourism is trendy, and refreshing. *Herald.* Retrieved from: 10.30892/gtg.46137-1031

Harrigan, P., Evers, U., Miles, M. and Daly, T. (2017). Customer engagement with tourism social media brands. *Tourism Management*, 59, 597–609.

Hofacker, C., Golgeci, I., Pillai, K.G. and Gligor, D.M. (2020). Digital marketing and business-to-business relation- ships: A close look at the interface and a roadmap for the future. *European Journal of Marketing*, 54(6), 1161–1179.

Jęczmyk, A., Uglis, J., Graja-Zwolińska, S., Maćkowiak, M., Spychała, A. and Sikora, J. (2014). Economic benefits of agritourism development in Poland - An empirical study. *Tourism Economics*, 21(5). 10.5367/te.2014.0

Jones, M., Kaminski, J., Christians, N. and Hoffmann, M. (2011). Using blogs to disseminate information in the turfgrass industry. *Journal of Extension*, 49(1), Retrieved from: http://www.joe.org/joe/2011february/rb7.php

Joyner, L., Kline, C., Oliver, J. and Kariko, D. (2018). Exploring emotional response to images used in agritourism destination marketing. *Journal of Destination Marketing and Management*, 9, 44–55.

Karjaluoto, H., Kefi, H., Krishen, A.S., Kumar, V., Rahman, M.M., Raman, R., Rauschnabel, P.A., Rowley, J., Salo, J., Tran, G.A. and Wang, Y. (2021). Setting the future of digital and social media marketing research: Perspectives and research propositions. *International Journal of Information Management*, 59, 102168.

Khanal, A., Honey, U. and Omobitan, O. (2019). Diversification through 'fun in the farm': analyzing structural factors affecting agritourism in Tennessee. *International Food and Agribusiness Management Review*, 23, 1–16.

Liang, A.R.D., Nie, Y.Y., Chen, D.J. and Chen P. (2020). Case studies on co-branding and farm tourism: The best match between farm image and experience activities. *Journal of Hospitality and Tourism Management*, 42, 107–118.

Matarise, R. (2023). Agritourism in Zimbabwe, Herald, April, 20230.

Mathew, V. and Soliman, M. (2021). Does digital content marketing affect tourism consumer behavior? An extension of technology acceptance model. *Journal of Consumer Behaviour*, 20(1), 61–75.

Miller, K., Settle., Q., King, A.E.H. and Kisling, B. (2021). How agritourism operators make marketing and promotion decisions. *Journal of Human Sciences & Extension*, 11(3), 1–16.

Mpiti, K. and Harpe, A. (2015). ICT factors affecting agritourism growth in rural communities of Lesotho. *African Journal of Hospitality, Tourism and Leisure*, 4(2), 1–11.

Mutami, C. (2015). Smallholder agriculture production in Zimbabwe: A survey. *The Journal of Sustainable Development*, 14(2), 140–157.

Ndlovu, J. and Heath, E. (2013). Rebranding of Zimbabwe to enhance sustainable tourism development: Panacea or villain. *Academic Journals*, 1(12), 947–955.

Nyahunzwi, D.K. (2019). The online marketing strategies of the Zimbabwe Tourism Authority (ZTA) and South Africa Tourism (SAT): A comparative study. *Journal of Tourism and Hospitality*, 8(3), 1–10.

Ollenburg, C. and Buckley, R. (2007). Stated economic and social motivations of farm tourism operators. *Journal of Travel Research*, 45(4), 444–452.

Orias, M.J.S. and Borbon, N.M.D. (2022). Adoption of digital marketing among farm tourism sites in the province of Quezon, Philippines. *International Journal of Research Studies in Management*, 10(1), 29–40.

Runyowa, D. and Nyaruwata, S. (2023). The impact of COVID-19 on rural tourism enterprises and their future growth prospects in Zimbabwe. In K. Dube, O.L. Kupika and D. Chikodzi (eds.), *COVID-19, Tourist Destinations and Prospects for Recovery*. Cham: Springer, pp. 97–111.

Sanderson, S., Nyamadzawo, J., Nyaruwata, S. and Moyo, C. (2013). *USAID Strategic Economic Research And Analysis – Zimbabwe (Sera) Program Positioning The Zimbabwe Tourism Sector For Growth: Issues and Challenges*. Harare: Nathan Associates Inc.

Shin, H. and Perdue, R.R. (2022). Hospitality and tourism service innovation: A bibliometric review and future research agenda. *International Journal of Hospitality Management*, 102, 103176.

STATSSA (2023). *How important is tourism to the South African Economy*. Retrieved from: http://www.statssa.gov.za/?p=11030

Stelzner, M.A. (2018). *2018 Social media marketing industry report: How marketers are using social media to grow their business*. Retrieved from: https://www.socialmediaexaminer.com

Suhartanto, D., Dean, D., Chen, B.T. and Kusdibyo, L. (2020). Tourist Experience With Agritourism Attractions: What leads to loyalty? *Tourism Recreation Research*, 45(3), 364–375.

Tew, C. and Barbieri, C. (2012). The perceived benefits of agritourism: The provider's perspective. *Tourism Management*, 33(1), 215–224.

Veeck, G., Che, D. and Veeck, J. (2006). America's changing farmscape: A study of agricultural tourism in Michigan. *The Professional Geographer*, 58(3), 235–248.

Vuori, V. and Okkonen, J. (2012). Refining information and knowledge by social media application: Adding value by insight. *VINE*, 42, 117–128.

Wiig, A. (2003). Developing countries and the tourist industry in the Internet age: The Namibian case. *Forum for Development Studies*, 30(1), 59–87.

Wijijayanti, T., Salleh, N.H.M., Hashim, N.A., MohdSaukani, M.N. and Abu Bakar, N. (2023). The feasibility of rural tourism fostering real sustainable development in host communities. *Journal of Tourism and Geosites*, 46(1), 336–345.

Wong, G., Westhorp, T., Westhorp, G., Buckingham, J. and Pawson R. (2013). RAMESES publication standards: Meta-narrative reviews. *BMC Medicine*, 11(12).

Yasmin, A., Tasneem, S. and Fatema, K. (2015). Effectiveness of digital marketing in the challenging age: An empirical study. *The International Journal of Management Science and Business Administration*, 1(5), 69–80.

Zheng, C., Yuan, J., Zhu, L., Zhang, Y. and Shao, Q. (2020). From digital to sustainable: A scientometric review of smart city literature between 1990 and 2019. *Journal of Cleaner Production*, 258, 120689.

Zvavahera, P. and Chigora, F. (2023). Agritourism: A source for socio-economic transformation in developing economies. *Qeios*. doi:10.32388/DXTYIG

11 Exploring How Disruptive Technological Innovation Can Positively Impact Agritourism in Developing Countries

Echoes from Zimbabwe

Samuel Musungwini, Edward Mudzimba, Mercy Dube, and Sinothando Tshuma

Introduction

In recent years, the African tourism sector has expanded significantly, making a considerable economic contribution to individual countries on the African continent and to the world at large. For many years, tourists have flocked to Zimbabwe, which is famed for its diverse wildlife, fauna, flora, and rich cultural heritage. However, since the fast-track land reform initiative of 2000, the number of tourists has dropped sharply. The development of COVID-19 aggravated the already grave position as it led to a global loss in income of more than 50% worldwide, as reported by Abbas et al. (2021). However, disruptive technological innovations have brought some renewed impetus, and the traditional tourism landscape is transforming the world (Buhalis et al. 2019). Jafari-Sadeghi et al. (2021) suggest that technological entrepreneurship and technological market expansion have been taking place in the tourism sector in recent years and this is important since these concepts are part of the dynamic capabilities that help in embracing digital innovation in the country's national level. This book chapter investigated how disruptive technological innovation could be leveraged to positively influence Zimbabwe's travel and tourism sector. The advent of disruptive technologies that radically alter how companies conduct business and provide goods and services has brought with it mixed fortunes for different contexts. Disruptive technologies could completely transform the tourist industry, from pre-trip preparation and reservation through on-site experiences and post-trip feedback (Musungwini et al., 2022; Woodhead et al., 2018). The development of digital platforms, mobile applications, and virtual reality has provided new prospects for Zimbabwean tourism enterprises to expand their markets and improve the overall travel experience.

This is because disruptive technological innovations have widely been embraced in many sectors across a broad spectrum of industries. These innovations have a great capacity to disrupt the status quo, change the way people live and work, reorganize value chains, and promote the emergency of

DOI: 10.4324/9781032696188-11

entirely new products and services as posited by Dube et al. (2023). According to McKinsey and Company (2013), the Print media industry today is in a life-and-death struggle to remain relevant in a world of instant, online news and entertainment. The Transport and Entertainment industries have and continue to experience significant disruptions from digital technology, leading to redefining every stage of consumer experiences. Indeed, the competition landscape has now been tilted in favor of corporates that have embraced technological advancements and race to create next-generation solutions. Within the Tourism industry, for instance, Benckendorff and Shu (2019) claim that new technologies have the potential to disrupt existing service offerings and force a re-imagining of the visitor experiences and the end-to-end visitor journey. In this chapter, the authors look at the various disruptive technological innovations and explore how these innovations can positively impact tourism in Zimbabwe. We make the case that business executives in the Tourism industry in emerging nations like Zimbabwe shouldn't wait until disruptive technological innovations negatively affect their operations. Instead, they should adopt a proactive strategy while fully cognizant of the possibility that distinctive competencies upon which they may have based their strategies could at any time be undermined or strengthened.

The main objective of our book chapter is to examine the role of disruptive technological innovation in the Agritourism context, with a specific focus on Zimbabwe. Our chapter aspires to achieve the following objectives:

a To provide an overview of the current state of the tourism industry in Zimbabwe and its potential for growth through agritourism.
b To identify and analyze the disruptive technological innovations that are transforming the tourism sector globally.
c To explore the impact of these disruptive technologies on different stakeholders in the tourism industry, including tourists, businesses, and the government.
d To highlight the challenges and opportunities presented by disruptive technological innovation in the Zimbabwean tourism industry.
e To propose recommendations for tourism businesses and policymakers to effectively leverage disruptive technologies for sustainable tourism development in Zimbabwe.

Theoretical Development

The concept of disruptive technological innovation owes its roots to the pioneering work of Christensen (1997) in the disruptive innovation theory. Christensen coined the term disruptive technology and later replaced it with disruptive innovation Christensen and Raynor (2003) to widen the applicability of the theory to encompass services and business model innovations. In this chapter, we take disruptive technological innovations to encompass the new emerging technologies that provide different values from mainstream technologies and are initially inferior to mainstream technologies along the dimensions of performance that are most important to mainstream customers. These technological innovations are capable of changing performance metrics or consumer expectations of the market by providing radically new functionality, and discontinuous technical standards. We explore some of the main disruptive technological innovations including artificial intelligence, mixed reality, virtual reality, augmented reality, metaverse, emerging payment platforms as well as the Internet of Things, and examine how players in the Zimbabwean tourism industry could embrace these emerging technologies to their advantage.

Tourism is a key industry contributing immensely to economic development across many nations globally. If investments are channeled towards the efforts targeted at amending any potential mishaps, Mzobe et al. (2023) allege that this industry can turn around economies. This industry has been cited as the largest in the world and rated as one of the most developed industries. This is demonstrated by the industry's annual average growth rate, which ranges between 4–5% (Nyagadza et al., 2022), its 10% contribution to global employment, and an 8% contribution to GDP at a global level (Nyagadza et al., 2022). The appropriateness of this industry as an ideal target for economic transformation lies also in its diverse nature. Meschede (2020) asserts that this industry has an integrating effect that permeates all areas of the economy. Therefore, we contend in this chapter that through the adoption of our recommended strategies, not only will the players in the Tourism industry of Zimbabwe reap the positives from disruptive technological innovations but all value chain players and, ultimately, the economy in its entirety.

Past researchers, among them Abbas et al. (2021), Fernandes (2020) and Novelli et al. (2018) concur on the fact that, had it not been for the deadly impact of the COVID-19 pandemic which curtailed the performance of this industry, its growth and promise would have transcended into the current and future periods. Sadly again, this industry is very slow in recovering from pandemics of any kind (Novelli et al., 2018). Countries whose economies rely on tourism like Zimbabwe are not immune to the challenges currently affecting the industry at the Global level. In Zimbabwe, for instance, the performance of the Tourism industry in terms of tourist arrival fell by 504% in the year 2021 compared to 2019 before the COVID-19 pandemic (Baipai et al., 2023). This industry however remains a significant player in the Zimbabwean economy and in 2019 contributed to 5.8% of the country's GDP and 8.6% of the country's total employment (Fernandes, 2020). Due to the negative impact of the COVID-19 pandemic, the contribution to employment and the country's GDP fell to 7.1% and 4.1%, respectively (Monitor, 2020).

It's important to note however that Zimbabwe is blessed with tourist attractions that attract visitors from across the world. More importantly, the country is home to the Mighty Victoria Falls and Great Zimbabwe Ruins, popular tourist attractions that have been accorded world heritage status by the UNESCO World Heritage Organization. The country's tourism industry enjoyed tremendous growth since gaining independence in the year 1980 to 1999 with average tourist arrival growth and tourism receipts averaging around 17.5% and 18%, respectively. Since the year 2000, the Zimbabwean tourism industry witnessed a huge decline in tourist arrivals owing to political instability and poor country image which had been triggered by the country's land reform programme (Ngarava, 2020). Apart from the land reform programme, the government of Zimbabwe blamed the decline on illegal sanctions that were imposed on the country by Britain (Government, n.d.). The situation was further aggravated by the issuance of travel warnings against the country by the United Kingdom and the United States of America. This presented a significant blow to the industry considering that these countries are the major source markets for Zimbabwean tourism. In the year 2005, for instance, the country witnessed a total of 112,608 tourist arrivals from Europe and the USA markets compared to a total of 400 million who arrived in the year 1991.

The plight of the Zimbabwean industries, in general, would be incomplete without mentioning the never-ending economic crises that continue to make headlines across the world. The annual Inflation figure is currently pegged at 473%, becoming the second highest in the world after Venezuela (Hanke, n.d.). We argue in this chapter that Zimbabwe, through its diverse tourism product which encompasses its people, culture, history, and natural resources if proactively managed, considering a myriad of benefits that can be brought by harnessing disruptive technological innovations, can significantly assist in the transformation of the economic fortunes of the country. Zimbabwe is an agrarian economy, with 70% of its population receiving direct income and employment from agriculture. Our chapter zeroes in on the potential of agritourism to transform the fortunes of the country by leveraging the emerging disruptive ICT technologies. Agritourism is a segment of tourism focusing on visiting farm operations for recreation, leisure, education, or involvement in farm activities (Chatterjee and Prasad, 2019; Yamagishi et al., 2021). Past studies support the idea that, in contemporary tourism, there is a growing segment of the tourist market now tired of the same old sun, sea, and vacations and now crave meaningful journeys fostering people-to-people connections with local people working and living in destinations, as submitted by Chikuta and Makacha (2016) and echoed by Barbieri (2020) and Baipai et al. (2023).

Past studies have confirmed that agritourism can be an alternative tourism product in Zimbabwe as a follow-up to its agrarian reform, which has seen land ownership transferred from a minority to the hands of a significant majority of its local people (Scoones et al., 2011). The authors are however aware of the fact that the concept has not yet gained much traction in the country owing to several challenges revolving around the lack of capacity-building programs (Baipai et al., 2023; Guvamombe, 2019). We believe that for a country that is a popular destination for tourists and supported by an agro-based economy, agritourism can be a viable tourism product especially if players find ways of harnessing disruptive technological innovations to their advantage. We are however cognizant of the fact that disruptive technological innovations, by their very nature, are difficult to predict in so far as their likely impact on industries. However, considering evidence from other industries, particularly the entertainment industry, we believe that these technological innovations if proactively managed and harnessed can provide a springboard for the Agritourism sector of Zimbabwe to become a significant player in the country's quest to turn around its economy.

Methodology

Methodologically, a qualitative approach was used to construct this piece of work (Lub, 2015) and thus, our book chapter is based on a comprehensive review of existing literature on the evolution of agritourism, disruptive technological innovation, and its potential impact on the tourism industry. The primary data was collected through interviews with key stakeholders in the Zimbabwean tourism industry, including government officials, tourism business owners, tourists, and ICT experts. Table 11.1 on the next page provides demographic data profiles of the research participants. The research is qualitative in nature; hence, the data was analyzed using content analysis to identify key themes and patterns.

Literature Review

This section of the chapter focuses on a review of literature relevant to how disruptive technological innovations can positively impact agritourism in Zimbabwe. Literature related to agritourism and disruptive innovations will be explored. There will also be a review of the literature on how disruptive technological innovations in the agritourism sector can enhance consumer interaction. Challenges and opportunities for leveraging disruptive technologies in the agritourism sector will also be explored in this section. To come up with the research the researchers searched the major databases for relevant articles namely Elsevier, Science Direct, Thompson Reuters, Google Scholar and Scopus. This was done using the following keywords 'Disruptive/ Emerging technology and Innovation in Agritourism', 'Digital technology impact on Agritourism in Developing countries' and all of these threads were suffixed with 'Sub-Saharan Africa'.

Agritourism: An Emerging Field in the Tourism Sector

From the background section of this chapter, it is clear that agritourism is an emerging wing of the tourism sector that has gained prominence in both developed and developing countries. This growth has been motivated by an urge to re-create farms as agritourism destinations to fulfill intrinsic desires of achievement and locus of control and seize the opportunities brought by

Table 11.1 Demographic data of research participants

Sector	Representatives	Age	Gender	Highest academic qualifications attained	Date of interview	Interview method	Duration of Interview	Experience
Tourism Experts	Tourism academic professor	56	Male	MBA	21 August 2023	Face-to-face	48 Minutes	27 Years
	Tourism operator	36	Female	BCom Tourism	29 August 2023	Face-to-face	44 Minutes	12 Years
	Tourism practitioner	32	Male	MCom Business Management and Corporate Strategy	30 August 2023	Face-to-face	46 Minutes	10 Years
	Tour guide	27	Male	BCom Business Management	19 August 2023	Face-to-Face	40 Minutes	5 Years
	Tour guide	25	Female	Bachelor of Business Administration	19 August 2023	Face-to-face	49 Minutes	6 Years
Marketing professionals	Marketing Manager	42	Female	MCom Marketing and Corporate Strategy	21 September 2023	Face-to-face	45 Minutes	18 Years
	Marketing Manager	32	Female	BCom Marketing	21 September 2023	Face-to-face	48 Minutes	9 Years
Agricultural development experts	Agricultural development officer	34	Female	BSc Agriculture	25 September 2023	Face-to-Face	39 Minutes	13 Years
	Rural agriculture planning officer.	42	Male	MSc Agriculture and Wildlife Management	25 September 2023	Google Meet	45 Minutes	17 Years
ICT Experts	ICT Academic professor	52	Male	Ph.D. in Information Systems	31 October 2023	Face-to-face	49 Minutes	27 Years
	ICT Engineer	39	Male	MTec Computer Systems Engineering	31 October 2023	Face-to-face	46 Minutes	12 Years
	Digital Transformation expert	31	Female	MSc Information Systems Management	31 October 2023	Google Meet	42 Minutes	8 Years

extrinsic factors of market-driven, profitability, as well as economic impact. As a fusion concept uniting the fundamentals of the tourism and agriculture industries, agritourism can be exploited as an organization of special interest tourism, emphasizing the distinctive travel experiences and activities that people can have in rural settings. Agritourism is a subsector of the tourism industry and a subset of rural tourism that uses a diverse array of farms as tourist destinations (Philips et al., 2010). It supports farm activities targeting tourists, including farm demonstrations, farm visits, farm training, as well as on-site value-adding and sales of farm products. Agritourism occurs on a working farm or agricultural plant and is conducted for the visitors' enjoyment and to generate supplemental income for farmers. Agritourism and nature tourism farms might include a diverse array of activities such as outdoor recreation, farming experiences, cultural entertainment, hospitality services, and on-farm direct sales.

The term agritourism has often been used alternatively with farm tourism, farm-based tourism, and rural tourism (McGehee and Kim, 2004; Clarke, 1999; Ilbery et al., 1998; Roberts and Hall, 2001; Barbieri and Mshenga, 2008). Agritourism may be characterized as "rural enterprises that integrate both a working farm environment and a commercial tourism element" (McGehee et al., 2007). Barbieri and Mshenga (2008) referred to agritourism as any practice developed on a working farm to attract visitors. Farm enterprise diversification has become an approach for small farms to remain practical, particularly during high risks facing situations of modern-day farming. Agritourism is a set of tourism-related activities operating on agriculture, applied in rural areas either in groups or individually (Fahmi et al., 2013). With farmers becoming more market-orientated in response to declining farm incomes and rural restructuring (Meert et al., 2005; OECD, 2009), the reorientation of farming away from production to more entrepreneurial models of agriculture (Phillipson et al., 2004) emerged. Thus, as diversification becomes an almost anticipated practice, farmers are being accepted as entrepreneurial, having to develop new skills and competencies to remain sustainable. In both developing and developed countries, agritourism is now dubbed the new economic game-changer and one of the propellers of economic expansion, especially in rural areas.

Agritourism Trends and Developments in Zimbabwe

Agritourism, the practice of inviting tourists to rural areas to experience and engage in farming and agricultural activities, has gained popularity world-wide as a means to diversify income sources for farmers and promote rural development. In the case of Zimbabwe, the concept of agritourism has been explored in both academic literature and government initiatives. Agritourism in Zimbabwe often integrates cultural and culinary experiences. Traditional food, music, and cultural practices play a crucial role in attracting and engaging tourists.

Disruptive Technological Innovations in the Agritourism Sector

Agritourism is a niche within the broader tourism industry that involves visitors engaging in agricultural activities or experiencing rural life. Disruptive technological innovations have had a significant impact on this sector, enhancing the visitor experience, increasing the efficiency of operations, and expanding the reach of agritourism businesses. The most common disruptive technological innovations that have had a great impact on the sector include digital marketing and social media, mixed reality technologies, online booking platforms, and mobile applications, among others. These are discussed below:

Digital Marketing and Social Media: Agritourism operators have increasingly harnessed the power of digital marketing and social media to reach a wider audience. Online platforms, such as Facebook, Instagram, and Twitter, have enabled farms and rural destinations to promote their offerings, attract visitors, and engage with their target market effectively. These tools have disrupted traditional advertising methods and expanded the scope of agritourism.

Virtual Reality (VR) and Augmented Reality (AR): VR and AR technologies have transformed the agritourism experience. They allow visitors to take virtual farm tours, interact with 3D models of agricultural processes, and experience rural life without actually being present. This disruptive innovation enhances the accessibility and engagement of agritourism destinations for a broader range of people. Mixed reality (MR) is a technology that blends elements of both virtual reality (VR) and augmented reality (AR) to create immersive and interactive environments (Dube et al., 2023). The application of MR to agritourism can enhance the visitor experience and offer new opportunities for education, entertainment, and engagement.

Online Booking Platforms: The emergence of online booking platforms, such as Airbnb, Booking.com, and specialized agritourism booking websites, has made it easier for tourists to discover and book farm stays and other rural experiences. This innovation has increased the visibility of agritourism businesses and streamlined the reservation process, thereby disrupting traditional booking methods.

Mobile Apps: Agritourism mobile apps offer information on farm locations, events, and activities, as well as provide interactive maps and navigation to rural destinations. These apps have disrupted the way visitors access information and navigate through agritourism sites, enhancing their experience.

Smart Farming Technologies: Smart farming technologies, including sensors, drones, and data analytics, have improved the efficiency of agricultural processes. Agritourism businesses have adopted these innovations to provide educational tours and interactive experiences related to precision farming, thereby enhancing the educational aspect of agritourism.

E-commerce and Farm-to-Table Sales: E-commerce platforms have disrupted the traditional farm-to-table supply chain by allowing consumers to purchase fresh produce directly from farms. Agritourism operators have leveraged these platforms to sell their products to a wider customer base, cutting out intermediaries and increasing their revenue.

Educational Platforms: Online courses, webinars, and educational content platforms have allowed agritourism destinations to provide educational experiences to a global audience. This innovation has disrupted the traditional model of on-site education, expanding the reach and impact of agritourism education.

Blockchain for Transparency: Some agritourism businesses have adopted blockchain technology to provide transparency in the supply chain. This technology allows visitors to trace the origin of the products they consume, fostering trust and authenticity in agritourism experiences.

Sustainable Practices through Technology: Agritourism destinations have integrated sustainable technologies such as renewable energy, recycling systems, and eco-friendly practices. These technologies not only enhance the sustainability of agritourism but also attract environmentally conscious tourists.

Robotic Farm Assistants: The use of robots in agriculture has created opportunities for agritourism businesses to showcase cutting-edge technology and educate visitors on the future of farming. Robotic assistants for tasks like harvesting and weeding are increasingly common on farms and are incorporated into agritourism experiences. As a result, disruptive technological innovations have transformed the agritourism sector by enhancing the visitor experience, expanding the reach of agritourism businesses, and increasing operational efficiency in developed economies. These innovations have enabled agritourism operators to adapt to changing consumer preferences and offer unique and engaging experiences to tourists interested in rural and agricultural experiences.

Disruptive Technologies and Consumer Interaction in the Agritourism Sector

Agritourism, the intersection of agriculture and tourism, has gained significant attention as a sustainable and economically viable practice in recent years. Disruptive innovations in agritourism play a pivotal role in transforming the industry to be more sustainable, technologically advanced, and consumer-centric. The thrust of all these innovations is to enhance consumer experiences in the agritourism sector. Globally, mobile cellular subscriptions have been growing over recent years. Between 2013 and 2018, there were 1 billion new mobile subscribers and 67% of the world population is now subscribed to mobile services. Much of this recent growth has been driven by countries in Africa, Asia, and the Pacific. Access to computers and the internet has also been increasing in LDCs and developing economies. This increase in digital transformation has promoted the adoption of disruptive technological innovations in every sector, and the agritourism sector has not been left behind. The adoption of disruptive technological innovations has made it possible for the sector to gain a competitive advantage and be able to give the best experiences to tourists. While disruptive innovations in agritourism are still evolving, they offer exciting opportunities for sustainable growth in the industry. The consumer experience can be enhanced through the following:

Digital Platforms and AgTech

Digital innovations can assist developing countries in overcoming global poverty and hunger quickly in rural areas (Padhy et al., 2022). Digital platforms and agricultural technologies (AgTech) have revolutionized agritourism. These innovations encompass various aspects, such as farm management software, online booking systems, and mobile apps (Musungwini et al., 2023). They connect farmers directly with consumers, making it easier for tourists to access and experience agricultural activities. The literature discusses the benefits of these digital platforms in improving the visitor experience and the efficiency of agritourism operations. AR and VR technologies have been used in agritourism to provide consumers with immersive experiences. For instance, virtual farm tours or augmented reality apps can enhance the visitor experience by providing additional information or interactive features during tours.

Farm-to-Table Experiences

Disruptive innovations in the food industry, such as the farm-to-table movement, have had a significant impact on agritourism. Consumers are increasingly interested in the origin of their food, leading to a demand for authentic, locally sourced experiences. Farmers who can provide farm-to-table experiences through tours and dining options are often at the forefront of sustainable agritourism.

Eco-Friendly Practices

Sustainability is a central theme in agritourism, and disruptive innovations in sustainable farming practices are crucial. There are various eco-friendly innovations, including organic farming, permaculture, and the use of renewable energy sources. These practices not only reduce the environmental impact but also attract tourists seeking authentic and sustainable experiences.

Education and Engagement

Innovations in educational agritourism experiences are gaining prominence. Hands-on workshops, classes, and interactive tours provide tourists with insights into farming practices and environmental stewardship. These initiatives not only engage visitors but also promote a deeper understanding of sustainable agriculture.

Collaborative and Sharing Economy

The sharing economy model is also making inroads into agritourism. Peer-to-peer platforms allow farmers to share their land or resources for agritourism activities, opening up new possibilities for rural communities and encouraging sustainable land use. Different farmers can collaborate and share resources, enhancing user engagement and experiences.

Sustainable Accommodation

Agritourism often involves overnight stays on farms, and innovative accommodation options are emerging. Sustainable lodging, such as eco-friendly cabins or glamping sites, adds to the overall appeal of agritourism destinations. Most consumers are now more enlightened and hence require environmentally friendly services.

Storytelling and Branding

Effective storytelling and branding strategies are crucial to attracting tourists. Disruptive innovations in marketing and communication help farms create compelling narratives and build strong brands, which can differentiate them in the competitive agritourism market. Agritourism entrepreneurs can capitalize on different social media platforms that allow consumers to interact and share their experiences at the different farms that they have visited. The sharing of stories and experiences on different platforms will thus attract more visitors to the destinations.

Data Analytics and Customer Insights

The use of data analytics and customer insights is another disruptive innovation in agritourism. Digital platforms enable the tourism sector to understand consumer preferences and behaviors, such that farms can tailor their offerings to meet specific demands, enhancing the visitor experience and overall sustainability. Tailoring offerings to individual preferences through data analytics or offering apps that provide personalized recommendations can enhance consumer engagement. Consumer insight can be obtained by tracking the online activities of the visitors using cookies.

Opportunities for Disruptive Technologies Adoption in the Agritourism Sector

Disruptive technologies have redefined the tourist's lifestyle, as suggested by Rai and Pinto (2022), leading to the creation and rise of the agritourism sector as a new model in the digital arena in times of disruptions Adom et al. (2021). This has resulted in the development of new innovative tourism products as a result (Colombo and Baggio, 2017), effective real-time booking (Chang et al., 2019), and increasing customer satisfaction as the customer journey can be tracked and feedback on experience examined (Fragnière et al., 2022). According to Ranganai et al. (2023), disruptive technologies in the tourism sector in South Africa and Zimbabwe have created more opportunities for the tourism industry, such as anticipating visitor tour needs and wants, creating beneficial technological-based artificial product exchanges such as online bookings and e-payments, and e-service delivery effectively and efficiently beyond the customers' expectations. On another dimension, one of the reasons that has led to the continued struggles in the revival of the tourism sector in Zimbabwe is the lack of technological support innovations (Musasa and Mago (2014), and as per the recommendations by Baipai et al. (2023), the need for ICT-based investments as a sustainable development strategy for agritourism sector revival. When the system has technological innovation support, the examination of customer buyer behavior is made easy and competitor analysis effective (Fragnière et al., 2022; Kolajo et al., 2019).

The major opportunities the disruptive technologies have brought in the sector are their ability to create a stakeholder community through various evolving digital platforms, which are mainly geared toward improving the service outcome in the tourism sector (Stylos et al., 2021). Past delighted tourists, through their positive reviews, have created an increased platform of new subscribers and new future tourists for the destination. The rise in technological innovations has also made marketing research easy and corporations to manage their brands online (Pereira et al., 2022), improve tourist retention (Fan et al., 2015), analyze potential visitor expectations before they visit the tourism site (Zhang et al., 2019), create an opportunity to make predictions of bookings to avoid underbooking or overbooking (Duft and Durana, 2020), lead to real-time decision-making before the problem is acute (Ranganai et al., 2023), facilitate touring route planning, lower customer tracking and management costs, and create a lucrative investment opportunity for sustainable development of the sector (Baipai et al., 2023).

The disruptive technologies have helped in creating a less congested and flexible virtual tourism platform for many tourists, as they can be involved in different activities depending on their likes. In agritourism, some can be doing virtual farm tours, virtual food preparation, packaging and tasting (Garibaldi and Pozzi, 2020). It has also been noted that many individuals who subscribe to virtual tours and later physical agritourism tours are mainly digitally literate consumers who would have been exposed to the farms' social media sites or websites, among other platforms (Liu et al., 2020; Hardesty and Leff, 2020). Many consumers track consumer reviews before visiting a destination (Sigala, 2020), and farms have used these social media platforms to correct tourists' misconceptions about the destinations and build their agritourism sites' reputations (Sigala, 2020). Many tourists are convinced that agritourism is safer than just other tourism ventures.

According to Chin and Musa, the use of the Al-powered app has helped in the easy management of tourists when they visit by checking their health status to see whether they do not have contaminable diseases that may risk other tourists' lives. The future benefits of evolving technologies in the tourism sector are still being explored with many borrowing from the lessons learned during the COVID-19 pandemic; this is because the pandemic created technologies for remote management of sites, remote tour guiding and digital home-based tourism adventures for many.

Challenges of Disruptive Technological Innovations in the Agritourism Sector

The rise in COVID-19 led to the fatal collapse of many businesses, especially in the tourism sector (Chin and Musa, 2021), creating perennial challenges that some economies are still crafting strategies to alleviate. In the agritourism sector, disruptive technological innovations did not only create opportunities but also challenges that the sector should address to survive the inevitable economic restructuring phase. The rise in the use of digital technologies in the tourism sector has intensified machine-based competition between tourist resort centers (Vanhove, 2017), and the rise in digital automation can lead to the extinction of some firms in the industry (Vanhove, 2017). The new digital models created have led to the new customer tourists' needs, with the rise in customer e-services that many resort centers cannot adopt and commercialize (Kenney, 2017) due to the country-based technological levels. The tourism sector invested in many digital technology platforms to ensure tourists still enjoyed tourism artefacts during and after the peak of COVID-19 in the onset and offset of travel restrictions (Chin and Musa, 2021). The digital platforms ensured that tourists embarked on virtual tours, remote social media-based team touring, remote partying, and farm touring (Garibaldi and Pozzi, 2020). Some online activities reminded tourists of the agritourism sites that they should visit after the pandemic for mental and physical healing through nature tours (Chin and Musa, 2021).

Many tourist resort areas installed contact tracing apps to track the tourists from the time they visit and the time they check out of the resort area, in the quest to provide up-to-date data about the health status of the client during the COVID-19 peak. These platforms were not abandoned when COVID-19 started to surge, as they proved effective for online customer database tracking. The application was however criticized for intruding into the tourists' privacy and making the person vulnerable to online digital cyber crimes in the future, as the application permitted access by many sectors of the COVID-19 response team (Gasser et al., 2020).

The usage of disruptive technologies that were mainly invested in during the COVID-19 era in the agritourism sector created more challenges than anticipated opportunities, as their ethical, social, political, and legal implications still need to be revisited. Location tracker apps have always been criticized for personal life intrusion (Toh and Brown, 2020). The longevity and future reliability of these applications are questionable as applications are now being modified to suit the agritourism industry marketing activities that they were not solely designed to execute. On the other hand, human rights activism groups have for long queried the usage of private data for other purposes besides what it has been gathered for without the customer's consent (Human Rights Watch, 2023). The virtual e-guided tours create flexible off-campus tours for a destination that can be physically toured in the future; however, some tourists have resorted to online touring rather than physical tours, reducing the anticipated future revenues (Garibaldi and Pozzi, 2020).

Musa and Chin articulate that virtual reality tours are bound to replace traditional physical tours in the future and are bound to make many future tourists skeptical of visiting the agritourism site before a digital tour of the site. This, in turn, forces every agritourism firm to subscribe to digital marketing platforms as their future survival strategy.

The main challenges of disruptive technologies also include network issues and technical complications, loss of the competitiveness of ancient tourism attractions (Zaman et al., 2017) due to technological modification of the tourism physical and virtual destinations (Fragnière et al., 2022), redefining the tourism industry staff roles and some being laid off as their duties are possessed by the digital technologies (Zheng et al., 2018). On the other hand, they are expensive to maintain from a company perspective; they create a permanent financial expense with the need to update the system regularly. The disruptive technologies also require staff training on how to use the technology for resort benefit; also deceptive marketing with many artificial animations of the agritourism sites havs led to an outcry of many tourists after physically visiting the destination (Preko and Anyigba, 2022; Fragnière et al., 2022).

It can also not be undermined those digital disruptions in the tourism sector dominated by digital accommodation, digital transportation, digital shared food economy services, digital travel planning, and activities have led to the death of some internal sectors within the tourism companies as tourists can now book resorts and digitally outsource food during their touring days. These technologies have also led to the loss of government tax revenue as they do not have frameworks to tax the outsourced services (Adeyinka-Ojo and Abdullah, 2019).

How Can Emerging Digital Technologies Positively Impact Agritourism in Zimbabwe?

Since agritourism is a fusion of agriculture and tourism together, it therefore offers inimitable experiences for tourists who may be seeking rural charm and authentic connections with nature which is abundantly offered by Zimbabwe. Choi et al. posit that Information Technology provided a buffer for the tourism industry during the COVID-19 pandemic. It was like a helmet for those who had invested in ICTs in their operations, which is why disruptive technological innovations are deemed to have the potential to revolutionize this sector by enhancing such aspects as marketing, efficiency, sustainability, and overall visitor engagement to the rural areas of Zimbabwe. The literature analysis suggests that rural farmers can integrate and employ disruptive technologies by implementing smart farming techniques, which include the Internet of Things, precision agriculture practice, data analytics, optimizing resource allocation, reducing environmental impact and increasing productivity. These practices can then be promoted through digital electronic platforms, including social media platforms like Facebook, Twitter and YouTube, among others. We, therefore, believe that the digital platforms can also facilitate direct sales of local produce, handicrafts, and tourist experiences thereby expanding the agritourism operators' market reach and this may result in the generation of additional revenue streams for operators, the tourism industry and the country at large. The adoption and use of online booking systems can also improve convenience for tourists, tour operators and businesses, thereby enabling seamless planning and reservations 24/7. Disruptive technology can further enable Zimbabwean agritourism players to automate and streamline their agritourism activities and tasks, allowing operators to focus on value-added activities and unique personalized tourist experiences for every visitor.

The literature also suggests that disruptive emerging technologies can be a vehicle for agritourism in Zimbabwe enabling the usage of Artificial Intelligence (AI) to facilitate the use of Virtual Reality (VR) and Augmented Reality (AR), thereby affording tourism operators with possible innovative solutions to afford Immersive Virtual Reality experiences to transport visitors to virtual agritourism farms and showcasing their agricultural practices in their farms and villages and rural life experiences. This innovation enables the tourists to have a sense of feeling, generate curiosity, and appeal to many international and local tourists even before their physical arrival. This can result in more and more bookings being realized by the agritourism operators, the tourism industry, and the country at large. AR overlays digital information onto real-world environments, providing interactive learning opportunities about crops, livestock, and ecosystems to Zimbabwean urban dwellers and international tourists.

There are a good number of smallholder farmers who practice agriculture using organic manure and these are an exceptional breed of farmers who may turn to digital disruptive technologies like blockchain for traceability. Such farmers may venture into the agritourism business and create virtual farms for educational purposes and promote sustainable agritourism practices. This is because blockchain technology safeguards transparency and traceability of production throughout the agricultural supply chain. This has the effect of assuring consumers about the authenticity of origin and quality of agricultural produce from farms through the value chain up to the table. This has the effect of fostering trust and encouraging conscious consumption, benefiting both farmers and agritourism businesses. We believe that this data-driven approach may contribute to sustainable business practices and informed decision-making. Therefore, embracing these emerging disruptive digital technological innovations can be a boon to the agritourism business; hence, it can become more resilient, sustainable, and appealing to international tourists and urban dwellers in Zimbabwe. This can ultimately lead to a more positive impact on rural communities and the overall tourism industry in the country.

Empirical Data from Zimbabwe

We interviewed 12 people in total. Of these, there is one tourism academic, two tourism experts, two tour guides, two marketing professionals, two ICT experts, and two rural agricultural experts, one of them operating around Tokwe Mukosi. The demographic profiles of the research participants have been provided in Table 1. The research participants were required to provide an overview of the current state of the tourism industry in Zimbabwe and its potential for growth through the agritourism branch. According to the interviewees, the Zimbabwe tourism sector has historically made a significant economic contribution to the country's fiscus over the years until the fast-track land reform program of 2000. These events led to a sharp decline in tourist arrivals and the subsequent collapse of the country's economy. However, the research participants were positive that things were beginning to show positive signs, and on the horizon, growth opportunities were looming over the sector, and it may be a matter of time before it reclaims or even better its past glory.

The interviewed experts believe that the government and the tourism industry players should take advantage of opportunities presented by the situation like the completion of the Tokwe Mukosi Dam. They believe that this presents the opportunity to develop and grow tourism variants like agritourism and growth offers even more opportunities. The thawing of Zimbabwe's relations with the international community presents an opportunity to be exploited by the government and industry to awaken this sleeping giant. Historically, Zimbabwe was a popular travel destination because of its abundance of natural resources, varied animals, and dynamic culture, and this needs to be rekindled for the betterment of the country. Notwithstanding the challenges that may be faced by the sector, agritourism is believed to have a bright future in Zimbabwe, and it could complement traditional tourist types, including wildlife safaris, adventure travel, and cultural tourism, and equally play a major role in Zimbabwe's tourism economy. Over the years, the traditional variants of tourism have been effective in drawing tourists to the nation, but to maintain and expand the sector, diversification is required.

The participants believe that visiting farms, ranches, and other agricultural enterprises to experience rural life and take part in agricultural activities may just be promoted to drive agricultural tourism. With most people working in agriculture, Zimbabwe boasts a robust agricultural economy, and the completion of the Tokwe Mukosi dam is bound to drive the agricultural revolution in Masvingo province. The dam itself presents a lot of opportunities for tourism-inclined activities like tiger fishing tournaments, canoeing, and fish farming, among others. As a result, this offers the country an opportunity to establish agritourism as a novel and supplementary kind of travel. The ability of agritourism to provide employment and strengthen the local economy is one of its key benefits. Zimbabwe may take advantage of the enormous potential of its agricultural industry and generate jobs for the local populace by fostering and growing agritourism. This will help the travel and tourism sectors as well as the general prosperity of the country.

Research participants suggested that the enactment of agritourism may attract a peculiar visitor to the sites, as agritourism can be attractive to domestic travelers, even if traditional tourism tends to draw more foreign visitors. This will lessen the nation's dependency on foreign tourists and assist in diversifying the tourism industry. Moreover, agritourism may benefit the environment in other ways, as has been seen with the promotion of tried and tested traditional herbs like the popular 'muchemedza mbuya' found at Great Zimbabwe village. Agritourism may contribute to the preservation of Zimbabwe's natural resources and wildlife by encouraging sustainable agricultural methods and teaching tourists about the value of conservation. In addition to helping the environment, this will improve the nation's standing as an ethical and sustainable travel destination.

However, participants were quick to point out that for agritourism in Zimbabwe to realize its full potential, there are a few issues that must be resolved. They decried the absence of amenities and infrastructure in rural areas, where most agricultural activities are conducted, as one of the major problems bedeviling the country. This covers lodging, travel, and other services that are essential for visitors. They suggested that to develop these sites and increase their tourist accessibility, there is a need for collaboration efforts from the government, private sector, non-governmental organizations, and communities to come together and find common ground. There is also a need for agritourism stakeholders, including farmers, to receive training and capacity building to guarantee that they possess the abilities and know-how required to give visitors a first-rate experience and successfully run their enterprises. Agritourism has an immense opportunity to boost the nation's economy while preserving its natural resources by diversifying the travel industry, generating jobs, and encouraging sustainable practices. Therefore, both the private and public sectors must collaborate to address the issues and fully realize Zimbabwe's agritourism potential.

However, since the industry has been facing impediments in recent years including political unrest, economic downturns, and natural disasters we believe that the country needs to turn to ICT and ride on the disruptive emerging technologies to turn the wheels of fortune in motion for the tourism industry in Zimbabwe.

Discussion

As the literature review has shown, many developed countries have turned to technology to turn the tides of economic fortunes in various sectors including tourism. The research participants equally pointed out that while disruptive technical innovations present an opportunity to be seized, several issues need to be addressed first to ride the tide of technology in Zimbabwe. Therefore, our chapter contends that disruptive emerging technologies present numerous opportunities to the tourism sector in terms of promoting and igniting agritourism activities. Thus, there is a need to address such issues as limited access to technology for many people. The constrained accessibility of technology is one of the primary concerns facing Zimbabwe's tourism sector. The infrastructure of the nation is under-developed, and many places lack access to even the most basic amenities like power and the internet. This makes it more difficult for companies in the tourism industry to accept and use new technologies, especially the purported agritourism players, and this reduces their capacity to compete with other locations. This is coupled with the high price of technology and Micro, Small and medium-sized firms (MSMEs) that lack the funds to engage in pricey technology comprise a large portion of Zimbabwe's tourism industry. This hinders their capacity to implement new technology and maintain their position as market leaders.

There is also the absence of a specialized labor force, and knowledgeable staff with a focus on emerging technology due to brain drain because of poor performance of the economy. Because of this, it is challenging for organizations to fully employ and integrate technology into their operations. Because of this challenge, the remnants available lack understanding and sometimes there is bound to be resistance to change. Many Zimbabwean tourism-related enterprises especially the targeted traditional ones may be reluctant to embrace new technology and are resistant to change. This is brought on by a lack of awareness of the advantages of technology. The development and competitiveness of the sector may be hampered by this resistance to change; hence, there is a need to consider borrowing the concept of a digital transformation strategy by Furusa et al. (2022) and applying it to the agritourism sector.

However, despite the challenges identified above, the participants were very livid about the potential presented by the disruptive emerging technologies to the tourism industry. They indicated that the emerging technologies may bring improved efficiency and productivity to the industry. Augmented reality, virtual reality, mixed reality, Artificial intelligence, big data analytics, and cloud computing are examples of disruptive technological advancements that can boost productivity and efficiency in Zimbabwe's tourism industry. These technologies can speed up and improve the precision of operations while automating procedures and cutting costs. The adoption and implementation of disruptive technology can lead to improved customer experience. In the travel and tourism sector, technology could improve the general customer experience as travelers can have a more immersive and participatory experience and create more lasting memories by utilizing virtual and augmented reality. This has the potential to boost business revenue and bring in more tourists to the nation. This is in line with the research by Yamagishi et al. (2021), who suggested that the future of agritourism lies in identifying and understanding the contextual challenges faced and fashioning strategies to overcome the identified challenges.

This work suggests that increased usage of emerging disruptive technology can lead to enhanced marketing and promotion of Zimbabwe's agritourism activities; hence, the players can now have an easier time reaching a larger audience by marketing and promoting their services thanks to technology. Agritourism actors can contact potential clients worldwide by utilizing social media platforms, websites, and online booking systems, which enhances their visibility and growth potential. This posits that disruptive technology also promotes product diversification for agritourism, enhances cross-sector collaboration, thereby enhancing value creation and generating employment opportunities leading to socio-economic transformation. Zimbabwe's agritourism business operators can ride on disruptive technology innovations and afford more opportunities and accessibility avenues to their clientele. This is in line with the research by Njerekai who proposed the need for the production of virtual videos showing the spherical images of the Great Zimbabwe monument by using a 360° camera. Such videos can then be run on software that improves the resolution quality to get high-resolution videos. The generated videos can then be marketed and accessed on the internet to provide agritourism virtual tours of farm and agricultural activities. These may also provide historical and contextual well as listen to highly educational and informative commentaries about the traditional farming systems.

This enables companies to provide tourists with distinctive and cutting-edge experiences, drawing in a larger clientele from the international markets as well as urban dwellers in the country. Disruptive digital technologies can provide the ability for sustainable agritourism development as this cutting-edge technology has the potential to promote sustainable practices in Zimbabwe. Businesses can lessen their carbon footprint and help preserve the nation's natural resources by using eco-friendly procedures and renewable energy sources. This tallies with the recommendations made by Brune et al., who identified that technology provided coping strategies to agritourism operators in the face of COVID-19.

The growth and development of Zimbabwe's agritourism activities through the leveraging of disruptive technological innovation has the potential to afford resilience to the rural economies and tourism at large. The research participants also recommended that a variety of disruptive technologies be harnessed, such as AI, robotics, digital finance, augmented reality, and virtual reality. These technologies are currently being employed, even though to varying levels, in a diversity of industries and geographical locations worldwide, and they are at the core of driving several disruptive changes. Web 3.0, which is also known as the semantic web, was made possible by formatting massive data so that software agents could understand it and allow computer-to-computer inter-operability. Big data linking and integration from a variety of data sets enhances data management, facilitates seamless integration, fosters innovation and creativity, and promotes cooperation on social media platforms. This resonates with the research by Nyagadza et al. (2022).

The way individuals interact and communicate has changed because of smartphones and other mobile devices, which can be leveraged to mediate the tourist experience. To provide the infrastructure for value co-creation for the tourism sector gave rise to smart tourism. Therefore, interconnectedness and interoperability of integrated technologies are exploited by intelligence which aims to maximize stakeholder value by reengineering processes and data to create novel services, products, and procedures. The public sector, suppliers, intermediaries, and consumers are all becoming dynamically networked, which co-produces value for all parties involved in the ecosystem. Through the assistance of travelers with mobility, visual, auditory, and cognitive disabilities in navigating physical and service barriers, smartness promotes inclusivity as well as accessibility for travelers. By rewarding interactions and raising visitor pleasure and engagement levels technology can facilitate innovative activities like gamification which can also have a positive impact. Everybody is connected and processes are integrated to create value through dynamic co-creation, personalization, and adaptation thanks to the universality and ubiquity of computing.

We think that the development of info structures and smart digital grids can be made possible by these disruptive technologies, and that, in turn, facilitates the seamless cooperation of all parties involved. Therefore, linking all stakeholders in the tourism ecosystem is essential because it nurtures fluidity in digital and physical interactions, which empowers dynamic networks that are constantly evolving and drive change within the firmly established tourism industry. Other elements, such as ambient intelligence may infuse ecosystems related to tourism with intelligence and make them responsive, adaptable, and adjustable to the demands of agritourism stakeholders. Thus, virtual reality, augmented reality mixed reality, robotics, and autonomous gadgets are all supported by these technologies, thereby improving the tourism experience. Hence, real-time service is supported by growing intelligence and ambient intelligence, which enables the co-creation of value for all agritourism stakeholders across many platforms. Therefore, real-time interactions in the tourism sector can happen when tourists are ready and eager to relate with brands.

Recommendations for Leveraging Disruptive Innovations for Sustainable Agritourism in Zimbabwe

We recommend that there is a need for the digital transformation of the tourism industry in Zimbabwe which encompasses all aspects including agritourism. This would enable the sector to adopt digital marketing and online booking platforms in their operational processes. Agritourism enterprises in Zimbabwe must adopt digital marketing and online booking platforms to reach a larger audience and simplify holidaymakers' agritourism experience bookings. This will improve their exposure and facilitate travelers' ability to plan and schedule their travels. Implement sustainable practices. Aspiring Zimbabwean agritourism players must implement sustainable practices into their operations given the growing concern for environmental sustainability. This can involve putting waste management systems in place, encouraging eco-friendly tourism activities, and utilizing renewable energy sources.

There is a need for collaboration in rural communities and this can be done by fostering job opportunities and promoting the genesis of an agritourism ecosystem that ensures that family communities and the birth of many micro, small, and medium enterprises (MSMEs) in the tourism sector benefit the nearby areas. Agritourism players may give visitors a more authentic and immersive experience while also promoting the community's sustainable growth by collaborating with local communities. There is a need to provide tourists with unique and immersive experiences. Providing tourists with unique and immersive experiences can be one of the disruptive developments in agritourism. This can involve things like farm-to-table meals, practical farming, and cross-cultural interactions with nearby people. These experiences have the power to draw in a specific type of visitor and set agritourism companies apart from more conventional travel options.

The usage of technology to manage agricultural activities is vital as this can help agritourism firms become more productive and efficient. This can involve applying precision farming methods, employing data analytics to improve decision-making, and using drones to monitor crops. By so doing, farmers can then promote their agricultural activities for educational tours, thereby boosting the agritourism business. This can subsequently improve their operations and give visitors a more cutting-edge, contemporary experience by utilizing technology. Working together with other industries may assist agritourism to gain from such fostered partnerships with the food and beverages, hospitality, and transportation sectors, among others. Agritourism companies may provide visitors with a more complete and seamless experience by collaborating with various sectors of the economy, opening new avenues for development and innovation.

There is a need to emphasize hands-on learning activities, and agritourism can serve as a platform for hands-on learning. Through educational programs and workshops, visitors can gain knowledge about environmentally conscious farming methods, regional customs and culture, and the value of protecting the environment and preserving traditional herbs and practices. This can help create a more informed and responsible society in addition to drawing in a specific type of tourist. There is a need to invest in the development of facilities and infrastructure. Agritourism enterprises must make these investments to give visitors a relaxing and pleasurable experience. This can involve creating leisure spaces, renovating lodgings, and enhancing transit choices. Agritourism companies can improve the entire traveler experience and eventually draw in more guests by making these kinds of investments.

The Implications of This Research

This research has implications for theory, policy and practice for agritourism in Zimbabwe and potentially for sub-Saharan Africa and the developing world at large. This chapter established that the adoption, application and implementation of disruptive emerging ICT technologies in agritourism have significant implications for innovation theory, policy, and practice in agritourism business. These new and innovative technologies have the potential to meaningfully transform or disrupt the existing operational landscape of the operators or the general practices. The research has shown that these emerging technologies can foster an enabling environment that inculcates a culture of innovation within the context of agritourism. This research suggests that these disruptive emerging ICT technologies can provide new tools, approaches and opportunities for agritourism businesses to enhance their offerings and reach new clients from the international market as well as on the local front from the urban dwellers and this can significantly improve their operations.

The implications for Innovation Theory - The adoption and implementation of disruptive technologies in agritourism challenges the status quo of traditional models of innovation in agritourism. This therefore requires that the researchers explore and fashion innovation frameworks that can account for the unique characteristics of these emerging, disruptive ICT technologies. The study and analysis of these emerging technologies and their impact on agritourism contribute to the broader field of innovation theory. The research findings in our context can inform a new understanding of how disruptive technology can facilitate innovative activities in business and society as a whole.

The implications for policy and regulation Our research suggests that the Zimbabwean government should craft policies that support the integration of emerging technologies in agritourism. Therefore, it would be critical for the government to ensure that the issues of the ICT domain ensure that the digital ecosystem is thriving. This will result in other areas like agritourism easily tapping into these technologies and generating innovative new products and services that improve and expand their operational activities. However, it is imperative that harnessing these disruptive technologies should be done while also addressing potential challenges related to data privacy, cybersecurity, and ethical considerations.

The implications for practice and implementation Our book chapter advocates that agritourism operators must embrace these emerging disruptive digital technologies to enhance their competitiveness and ultimately provide innovative experiences for tourists. The results show that there may be a need for training and support programs to ensure successful implementation and adoption of these technologies.

The research is an intersection of disruptive, emerging digital technologies, agritourism, and innovation theory, and the research findings suggest that these technologies have the potential to transform agritourism practices and contribute to our better understanding of innovation. That is why we believe that the integration of emerging disruptive technologies in agritourism presents exciting opportunities for innovation and growth in agritourism activities. By leveraging these technologies, agritourism players can enhance their efficiency, sustainability, and profitability while offering unique and immersive experiences for tourists.

There is a need to design an agritourism digital transformation model for Zimbabwe. The model should be developed considering the contextual factors that are found in Zimbabwe. This should consider looking at the agritourism ecosystem that constitutes the Zimbabwe landscape to establish the capacity, current utilization level and potential for innovation, development and growth. Secondly, there is a need to analyze the Zimbabwean digital ecosystem and establish the current establishment in comparison with the global digital ecosystem. This will facilitate the establishment of the current status quo and the existing gaps in the local digital ecosystem establishment.

Conclusion

We believe that our book chapter will contribute to the existing literature on disruptive technological innovation in the tourism industry, specifically in the African context. It will provide insights into the current state of the tourism industry in Zimbabwe and the potential for growth through the adoption of disruptive technologies, and with the ongoing evolution and revolution taking place in the ICT digital ecosystem developing countries like Zimbabwe have the opportunity to make great leap forward. Thus, they need to ride on the wave of these disruptive technologies and embrace the innovative opportunities that come with this technology. But to realize the full spectrum benefits of these disruptive technologies, there is a need for the government to craft policies that promote inclusive digitalization processes targeted at the informal sector and hopeful marginalized communities as proposed by Musungwini et al. (2022).

This suggests that if this is done properly within the inclusive spirit, it can potentially generate novel ways of doing things in the agritourism sector and this may generate new and more employment opportunities and augment the economy of the country. If the disruptive emerging digital technologies are left to adoption and implementation by individual operators have the potential to illuminate the existing gaps between the rich and the poor among agritourism operators. There is a need to categorize digital ecosystems and specific agritourism communities and subsequently invest in those digital ecosystems to ensure that there is increased adoption and implementation across the agritourism ecosystem. This will ensure that the country's economic and social sectors and actors can reap the digital dividends as suggested by Mukora et al. (2022). Hence, the findings will be valuable for agritourism business operators, policymakers, and researchers interested in understanding the impact of disruptive ICT technology on the tourism industry in developing countries. This book chapter is in sync with the key themes of the book hence it directly speaks to the title of the book.

References

Abbas, J., Mubeen, R., Iorember, P. T., Raza, S., and Mamirkulova, G. (2021). Exploring the impact of COVID-19 on tourism: Transformational potential and implications for a sustainable recovery of the travel and leisure industry. *Current Research in Behaviourral Sciences*, 2(February), 100033. 10.1016/j.crbeha.2021. 100033

Adeyinka-Ojo, S., and Abdullah, S. K. (2019). Disruptive digital innovation and sharing economy in hospitality and tourism destination. *IOP Conf. Series: Materials Science and Engineering*, 495(2019), 012006 IOP Publishing doi:10. 1088/1757-899X/495/1/012006

Adom, D., Alimov, A., and Gouthami, V. (2021). Agritourism as a preferred traveling trend in boosting rural economies in the post-COVID-19 period: Nexus between agriculture, tourism, art and culture. *Journal of Migration, Culture and Society*, 1(1). Retrieved from https://royalliteglobal.com/jmcs/article/view/671

Baipai, R., Chikuta, O., Gandiwa, E., and Mutanga, C. N. (2023). A framework for sustainable agritourism development in Zimbabwe. *Cogent Social Sciences*, 9(1). 10.1080/23311886.2023.2201025

Barbieri, C., and Mshenga, M. (2008). The role of the firm and owner characteristics on the performance of agritourism farms. *Sociologia Ruralis*, 48(2), April 2008, 166–183.

Barbieri, C. (2020). Agritourism research: A perspective article. *Tourism Review*, 75(1), 149–152. 10.1108/TR-05-2019-0152

Benckendorff, P., and Shu, M. (Lavender). (2019). Research impact benchmarks for tourism, hospitality and events scholars in Australia and New Zealand. *Journal of Hospitality and Tourism Management*, 38(xxxx), 184–190. 10.1016/j.jhtm.2018. 04.005

Buhalis, D., Harwood, T., Bogicevic, V., Viglia, G., Beldona, S., and Hofacker, C. (2019). Technological disruptions in services: Lessons from tourism and hospitality. *Journal of Service Management*, 30(4), 484–506. 10.1108/JOSM-12-2018-0398

Caroline Mzobe, S., Makoni, L., and Nyikana, S. (2023). Unlocking the potential of domestic tourism in uncertain times: Lessons from the COVID-19 pandemic in South Africa. *Studia Periegetica*, 38(2), 137–155. 10.5604/01.3001.0016.0570

Chatterjee, S., and Prasad, M. V. D. (2019). The evolution of agri-tourism practices in india: some success stories. *Madridge Journal of Agriculture and Environmental Sciences, 1*(1), 19–25. 10.18689/mjaes-1000104

Christensen, C. M. (1997). *The innovator's dilemma: When new technologies cause great firms to fail.* Harvard Business School Press.

Chang, S. E., Chen, Y.-C., and Lu, M.-F. (2019). Supply chain re-engineering using blockchain technology: A case of smart contract based tracking process. *Technological Forecasting and Social Change, 144*(July), 1–11.

Clarke, C. (1999). Marketing structures for farm tourism: Beyond the individual provider of rural tourism. *Journal of Sustainable Tourism, 7*(1), 26–47.

Colombo, E., and Baggio, R. (2017), Tourism distribution channels: Knowledge requirements. In Scott, N., De Martino, M., and Van Niekerk, T. (Eds.), *Knowledge Transfer to and within Tourism: Academic, Industry and Government Bridges* (pp. 741–754), Bingley, Emerald Publishing.

Christensen, C. M., and Raynor, M. E. (2003). Why hard-nosed executives should care about management theory. *Harvard Business Review, 81*(9).

Dube, M., Musungwini, S., Mudzimba, E., and Watyoka, N. (2023). Mixed reality in confronting consumer security and privacy issues in digital marketing: Integrating the best of both worlds for better interaction with users. In *Confronting Security and Privacy Challenges in Digital Marketing* (pp. 252–266). 10.4018/978-1-6684-8958-1.ch012

Duft, G., and Durana, P. (2020). Artificial intelligence-based decision-making algorithms, automated production systems, and Big data-driven innovation in sustainable industry 4.0. *Economics, Management and Financial Markets, 15*(4), 9–18.

Fernandes, N. (2020). Economic effects of coronavirus outbreak (COVID-19) on the world economy Nuno Fernandes Full Professor of Finance IESE Business School Spain. *SSRN Electronic Journal, ISSN 1556-5068, Elsevier BV*, 0–29.

Fahmi, Z., Samah, B. A., and Abdullah, H. (2013). Paddy industry and paddy farmers wellbeing: A success recipe for agriculture industry in Malaysia. *Asian Social Science, 9*(3), 177–181.

Fan, S., Lau, R. Y. K., and Zhao, J. L. (2015). Demystifying big data analytics for business intelligence through the lens of marketing mix. *Big Data Research, 2*(1), 28–32.

Fragnière E., Sahut J. M., Hikkerova L., Schegg R., Schumacher M., Grèzes S., and Ramseyer R. (2022). Blockchain technology in the tourism industry: New perspectives in Switzerland. *Journal of Innovation Economics & Management, 2022/1*(N° 37), pages 65 à 90, DOI 10.3917/jie.pr1.0111

Furusa, S. S., Musungwini, S., Gavai, P. V., and Maseko, M. (2022). A strategy for the digitalization of marginalized communities. In Zhou, M., Mahlangu, G., and Matsika, C. (Eds.), *IGI Global* (pp. 237–259). IGI Publishing. 10.4018/978-1-6684-3901-2.ch012

Garibaldi, R., and Pozzi, A. (2020). Gastronomy tourism and Covid-19: Technologies for overcoming current and future restrictions. *B. Federica Ed. Tourism Facing a Pandemic: From Crisis to Recovery*, 45–52.

Gasser, U., Ienca, M., Scheibner, J., Sleigh, J., and Vayena, E. (2020). Digital tools against COVID-19: Framing the ethical challenges and how to address them. *Computers and Society*, 1–15. https://arxiv.org/abs/2004.10236v1

Government, Z. (n.d.). *The 2011 National Budget Statement.*

Guvamombe. (2019). *The Herald – Breaking news.* https://www.herald.co.zw/farm-tourism-trendy-refreshing/

Hardesty, S., and Leff, P. (2020). California's agritourism operations expand despite facing regulatory challenges. *California Agriculture, 74*(3), 123–126. 10.3733/ca.2020a0026

Hanke, S. (n.d.). *Hanke's Inflation Dashboard*. Retrieved January 6, 2024, from https://www.investmentoffice.com/Observations/Macro_Observations/Hanke_s_Inflation_Dashboard.html

Ilbery, W., and Bowler, I. (1998). *From Agricultural Productivism to Post-productivism*. Longman.

Jafari-Sadeghi, V., Garcia-Perez, A., Candelo, E., and Couturier, J. (2021). Exploring the impact of digital transformation on technology entrepreneurship and technological market expansion: The role of technology readiness, exploration and exploitation. *Journal of Business Research, 124*(November 2020), 100–111. 10.1016/j.jbusres.2020.11.020

Kenney, M. (2017). Explaining the growth and globalization of silicon valley: The past and today

Kolajo, T., Daramola, O., and Adebiyi, A. (2019). Big data stream analysis: A systematic literature review. *J Big Data, 6*, 47.

Liu, Y., Chin, W. L., Nechita, F., and Candrea, A. N. (2020). Framing film-induced tourism into a sustainable perspective from Romania, Indonesia and Malaysia. *Sustainability, 12*(23), 9910. 10.3390/ su12239910

Lub, V. (2015). Validity in qualitative evaluation. *International Journal of Qualitative Methods, 14*(5), 160940691562140. 10.1177/1609406915621406

McGehee, N., and Kim, K. (2004). The motivation for agritourism entrepreneurship. *Journal of Travel Research, 43*(2), 161–170.

McGehee, N. G., Kim, K. M., and Jennings, G. R. (2007). Gender and motivation for agritourism entrepreneurship. *Tourism Management, 28*(1), 280–289.

McKinsey & Company. (2013). *How retailers can keep up with consumers*. McKinsey & Company, *October*, 1–10.

Meert, H., Van Huylenbroeck, G., Vernimmen, T., Bourgeois, M., and Van Hecke, E. (2005). Farm household survival strategies and diversification on marginal farms. *Journal of Rural Studies, 21*(1), 81–97.

Meschede, H. (2020). Analysis of the demand response potential in hotels with varying probabilistic influencing time series for the Canary Islands. *Renewable Energy, 160*, 1480–1491. 10.1016/j.renene.2020.06.024

Monitor, I. L. O. (2020). *Current situation: Why are labour markets important? March*, 1–15.

Mukora, F. N., Matekenya, T., Mahlangu, G., Chirisa, I., and Muparutsa, C. H. S. (2022). Zimbabwe's digitalisation agenda from digital divides to digital dividends. In Zhou, M., Mahlangu, G., and Matsika, C. (Eds.), *IGI Global* (pp. 123–139). IGI Publishing. 10.4018/978-1-6684-3901-2.ch006

Musasa, G., and Mago, S. (2014). Challenges of rural tourism development in Zimbabwe: A case of the Great Zimbabwe-Masvingo district. *African Journal of Hospitality, Tourism and Leisure, 3*(2), 1–12.

Musungwini, S., Furusa, S. S., Gavai, P. V., and Gumbo, R. (2022). *Inclusive digital transformation for the marginalized communities in a developing context*. IGI Global, 95–122. 10.4018/978-1-6684-3901-2.ch005

Musungwini, S., Gavai, P. V., Munyoro, B., and Chare, A. (2023). Emerging ICT technologies for agriculture, training, and capacity building for farmers in developing countries: A case study in Zimbabwe. *Applying Drone Technologies and Robotics for Agricultural Sustainability, January*, 12–30. 10.4018/978-1-6684-6413-7.ch002

Ngarava, S. (2020). Land use policy impact of the fast track land reform programme (FTLRP) on agricultural production: A tobacco success story in Zimbabwe? *Land Use Policy, 99*(February 2019), 105000. 10.1016/j.landusepol.2020.105000

Novelli, M., Gussing Burgess, L., Jones, A., and Ritchie, B. W. (2018). 'No Ebola … still doomed' – The Ebola-induced tourism crisis. *Annals of Tourism Research, 70*, 76–87. 10.1016/j.annals.2018.03.006

Nyagadza, B., Chuchu, T., and Chigora, F. (2022). Technology application in tourism events: Case of Africa. In *Digital Transformation and Innovation in Tourism Events* (pp. 107–116). Taylor and Francis. 10.4324/9781003271147-12

Oliver Chikuta, and Caroliny Makacha. (2016). Agritourism: A possible alternative to Zimbabwe's tourism product? *J. of Tourism and Hospitality Management, 4*(3). 10. 17265/2328-2169/2016.06.001

OECD (2009). *The role of agriculture and farm household diversification in the rural economy: Evidence and initial policy implications.* Paris: Organisation for Economic Co-operation and Development.

Padhy, C., Reddy, M., Raj, R. K., and Pattanayak, K. P. (2022). Role of digital technology in agriculture. *Indian Journal of Natural Sciences, 13* / Issue 71 / April / 2022. International Bimonthly (Print) ISSN: 0976–0997.

Pereira, C. S., Durao, N., Moreira, F., and Veloso, B. (2022). The importance of digital transformation in international business. *Sustainability, 14*, 834.

Philips, S., Hunter, C., and Blackstock, K. (2010). A typology for defining agritourism. *Elsevier Publications, Tourism Management, 31*, 754–758.

Phillipson, J., Gorton, M., Raley, M., and Moxey, A. (2004). Treating farms as firms? The evolution of farm business support from productions to entrepreneurial models. *Environment and Planning C: Government and Policy, 22*(1), 31–54.

Preko, A., and Anyigba, H. (2022). The tourism and hospitality career progression pathway. *International Hospitality Review*, (ahead-of-print).

Rai, V. S. R., and Pinto, V. S. (2022) Role of disruptive technology in tourism industry, leading to lifestyle transforming -A study with special reference to COORG, Conference: Transforming Business Practices through Disruptive Technologies At Bangalore

Ranganai, N., Basera, V., and Muwani, T. (2023). Impact of big data and analytics on quality management in rural tourism in southern Africa - Zimbabwe: A systematic literature review. *Journal of Tourism, Culinary, and Entrepreneurship, School of Tourism, Universitas Ciputra Surabaya, Indonesia.*

Roberts, L., and Hall, D. (2001). *Rural Tourism and Recreation: Principles to Practice.* CABI Publication.

Sigala, M. (2020). Tourism and COVID-19: Impacts and implications for advancing and resetting industry and research. *Journal of Business Research, 117*, 312–321. 10.1016/j.jbusres.2020.06.015

Scoones, I., Marongwe, N., Mavedzenge, B., Murimbarimba, F., Mahenehene, J., and Sukume, C. (2011). Zimbabwe's land reform: Challenging the myths. *Journal of Peasant Studies, 38*(5), 967–993. 10.1080/03066150.2011.622042

Stylos, N., Zwiegelaar, J., and Buhalis, D. (2021). Big data empowered agility for dynamic, volatile, and time-sensitive service industries: The case of the tourism sector. *International Journal of Contemporary Hospitality Management.*

Toh, A., and Brown, D. (2020, June 4). *How Digital Contact Tracing for COVID-19 Could Worsen Inequality.* Human Rights Watch. https://www.hrw.org/news/2020/06/04/how-digital-contact-tracing-covid-19-could-worsen-inequality

Vanhove, N. (2017). *The Economics of Tourism Destinations: Theory and Practice.* Routledge.

Watch, H. R. (2023). *Joint Civil Society Statement: States use of digital surveillance technologies to fight pandemics must respect human rights.* Human Rights Watch. https://www.hrw.org/news/2020/04/02/joint-civil-society-statement-states-use-of-digital-surveillance-technologies-fight

Wei Lee Chin, and Siti Fatimahwati Pehin Dato Musa (2021). Agritourism resilience against Covid-19: Impacts and management strategies. *Cogent Social Sciences, 7*(1), 1950290, DOI: 10.1080/23311886.2021.1950290

Woodhead, R., Stephenson, P., and Morrey, D. (2018). Digital construction: From point solutions to IoT ecosystem. *Automation in Construction*, *93*, 35–46. 10.1016/J.AUTCON.2018.05.004

Yamagishi, K., Gantalao, C., and Ocampo, L. (2021). The future of farm tourism in the Philippines: challenges, strategies and insights. *Journal of Tourism Futures*. 10.1108/JTF-06-2020-0101

Zaman, M., Botti, L., Boulin, J., and Corne, A. (2017), Intégration des Innovations Web: défi relevé pour les OGD français?, *Management & Avenir*, *91*(1), 163–184.

Zhang, K., Chen, Y., and Li, C. (2019). Discovering the tourists' behaviours and perceptions in a tourism destination by analyzing photos' visual content with a computer deep learning model: The case of Beijing. *Tourism Management*, *75*, 595–608.

Zheng, Z., Xie, S., Dai, H. N., Chen, X., and Wang, H. (2018). Blockchain challenges and opportunities: A Survey. *International Journal of Web and Grid Services*, *14*(4), 352–375.

12 Conclusion

The Future of Agritourism Development in Africa

Brighton Nyagadza, Farai Chigora, and Azizul Hassan

Exploring the Importance of Future Agritourism Development in Africa

Insights presented in this volume give rise to the continuing discussion on future effective agritourism development policies in Africa and the world at large. The current volume presents an understanding of the potential role of agritourism as a driver of economic growth and the improvement of agribusiness through meticulous analysis of findings from a wide range of economies (Nyagadza et al., 2024). Given the multi-faceted and multi-layered nature of agritourism, how the antecedents have framed and contoured past agritourism future research generations and societies might not be comparable, and transposable to how they will curve the current and future generations and societies.

The effective application of agritourism development in African economies is poised to stimulate higher professional levels of business productivity, greater forms of diversification, technological upgrading, and production of high-value added products (goods and services) (Guruwo, Nyagadza, and Manhimanzi, 2023). This is forecasted to provide more decent work opportunities, priority driven leadership (Gwiza et al., 2022), which increase social empowerment and wider choices of diverse people in meeting their life necessities and capabilities.

The need to encourage future sustainable agritourism in Africa has been urged by the increase in the general acceptance of global climate change. Informed by the 2030 Agenda for Sustainable Development and Sustainable Development Goals (SDGs) framework that emphasises the fostering of sustainable economic growth per capita according to national circumstances and higher gross domestic product growth per annum in the least developed countries, the current volume provides theoretically informed and practically grounded solutions to addressing sustainable development for humans, through enhanced agritourism development (Nyagadza et al., 2024). In the near future, it is important for Africa tourism stakeholders to adopt and implement agritourism aspects, which uplift the vulnerable, women, address inequality, and job creation, in order for agritourism to be significant at both

DOI: 10.4324/9781032696188-12

community and national levels when developing the agritourism sector. Evidence will show that though agritourism and hospitality can benefit local communities, in Africa they may still face marginalisation, exploitation, poverty, insecurity, political instability, economic downturn, and inequalities since unemployment is still a major hindrance to human capital development, hence affecting the achievement of SDG 8.

The scientific findings show that the relationship between emerging AI technologies and agritourism, is critical for economic development and growth in Africa. The results are necessary in strategically leveraging the potential of AI and addressing underlying challenges. The findings benefit policymakers, tourism professionals, agricultural practitioners, entrepreneurs, and researchers. They clearly indicate the significant importance related to this as it explores the potential of AI in promoting sustainable agritourism in Africa. AI-enhanced agritourism in the future will increase the opportunities for decent work for the employable population (Buhalis et al., 2023). This may offer valuable future digital transformative insights and recommendations for agritourism development policy (Hassan, 2022). The integration of AI in agritourism might have a ripple effect on tourism value chains thereby prompting the formulation of policies that bring about positive socio-economic impact. By addressing digital and social media marketing inefficiencies, environmental sustainability concerns, and limited access to information, AI can contribute to sustainable agritourism practices that preserve natural resources and promote responsible tourism (Ramos and Hassan, 2022; Nyagadza et al., 2022). However, the AI technologies are not readily available and accessible to agritourism operators in most parts of Africa, and there are many barriers and challenges to their adoption and deployment, such as lack of infrastructure, skills, data, funding, regulation, and awareness.

The volume shows the future deployment of precision agritourism can be facilitated through farmers' collective action, where a group of individuals invests in precision agritourism, and the resulting profits are distributed among the members. It shows the future ethical dilemmas probably faced by various agritourism actors, including the challenge of balancing economic gains with environmental conservation and preserving cultural authenticity while meeting the demands of tourists (Chigora et al., 2023b). Balancing these diverse tasks and allocating time effectively can be a challenge, especially during peak seasons when both ethical agricultural and hospitality activities demand attention. The investigation into the ethical aspects in this volume provides valuable insights for policymakers, entrepreneurs, and communities engaged in sustainable agritourism in African regions with similar aspirations. In addition, agritourism creates touristic jobs in farms. On the other hand, the tourism sector benefits from agritourism by having new tourist destinations and attractions. This implies innovative ways for building agricultural practising capacity to engage in transforming their farms and farming activities into agritourism ventures.

Based on the findings, recommendations are proffered with regard to how the African agritourism and hospitality experience can be continuously improved to retain its pole position within the global space, and how sustainability could be achieved from both policy and operational points of view. Related areas of further study in the future are also identified and recommended as part of supporting the continuous growth and development of the African tourism and hospitality industry. In this current volume, we found out that agritourism is an emerging phenomenon in Africa, but is gaining importance in rural development because it contributes to farm diversified economy in many countries. The findings indicate that agritourism in future can bring economic benefits both at the farm and national level. However, developing agritourism is constrained by inadequate entrepreneurial education (Maziriri et al., 2023), rural infrastructure and government enabling environment. Both the agriculture and tourism sectors in Africa, will have serious potential to grow, diversify products and services offered, and improve the economic welfare of farming communities surrounding tourist sites. While agritourism entrepreneurship appears to be lucrative in the contemporary tourism sector, it is worth mentioning that the success of agritourism entrepreneurship in the future African economy fully depends on the possession of robust knowledge related to contemporary agriculture, tourism and hospitality sector, and entrepreneurship (Khazami and Lakner, 2022).

The findings in the current volume are significant as they amplify on an alternative agritourism product often overlooked in Africa's traditional tourism packages at the global tourism stage. This provides insights into the need to reform the future of African regional agritourism policy frameworks, and regular reviews of agritourism marketing and promotional strategies at regional-level festivals. The future could be the establishment of African regional agritourism bodies focusing on coordinating the marketing of the region's agritourism products through impactful exhibition events and festivals (Chigora et al., 2023a). The recognition of the sector by farmers as a lucrative investment is likely going to intensify competition which can only be curbed through effective agricultural marketing (Nyagadza and Rukasha, 2023).

There is an argument that agritourism may require little investment yet high long-term returns, which will create room for agribusiness growth in future for many African farmers practising agritourism. Farmers-turned-entrepreneurs often find themselves juggling multiple responsibilities, such as crop cultivation, animal care, maintenance of farm infrastructure, and customer service. In future facilitating collaborative African governance through strategic multi-stakeholder committees and partnerships is recommended to foster open dialogue and cooperation, and regional bodies could help establish these governance mechanisms. Agritourism has great potential to support effective and sustainable economic growth for most African countries, Zimbabwe included, and can be instrumental in supporting recovery from the negative economic and socio-cultural effects of COVID-19. Further to this, the findings in the current volumes give vital views for African economies, directing future tactical choices in cultivating global trade connections, advancing economic expansion through agritourism, and moving through the intricacies of population-associated obstacles in agribusiness. The need by many African farmers practising agritourism to be sustainable in the future may lead to the growth of the sector as many farmers seek the need to increase profit margins and gain relevance to acquire state funding.

References

Buhalis, D., Leung, X. Y., Fan, D., Darcy, S., Chen, G., Xu, F., Wei-Han Tan, G., Nunkoo, R., and Farmaki, A. (2023). Tourism 2030 and the contribution to the sustainable development goals: The tourism review viewpoint. *Tourism Review*, 78(2), 293–313.

Chigora, F., Nyagadza, B., Katsande, C., and Zvavahera, P. (2023a). Building positive Zimbabwean tourism festival and event destination brand image and equity. In Kumar, J., Bayram, G. E. and Sharma, A. (Eds.), *Resilient and Sustainable Destinations After Disaster (Tourism Security-Safety and Post Conflict Destinations)*, pp. 63–74. Emerald Insight Publishers, Bingley, United Kingdom (UK). ISBN: 978-1-80382-021-7. 10.1108/978-1-80382-021-720231005

Chigora, F., Nyagadza, B., Katsande, C., and Zvavahera, P. (2023b). Destination marketing as an orienting tool in Zimbabwe's tourism image and publicity crisis. In Kumar, J., Bayram, G. E. and Sharma, A. (Eds.), *Resilient and Sustainable Destinations After Disaster (Tourism Security-Safety and Post Conflict Destinations)*, Emerald Publishing Limited, Bingley, pp. 75–85. Emerald Insight Publishers, Bingley, United Kingdom (UK). 10.1108/978-1-80382-021-720231006 ISBN: 978-1-80382-021-7

Gwiza, A., Hlungwani, P. M., Nyagadza, B., Masimo, P., and David. V. (2022). Developing priority-driven leadership and supervisory skills within public sector low resource settings. Book Chapter In Chiware, M., Nkala, B. and Chirisa, I. (Eds), *Transformational human resources management in Zimbabwe*, Springer Nature, Switzerland. 10.1007/978-981-19-4629-5_6

Guruwo, B., Nyagadza, B., and Manhimanzi, G. (2023). Assessing the influence of village savings and loan associations on climate resilience and food security: A case of Domboshava, Zimbabwe. In Nyagadza, B. and Rukasha, C. (Eds.), *Sustainable Agricultural Marketing and Agribusiness Development: An African perspective*, pp. 165–196. CABI Press, Wallingford, United Kingdom. ISBN: 978-1-80062-252-4. https://www.cabidigitallibrary.org/doi/abs/10.1079/9781800622548.0019

Hassan, A. (2022) Internet of things in the tourism industry. In Santos, J. and Sousa, B. (Eds.), *Promoting Organizational Performance Through 5G and Agile Marketing*, pp.165–178. IGI Global, Hershey, PA.

Khazami, N., and Lakner, Z. (2022). The development of social capital during the process of starting an agritourism business. *Tourism and Hospitality*, 3(1), 210–224.

Maziriri, E. T., Nyagadza, B., and Chuchu, T. (2023). Key innovation abilities on capability and the performance of women entrepreneurs: The role of entrepreneurial education and proactive personality, *Business Analyst Journal*, 44(2), 53–83. Emerald Insight, Bingley, United Kingdom (UK). https://www.emerald.com/insight/content/doi/10.1108/BAJ-02-2023-0044

Nyagadza, B., Chigora, F., and Hassan, A. (2024). Introduction: Agritourism for sustainable development. Nyagadza, B., Chigora, F., and Hassan, A. (Eds.), *Agritourism for Sustainable Development: Reflections from Emerging African Economies*, pp. 1–5. CABI Press, Wallingford, United Kingdom. ISBN-13:

9781800623699. https://www.cabidigitallibrary.org/doi/abs/10.1079/9781800623 705.0001

Nyagadza, B., Chuchu, T., and Chigora, F. (2022). Technology application in tourism events: Case of Africa. Chapter 9 In Hassan, A. (Ed.), *Digital Transformation and Innovation in Tourism Events*, Routledge, Taylor & Francis, Abingdon, United Kingdom (UK). ISBN: 9781032220963. https://www.taylorfrancis.com/chapters/ edit/10.4324/9781003271147-12/technology-application-tourism-events

Nyagadza, B., and Rukasha, C. (2023). Introduction: Envisioning the future of sustainable agricultural marketing and agribusiness development. In Nyagadza, B., and Rukasha, C. (Eds.), *Sustainable Agricultural Marketing and Agribusiness Development: An African perspective*, pp. 63–74. CABI Press, Wallingford, United Kingdom. ISBN: 978-1-80062-252-4. https://www.cabidigitallibrary.org/doi/abs/ 10.1079/9781800622548.0001

Ramos, C., and Hassan, A. (2022). The role of ICT applied to tourism and marketing in Asia. In Hassan, A. (Ed.), *Technology Application in Tourism in Asia: Theories, Innovations and Practices*, pp. 81–96. Springer, Singapore.

Index

Note: *Italicized* and **bold** page numbers refer to figures and tables.

Abbas, J. 212, 214
ABUAD *see* Afe Babalola University (ABUAD)
Accor 132
adult entrepreneurship education 48–49
AfCFTA *see* Africa Continental Free Trade Area (AfCFTA)
Afe Babalola University (ABUAD) 66
Africa Continental Free Trade Area (AfCFTA) 137
African Development Agenda 2063 57, 137
agricultural trade flows xxv, 2, 3, 5–6, 12–14
agritourism: agreements 9, 11; challenges to 5; definition of 1, 17, 77, 148; development, future of 240–243; entrepreneurship *see* agritourism entrepreneurship; necessary conditions for developing 173; outcomes of 4–5; precision xxvi–xxvii, 54–71, 241; products, marketing *see* marketing agritourism products; for rural development xxix; shift towards 194–195; sustainable *see* sustainable agritourism; *see also individual entries*
agritourism entrepreneurship xxvi, *50*; adult entrepreneurship education 48–49; awareness campaigns 49–50; central government support to 47; challenges to 44–46; conceptualization of 38–39; conducive legislative framework and robust policy 49; critical success factors for 37–52; financial resources of 48; global perspectives of 40–42; implications of

50–51; importance of 42–44; infrastructure development 49; limitations of 51; SDGs and 39–40
AgTech (agricultural technologies) 221
Ahmada, M. A. 58
AI *see* artificial intelligence (AI)
Ait-Yahia Ghidouche, K. 99
AI Virtual Assistant 25
Al-Baqura Village, Jordan: cottage farming agritourism 66–67
Algeria 99
Alignment-Interest and Influence Matrix (AIIM) 180, *181*
Alok, K. S. 151, 159
Alphonse, H. 100
Alternative Food Networks 113
alternative wetting and drying (AWD) 68
Anderson, J. 60–61
Anderson, W. 102
Angola Institute of Agricultural Development 109
AR *see* augmented reality (AR)
artificial intelligence (AI) xxvi, xxvii, 17–33, 80–81, 130, 241; agritourism activities 22, **23**; case studies of 21; challenges to 29, **30**; cultural implications of 28–29; definition of 18, 80; efficiency and effectiveness, improving 90–91; -enabled sustainable agritourism practices, empowerment and socio-economic development of local communities through 91, *91*; ethical considerations of 29, **30**; for human development 77–95; income streams 22–25, **23**; in knowledge transfer and training 25–26; limitations of 32; limited access to

information, addressing 89–90, *90*, **90**; practical implications of 29, 31; resource optimisation through, to bolster sustainable farming practices 26–28; stimulated agritourism 84–85; theoretical implications of 31–32; tools in knowledge transfer and training 26, **27**; utilitarian theory of 19, **20**, 31

ASEAN *see* Association of Southeast Asian Nations (ASEAN)

Ashley, C. 123

Association of Southeast Asian Nations (ASEAN) 6

Auboin, M. 3

augmented reality (AR) 219, 221, 226

AWD *see* alternative wetting and drying (AWD)

Back, R.M. 148, 157

Baiocco, S. 99

Baipai, R. 99, 192, 198, 215, 223

Bajona, C. 2

Barbieri, C. 5, 61, 215

Bastidas-Manzano, A. B. 135

Benckendorff, P. 213

Bergstrand, J. H. 3

Bertella, G. 101–102

Berthou, A. 3

Best Tourism Village 148

Best Western Hotels & Resorts 132

Bhagat, P. R. 84

Bhatta, K. 19

bibliometric analysis 167–188

blockchain technology 220

Boden: "AI: Its Nature and Origins" 80

Borbon, N. M. D. 195

Botswana 6

branding 222

Broccardo, L. 99

Butler's Model of Tourism Area Life Cycle 137

Caála Municipality: government authorities, as guardians of responsible tourism 112; indigenous communities in 111; sustainable agritourism in xxvii–xxviii, 97, 102–104, 106–108, 113–115

CAGR *see* compound annual growth rate (CAGR)

Campbell, J. M. 101

CAMPFIRE programme 135

Canada: agritourism entrepreneurship 37, 41

Caroline Mzobe, S. 214

cash flow management 46

Cawley, A. 26

CBT *see* Community-Based Tourism (CBT)

Central Angola, sustainable agritourism in 96–115, **105**; ethical considerations 106–109; ethical dilemmas faced by local actors 109–111; implications of 113–114; limitations of 114–115; local actors in shaping ethical agritourism, role of 111–112; recommendations for 115

central government support 47

Chase, L. C. 22

Chatbots 25

Chatterjee, S. 130

Chen-Tsang, T. 150

Chesvingo Cultural Village, Masvingo 135

Chigora, F. 39, 192, 195, 196

Chikuta, O. 156, 192, 198, 215

China: trade treaties 6

Choo, H. 146

Chouinard, H. H. 5

Christensen, C. M. 213

Circular Economy model 113

climate change mitigation 67

Cohesion Fund 41

collaborative economy 222

COMESA *see* Common Market of Eastern and Southern Africa (COMESA)

Common Market of Eastern and Southern Africa (COMESA) 6

Community-Based Tourism (CBT) 55, 61, 62, 66, 120, 121, 124, 137

community engagement 55, 81, 82, 113, 124, 197, 203; NGOs as advocates for 112; *vs.* resource management 108–109

compound annual growth rate (CAGR) 119, 125

conducive legislative framework and robust policy 49

Confederation of Farmer Associations and Agricultural Cooperatives of Angola (UNACA) 97

consumer interaction, in disruptive technological innovation 220–221

Corporate Social Responsibility (CSR) 62, 136

Costinot, A. 2
costs 5; marketing 45; promotion 45
cottage farming approach 54–71
COVID-19 pandemic 4, 18, 31, 54, 58,
   120, 125, 132, 138, 145, 195, 198, 212,
   224, 243
CPT *see* Cultural Tourism
   Programmes (CPT)
CSR *see* Corporate Social
   Responsibility (CSR)
culinary tourism 150
cultural authenticity 110
cultural preservation 5, 96, 97, 100, 103,
   111, 112, 115; economic aspirations
   vs. 107
Cultural Tourism Programmes (CPT) 62
customer insights 222

Dangi, T. B. 101
data analytics 222
de Crom, E. 155
degree of centrality 184–187
DeLay, N. D. 5
diffusion theory xxvii, 56, 57, 68
digital marketing, agritourism 219;
   consumer experience in 196, *196*, *197*,
   200; revolutionising agritourism
   through 195
digital platforms 221
disruptive technological innovation
   212–235; challenges of 224–225;
   impact of 225–227; implications of
   233–234; opportunities for 222–224;
   recommendations for leveraging
   232–233
diversified income 4
Donaldson, D. 2
drivers 149–150
Dube, K. 120, 153, 197, 213

East African Community (EAC) 6
eco-friendly innovations 221
e-commerce 220
economic aspirations *vs.* cultural
   preservation 107
ecosystems, bridging 80–81
eco-tourism 62
Ecotourism World: agritourism,
   definition of 17
education 221–222; educational
   agritourism 5, 24–25; educational
   platforms 220
Elkington, J. 102, *103*

environmental conservation: balancing
   economic gains with 110; *vs.* land use
   107–108
environmental sustainability xxvii, 78, 79,
   86, 87, 89, 92, 93, 96, 99–101, 103, 167,
   232, 241
EPA xxv, 2, 3, 6, 12
equitable revenue distribution 110–111
ERDF *see* European Regional
   Development Fund (ERDF)
ethical agritourism 97, 100–102, 108,
   110–112, 115; local actors, role of
   111–112
Ethiopia: agritourism 82
European Agricultural Fund for Rural
   Development (EAFRD) 41
European Regional Development Fund
   (ERDF) 41
European Union: on agritourism
   entrepreneurship 41–42
exchange rates 10
Extreme Learning Machine 83

Fair Trade Tourism (FTT) 100
Fanelli, R. M. 99
farmers, as stewards of the land 111
farm-to-table experience 37, 40–42, 82,
   104, 208, 220, 221
farm-to-table sales 220
Fernandes, N. 214
Ferreira, B. 158
financial resources 48
Finscope 60
Fleischer, A. 22, 148
Frąckiewicz, M. 28
FTT *see* Fair Trade Tourism (FTT)
future tourism picture 124–126

Galindo, I. M. 26
Gallemard, J. 25, 26
Gambia: ethical agritourism 100
gamification 24–25
Gandiwa, E. 192, 198
Gao, J. 148
GDP *see* gross domestic product (GDP)
Ghana: agritourism entrepreneurship 48;
   infrastructure development 49;
   Kumasi's cocoa farm experiences 7–8
Ghidouche, K. 21
Giller, M. 24
Gitau, D. K. 100
Global North 148
Global South 146, 148, 149, 155

Gohori, O. 135
government authorities, as guardians of responsible tourism 112
Grant, R. M. 161
gravity model of trade 1–2, 6, 12; specification of 8–9
Great Zimbabwe Monument 121
gross domestic product (GDP) xxvi, xxx, 1, 3, 10, 12, 13, 119, 124, 131, 132, 139, 149, 151, 167, 192, 214; Deflator and Exchange Rates xxv; trading partner 11

Halim, M. F. 57
Hall, C. M. 153
Haywood, L. K. 100
Heckscher-Ohlin theory 2–4, 12, 14
Hilton Hyatt International 132
hospitality 119–139; development 123–124; growth 123–124
Hospitality Global Report 119

IATA *see* International Aviation Transport Association (IATA)
ICTs *see* information and communication technologies (ICTs)
income poverty, alleviation of xxvi–xxvii, 54–71, 241
India: agritourism development 58
indigenous communities, as guardians of cultural heritage 111
Indonesia: agricultural trade flows 6
information and communication technologies (ICTs) xxx, 158, 196, 198, 215, 216, 223, 226, 233–235
infrastructure development 49
Ingram, J. 26
innovation diffusion theory 56
InterContinental Hotels Group 132
International Aviation Transport Association (IATA) 125
Italy 17; agritourism entrepreneurship 37, 40, 41; ethical dimensions of agritourism 99

Jafari-Sadeghi, V. 212
Jamal, T. 101
Janjua, Z. U. A. 101
J-Curve 3–4, 12
Jeou-Shan, H. 150
Jonas-Berki, M. 149

Karampela, S. 85
Kastenholz, E. 148

Kehoe, T. J. 2
Keisies Cottage 67
Kenya 6, 193–194; agritourism entrepreneurship awareness campaigns 50; Maasai Mara Serena Safari Lodge 81
Kenya-Kericho: agritourism 174
Khairabadi, O. 17
KICOTA *see* Kilimanjaro Indigenous Community Tourism Association (KICOTA)
Kilimanjaro Indigenous Community Tourism Association (KICOTA) 102
Kinigi Community Eco-Tourism Project 100
Kitsios, F. 18
Kizos, T. 85
knowledge transfer (KT): artificial intelligence in 25–26; definition of 25
KoMpisi Cultural Village, Victoria Falls 135
Krishna, D. K. 151, 159
KT *see* knowledge transfer (KT)
Kubickova, M. 101
Kupika, O. L. 197

Lak, A. 17
Lane, B. 148
Leontief paradox 2
Lesotho: infrastructure development 49
limited access to information, addressing 89–90, **90**, *90*
Lindstrom, K. 42–43
Lombard 82
Lowenberg-DeBoer, J. 67

machine learning (ML) 29
Madanaguli, A. 99
Makacha, C. 215
Malaysia: agricultural trade flows 6
Manalo, C. C. 99–100
Mangalis Hotel Group 132
marketing agritourism products xxix–xxx, 191–209
market insufficiencies 88, *89*, **89**
Marriott International 132
Marshall Lerner condition 3, 10, 12
Martinez-Zarzoso, I. 6
McCarthy, J. 18
McKinsey and Company 213
Melia Hotels & Resorts 132
Melubo, K. 102
Meschede, H. 214

micro, small, and medium enterprises (MSMEs) 232
Mitchell, C. J. A. 148–149, 156
Mitchell, J. 123
mixed reality (MR) 219
Mkwizu, K. H. 66
ML *see* machine learning (ML)
mobile apps 219
Morris, C. 26
Moyo, C. 196
Moyo, S. 84, 135
Mozambique 6
MR *see* mixed reality (MR)
Mshenga, P. M. 61
MSMEs *see* micro, small, and medium enterprises (MSMEs)
Mtandao wa Vikundi vya Wakulima Tanzania (MVIWATA) 63, 70
Mukora, F. N. 235
Muresherwa, G. 120
Musavengane, R. 119
Mutanga, C. 192, 198
MVIWATA *see* *Mtandao wa Vikundi vya Wakulima Tanzania* (MVIWATA)
Myer, S. 155

Namibia: agritourism development 58
natural language processing (NLP) 28–29
Neligwa, M. 63, 70
Nelson, F. 66
New Jersey: sustainable agritourism 83
Ngo, T. 99
NGOs, as advocates for responsible tourism and community engagement 112
Ngowi, P. 171
Nicolaides, A. 99
NLP *see* natural language processing (NLP)
Novelli, M. 214
Nyagadza, B. 231
Nyahunzwi, D. K. 199
Nyamadzawo, J. 196
Nyaruwata, S. 196, 197

Oates, B. J. 19
Ohe, Y. 19
Olofsson, P. 42–43
online booking platforms 219
opportunity-based theory of entrepreneurship 39
Orias, M. J. S. 195

Paniccia, P. 99
Park, J. 101
Paul, J. 47
peer-to-peer learning 26
Pérez-Olmos, K. N. 38
Perifanis, N.-A. 18
Petroman, I. 5
Philippines, The: agricultural trade flows 6
Piermartini, R. 2
Pinto, V. S. 222
population 10; trading partner 11
Pothipasa, P. 21, 83, 90
poverty: income, alleviation of xxvi–xxvii, 54–71, 241; reduction policy implications 167–188
PPT *see* Pro-Poor Tourism (PPT)
Prasad, D. M. V. 130
precision agritourism xxvi–xxvii, 54–71, 241; limitations of 60–68; methodology 59–60, **59**; recommendations for 68–71
pricing strategy 46
print media industry 213
Pro-Poor Tourism (PPT) 124, 137, 193
purchasing power parity theory 4

Quella, L. 22, 148

Radisson Hotel Group 132
Rai, V. S. R. 222
Rainbow Tourism Group 127
Ranganai, N. 223
RAPID *see* Research and Policy in Development (RAPID)
Rauniyar, S. 19
Raynor, M. E. 213
RBT *see* resource-based theory (RBT)
regulation 5
relative power parity 4
Research and Policy in Development (RAPID) 180
resource-based theory (RBT) 150, 161
revenue volatility 45
robotic farm assistants 220
Rogerson, C. M. 149
Rogerson, J. M. 149
Romagnoli, L. 99
Rotana Hotels 132
rural development xxix, 5, 39, 71, 97, 151, 152, 171, 177, 193, 242; AI for 92; promotion of 19, 38, 40, 96, 218;

sustainable xxvi, 41, 42, 168, 187; theory of 169
Rural Homestay Programme 104
rural transformation 173; agritourism nexus for *179*; implication of agritourism on 175–176; stakeholders, mapping 180–184, *181*
Ruta, M. 3
Rwanda: agritourism partnership 81; ethical agritourism 100

SADC *see* Southern African Development Community (SADC)
SADC-FTA *see* Southern African Development Community-Free Trade Area (SADC-FTA)
Sahebalzamani, S. 101–102
Sanderson, S. 196
Sarri, D. 67
Save Valley Conservancy 135
SDGs *see* Sustainable Development Goals (SDGs)
seasonality 5, 46
self-sustainability 136
sharing economy 222
Shen, C. C. 101
Shereni, N. C. 135, 136
Shinnon, M. 148–149, 156
Shu, M. 213
skill requirements 5
Skobkin, S. S. 131
smart farming technologies 219
SNA *see* Social Network Analysis (SNA)
SNV (Netherland Development Organizations) 62
social media marketing, in agritourism 219; challenges of 198–199, **204–205**; consumer experience **202**; opportunities for 199, **207**; sustainable agritourism 197
Social Network Analysis (SNA) 181–186
social sustainability xxiii, 103, 136
South Africa: adult entrepreneurship education 48; agricultural trade flows 6; agritourism activities 22; agritourism development 58, 173, **173**, 193–194; agritourism entrepreneurship 38, 48; cottage farming agritourism 67; Land and Agricultural Development Bank of South Africa 48; wine tourism, in Stellenbosch 7
Southern Africa, state of agritourism

development in 145–162; constraints 151–153; diversity 156–157; drivers of agritourism 149–150; frameworks 159; implications 160–161; limitations 162; marketing and promotion platforms 157–159; resource-based theory 150; socioeconomic benefits 149–150
Southern African Development Community (SADC) xxv, 2, 6, 153, 154
Southern African Development Community-Free Trade Area (SADC-FTA) 2, 3, 6, 12
staffing 45
storytelling 222
Suanpang, P. 21, 83, 90
sustainability 18; environmental xxvii, 78, 79, 86, 87, 89, 92, 93, 96, 99–101, 103, 167, 232, 241
sustainable accommodation 222
sustainable agritourism 68, 81, 161, 174, 199, 200, 221, 226, 230, 240, 241; AI-enabled xxvii, 77–95; in Caála Municipality xxvii–xxviii; in Central Angola 96–115; challenges of 84; disruptive innovations, recommendations for leveraging 232–233; ethical dimensions of 96–115; for human development 77–95; implications of 93–94; limitations of 94–95; opportunities of 82–84; social media marketing for 197, **202**
sustainable development: opportunities of 82–84
Sustainable Development Goals (SDGs) 21, 31, 101, 135, 151–153, 162, 168, 240; achievement of 84–85; and agritourism entrepreneurship 39–40; SDG 1 57, 100, 101, 152, 153, 160; SDG 2 57, 101; SDG 4 85; SDG 5 57; SDG 8 40, 51, 100, 122–124, 131, 136, 137, 152, 160, 241; SDG 10 101; SDG 11 100, 101; SDG 13 152, 160; SDG 13.3 57; SDG 15 100, 101
sustainable practices through technology 220
Sustainable Tourism Development Theory 137

TAMADI 63
Tanzania: 2013 National Agriculture Policy 176, 177; Agricultural

Marketing Cooperative Unions (AMCOs) 70; Agricultural Sector Development Lead Ministries (ASLMs) 62; Agricultural Sector Development Programme (ASDP II 2015/ 16–2025/26) 62; agriculture policy issue 177, *178*; agriculture-tourism interface 176, *178*; Build a Better Tomorrow (BBT) program model 70, 174; Community-Based Tourism 55, 61, 62, 66; Community Wildlife Management Areas Consortium (CWMAC) 63; cottage agritourism approach in poverty alleviation 66–68; CRDB bank 70; current status of agritourism 63–66, **64–65**; ethical agritourism 102; evolution of agritourism 61–63; GDP 54, 60; Housing and Human Settlement Development 63; income poverty alleviation among farming communities 54–71; Integrated Tourism Master Plan 62; Labor Force Survey 175; linkage effects of tourism **170**; Ministry of Agriculture 62; Ministry of Industries and Trade 63; Ministry of Land 63; Ministry of Livestock and Fisheries 62–63; Ministry of Natural Resources 62; Ministry of Natural Resources and Tourism (MNRT) 62; Ministry of Water 63; National Agricultural Policy 2013, Section 3.17 62; National Determined Contribution (NDC) of 2021 67; National Environment Management Council (NEMC) 63; National Tourist Policy 1999 62; Ngorongoro Conservation Area Authority (NCAA) 63, 68; policy issues 176, *177*; poverty levels of agricultural communities 60–61; precision agritourism xxvi–xxvii, 54–71, **69**, 241; President's Office, Regional Administration and Local Governments (PO-RALG) 63; Pro-Poor Tourism 54, 193; Responsible Tourism United Republic of Tanzania (RTTZ) 63; rural employment **176**; 'Rural Tourism Project' 63; Smart Machine for Agricultural Solutions High-tech (SMASH) project 67; Tanzania Association of Cultural Tourism Organizers (TACTO) 63;

Tanzania Horticultural Association (TAHA) 63; Tanzania Human Development Report 175; Tanzania Investment Center (TIC) 63; Tanzania National Parks Authority (TANAPA) 63; Tanzania National Road Agency (TANROAD) 71; Tanzania Rural and Urban Road Agency (TARURA) 71; Tanzania Tourist Board (TTB) 63; Tanzania Wildlife Policy 62; Tanzania Wild Life Research Institute (TAWIRI) 63; third Five Year Development Plan of 2021/22–2025/26 62; TIB development bank 70; Tourism Confederation of United Republic of Tanzania (TCT) 63; tourism policy issue 177–180; Tourism Strategic Plan 2021–2026 62; Tourist Master Plan 2002 62

TBL *see* triple bottom line (TBL) framework

technology diffusion theory 56

Teh, R. 2

TFTA *see* Tripartite Free Trade Area (TFTA)

Thailand: agricultural trade flows 6

Tichaawa, T. M. 120, 135

Tierhoek Cottages 67

time management 45–46

Tinbergen, J. 1

Togaymurodov, E. 149

Torabi, Z. A. 101

Tourism 2020 Vision 125

tourism in the future, shape of 129–130

trading partner: GDP 11; population 11

Transport and Entertainment industries 213

Tripartite Free Trade Area (TFTA) 6

triple bottom line (TBL) framework 79, 85, 93, 102–103, *103*, 112, 113, 124, 137

Tunisia: agritourism 174

2030 Agenda for Sustainable Development 240

Uganda: adult entrepreneurship education 48; agritourism 81–82

UNACA *see* Confederation of Farmer Associations and Agricultural Cooperatives of Angola (UNACA)

United Kingdom (UK): hospitality industry 120; sustainable agritourism 83

United Nations World Tourism
Organisation (UNWTO) 21, 119, 147;
Best Tourism Village 148
United States (US) 5, 17; agricultural
trade flows 6; agritourism
entrepreneurship 37, 40, 41, 51;
Agriturismo law of 1985 40
UNWTO *see* United Nations World
Tourism Organisation (UNWTO)
US *see* United States (US)
utilitarian theory xxvi, 18, 19, **20**, 31

van der Merwe, P. 22, 101, 135, 148, 157
Van Niekerk, J. 25
van Zyl, C. C. 22, 82, 101, 148, 157
Victoria Falls 132
virtual reality (VR) 219, 221, 226
virtual tourism 125–126, 223
VR *see* virtual reality (VR)

Wallach: "Artificial Intelligence: A Very
Short Introduction" 80
Wandschneider, P. R. 5
Wang. H. 156
Web 3.0 231
Wiltshier, P. 101
wine tourism, in Stellenbosch, South
Africa 7
WITS *see* World Integrated Trade
Solutions (WITS)
World Bank 135
World Integrated Trade Solutions
(WITS) xxv
World Tourism Authority (WTA) xxviii
World Travel and Tourism Council
(WTTC) 119, 124, 125
WTA *see* World Tourism
Authority (WTA)
WTTC *see* World Travel and Tourism
Council (WTTC)
Wu, B. 148

Xingping, L. 157
Xue, L. L. 101

Yamagishi, K. 130, 230
Yang, Q. 148
Yang, S. 6

Zambia 6
Zanzibar 58; agritourism 82
Zhou, Z. 119
Zimbabwe xxv–xxvi; agricultural trade
flows 2, 3, 6; agritourism 1–14, 174,
194; agritourism agreements 11;
agritourism entrepreneurship 38;
agritourism entrepreneurship
awareness campaigns 49; community
benefits of tourism and hospitality
135–138; digital and social media
marketing 191–209; disruptive
technological innovation 212–235;
employmentc creation 135–138; GDP
11; Hospitality Association of
Zimbabwe (HAZ) 129; hotels by city/
tourism destination **134**; marketing
agritourism products xxx; Ministry of
Environment, Tourism and
Hospitality Industry (METHI) 136;
population 11; tourism and hospitality
industry xxviii, 119–139; tourism
industry xxviii, 126–127; tourist
attractions in **133**; trends and
developments of agritourism 218;
UNESCO world heritage sites **134**;
wealth and uniqueness of tourism
industry 127–129; Zimbabwe Council
of Tourism (ZCT) 129; Zimbabwe
International Trade Fair (ZITF)
exhibition centre 154; Zimbabwe Sun
Hotels 127; Zimbabwe Tourism
Authority (ZTA), 135 121, 126, 128,
129, 154, 192, 194, 198, 199, 208;
Zimbabwe Tourism Development
Corporation (ZTDC) 126
Zvavahera, P. 39, 192, 195, 196

For Product Safety Concerns and Information please contact our EU
representative GPSR@taylorandfrancis.com
Taylor & Francis Verlag GmbH, Kaufingerstraße 24, 80331 München, Germany

www.ingramcontent.com/pod-product-compliance
Lightning Source LLC
Chambersburg PA
CBHW060236220326
41598CB00027B/3957

9 781032 696157